AF771560

Advances in Spacecraft Attitude Control

Edited by Timothy Sands

Published in London, United Kingdom

IntechOpen

Supporting open minds since 2005

Advances in Spacecraft Attitude Control
http://dx.doi.org/10.5772/intechopen.77574
Edited by Timothy Sands

Contributors
Brendon Smeresky, Alexa Rizzo, Matthew Cooper, Jonathan Lang, Henry Travis, Zachary A Lewis, Kyle Baker, Joshua Ten Eyck, Ranjan Vepa, Eryn Culton, Emanuele Calabrò, Elisa Capello, Matteo Dentis, Renato Bruni, Fabio Celani, Elżbieta Jarzębowska, Marcin Kłak, Peter Waswa, Sangram Redkar

© The Editor(s) and the Author(s) 2020
The rights of the editor(s) and the author(s) have been asserted in accordance with the Copyright, Designs and Patents Act 1988. All rights to the book as a whole are reserved by INTECHOPEN LIMITED. The book as a whole (compilation) cannot be reproduced, distributed or used for commercial or non-commercial purposes without INTECHOPEN LIMITED's written permission. Enquiries concerning the use of the book should be directed to INTECHOPEN LIMITED rights and permissions department (permissions@intechopen.com).
Violations are liable to prosecution under the governing Copyright Law.

(cc) BY-NC

Individual chapters of this publication are distributed under the terms of the Creative Commons Attribution - NonCommercial 4.0 International which permits use, distribution and reproduction of the individual chapters for non-commercial purposes, provided the original author(s) and source publication are appropriately acknowledged. More details and guidelines concerning content reuse and adaptation can be found at http://www.intechopen.com/copyright-policy.html.

Notice
Statements and opinions expressed in the chapters are these of the individual contributors and not necessarily those of the editors or publisher. No responsibility is accepted for the accuracy of information contained in the published chapters. The publisher assumes no responsibility for any damage or injury to persons or property arising out of the use of any materials, instructions, methods or ideas contained in the book.

First published in London, United Kingdom, 2020 by IntechOpen
IntechOpen is the global imprint of INTECHOPEN LIMITED, registered in England and Wales, registration number: 11086078, 7th floor, 10 Lower Thames Street, London, EC3R 6AF, United Kingdom
Printed in Croatia

British Library Cataloguing-in-Publication Data
A catalogue record for this book is available from the British Library

Additional hard and PDF copies can be obtained from orders@intechopen.com

Advances in Spacecraft Attitude Control
Edited by Timothy Sands
p. cm.
Print ISBN 978-1-78984-802-1
Online ISBN 978-1-78984-803-8
eBook (PDF) ISBN 978-1-83968-516-3

An electronic version of this book is freely available, thanks to the support of libraries working with Knowledge Unlatched. KU is a collaborative initiative designed to make high quality books Open Access for the public good. More information about the initiative and links to the Open Access version can be found at www.knowledgeunlatched.org

For EU product safety concerns:
IN TECH d.o.o., Prolaz Marije Krucifikse Kozulić 3, 51000 Rijeka, Croatia, info@intechopen.com or visit our website at intechopen.com.

We are IntechOpen,
the world's leading publisher of
Open Access books
Built by scientists, for scientists

4,500+
Open access books available

119,000+
International authors and editors

135M+
Downloads

151
Countries delivered to

Our authors are among the
Top 1%
most cited scientists

12.2%
Contributors from top 500 universities

WEB OF SCIENCE™

Selection of our books indexed in the Book Citation Index
in Web of Science™ Core Collection (BKCI)

Interested in publishing with us?
Contact book.department@intechopen.com

Numbers displayed above are based on latest data collected.
For more information visit www.intechopen.com

Meet the editor

Dr. Timothy Sands graduated from Columbia University, Stanford University, and the Naval Postgraduate School. He is an International Scholar Laureate of the Golden Key International Honor Society, a Fellow of the Defense Advanced Research Projects Agency, panelist of the National Science Foundation Graduate Research Fellowship program, and an interviewer for undergraduate admissions at Stanford University. He has published prolifically in archival journals, conference proceedings, books, and book chapters, in addition to giving plenary, keynote, and invitational presentations. He holds one patent in spacecraft attitude control. He is currently the Associate Dean of the Naval Postgraduate School's Graduate School of Engineering and Applied Science having previously served as a university chief academic officer, dean, and research center director.

Contents

Preface

Spacecraft attitude maneuvers comply with Euler's moment equations, a set of three nonlinear, coupled differential equations. Nonlinearities complicate the mathematical treatment of the seemingly simple action of rotating, and these complications lead to a robust lineage of research. This book begins with an introduction to the nature of attitude control before illustrating some recent advancements in the field, including optimal trajectory synthesis, parameter optimization, modern kinematics (questioning ubiquitously accepted practices), analysis and control of the nonlinear motion and utilization of precision control techniques, including application at relativistic velocities, and new developments in the use of control moment gyroscopes as actuators. Among the key facets of the book is the first-ever comprehensive treatment of mounting geometries of single-gimballed control moment gyroscopes, which will encourage readers to keep the book as a future reference text. The analysis is centered on performance and singularity-free operations.

The text is meant for basic scientifically inclined readers, and commences with a chapter on the basics of spaceflight and leverages this remediation to reveal very advanced topics to new spaceflight enthusiasts. The topics learned from reading this text will prepare students and faculties to investigate interesting spaceflight problems in an era where cube satellites have made such investigations attainable by even small universities. It is the fondest hope of the editor and authors that readers enjoy this book.

Dr. Timothy Sands
Columbia University (CVN),
Stanford University,
Naval Postgraduate School,
USA

Section 1

Overviews

Chapter 1

Introduction to Satellite Attitude Control

Henry Travis

Abstract

This chapter will introduce the space environment satellites must operate in, the motion they make in orbit, and their orientation while in orbit. The forces acting on the spacecraft will be considered, along with the implications of conservation of energy. The fundamentals of orbital mechanics will be presented, so common orbits can be visualized and discussed in terms of the six classical orbital elements. Perturbations impacting the orbit are covered for a better understanding of how orbits change over time. The inertial frame of reference will be defined and then transformed into body coordinates of the satellite using the direction cosine matrix and quaternions to describe the attitude of the spacecraft. A variety of modern attitude control techniques will be developed in the following chapters.

Keywords: satellite, space environment, gravitational force, conservation of momentum, orbital mechanics, classical orbital elements, orbital perturbations, frame of reference, LEO, MEO, GEO, HEO, direction cosine matrix, quaternion

1. Introduction

Controlling satellites begins with understanding the space environment they operate in and what forces are acting on them. Along with the solution to the two-body problem, the motion of satellites can be visualized quickly with a basic knowledge of the six classical orbital elements (COEs). Several common orbits are described in terms of their COEs. Lastly, the orientation of the satellite is described relative to an inertial frame of reference using the direction cosine matrix and quaternions.

2. Background

2.1 Environment

The space environment generally refers to the conditions existing above the earth's atmosphere. Since the atmosphere gradually dissipates as altitude increases, there is no fixed line of demarcation to define the edge of space. One convention defines the edge of the atmosphere where space begins at 100 km above the earth's surface, which will serve the purposes of this book. Once out of the atmosphere, the satellite will operate in a vacuum. There are several consequences to a vacuum environment, including outgassing, cold welding, and no heat transfer through convection. Without the atmosphere's protection, spacecraft are also susceptible

IntechOpen

micro-meteors. Similarly, the earth's magnetosphere protects the spacecraft from electromagnetic radiation and charged particles. There are a lot of advantages to staying in the atmosphere, but satellites need to leave the relative safety of the atmosphere to reach the altitudes their missions demand.

But of all the effects in the space environment, the greatest is arguably gravity. For reasons we will touch on shortly, the force due to gravity causes the spacecraft to move through space in a very specific and predictable way. All objects have mass, the amount of "stuff" that gravitationally attracts other objects. And significantly for this book, how that mass is arranged impacts how an object resists changes in motion. Weight is the force of gravitational attraction of two objects, and we are most familiar with this force when we step on a scale. As seen in Eq. (1), weight is dependent on the distance between the two masses.

$$F_{gravity} = a_{gravity} m_{satellite} = G\frac{m_{earth}}{R^2} m_{satellite} = \mu_{earth}\frac{m_{sat}}{R^2} \quad (1)$$

Gravity is then seen as the interaction of two masses at a given distance from each other. The earth's mass can be treated as a constant point mass, and the satellite's mass when at the surface is 6378 km above the center of the earth. A quick calculation of this situation shows the acceleration due to gravity on the earth's surface is the familiar 9.8 m/s^2.

$$F_{gravity@surface} = 3.986x10^{14}\frac{m_{sat}}{(6378000)^2} = 9.8m_{sat} \quad (2)$$

There are other forces acting on the spacecraft, and for more accurate results they need to be considered.

$$\Sigma F_{external} = F_{gravity} + F_{drag} + F_{thrust} + F_{3rd\ body} + F_{other} = m_{sat}a \quad (3)$$

Drag is the force the satellite feels as it passes through the atmosphere, similar to the force a hand feels when stuck out a window of a moving car. Even though the atmosphere dwindles off to nothing around 600 km above the earth's surface, its effects are still felt by satellites in low earth orbit (LEO). LEO is described in detail in Section 4.1. Thrust can be generated from rockets on the spacecraft for various purposes, but for now we will leave thrusters off. Other celestial bodies like the moon and Jupiter can also impart a gravitational force on the spacecraft. However, to understand the basic principles of satellite motion in space, the earth's force due to gravity is the most significant and sufficient. Adding a unit vector to Eq. (1) to show gravity pulling the satellite in the opposite direction, towards the center of the earth, yields:

$$\Sigma F_{external} = F_{gravity} = -\mu_{earth}\frac{m_{sat}}{R^2}\hat{R} = m_{sat}\vec{a} = m_{sat}\ddot{\vec{R}} \quad (4)$$

This results in the traditional two-body problem differential equation:

$$\ddot{\vec{R}} + \mu\frac{m_{sat}}{R^2}\hat{R} = 0 \quad (5)$$

With the solution

$$R = \frac{a(1-e^2)}{1+ecos\upsilon} \quad (6)$$

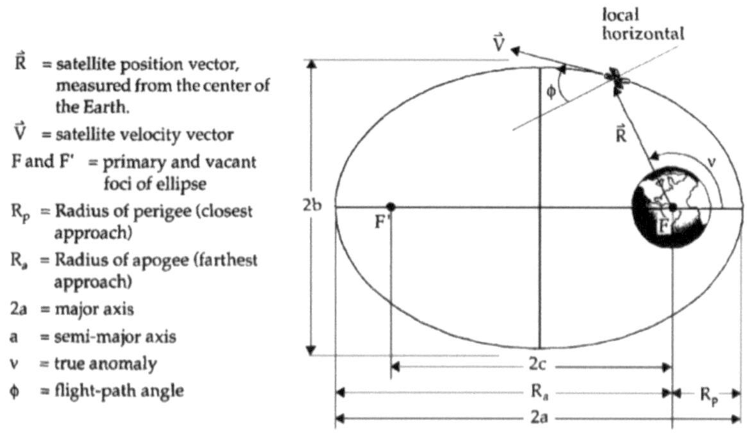

Figure 1.
Elliptical orbit [1].

where the variables are labeled in **Figure 1**. Please take time to familiarize yourself with the naming conventions used here. This solution gives the distance of the satellite from the center of the earth at all points in its orbit.

2.2 Conservation of energy

Intuitively, the more mass a satellite has, the more energy is required to achieve a particular orbit. When a satellite is launched, the chemical energy of the rockets is converted to kinetic energy to move the satellite. The higher the satellite goes, the more potential energy it has. At some point during launch, the rocket must turn over on its side to generate enough horizontal speed for the satellite to stay in orbit—otherwise it would just fall back down to the ground like a ball.

As a thought exercise, imagine a goalie throwing a soccer ball parallel to the ground. The harder the ball is thrown, the further it goes before it hits the ground. Keeping the initial trajectory parallel to the ground, throwing the ball with enough energy will cause the ball to go past the horizon of the earth. But gravity will still pull it back towards earth, or will it? With enough kinetic energy, the ball will move beyond the pull of the earth, beyond even the pull of the sun. The speed required to do so is called the escape velocity. Since the satellite is intended to orbit the earth, the energy imparted onto the satellite needs to be limited to what the earth's gravity can keep in orbit.

Through the law of conservation of energy, we can calculate how much chemical energy is required to put the satellite into the desired orbit. As you have undoubtedly seen, satellites require very, very large rockets to achieve orbit around the earth.

3. Classical orbital elements

To better understand the motion of the satellite in space, this section will elaborate on the location of the satellite in its orbital path. There are six orbital elements, collectively referred to as the classical orbital elements (COEs).

3.1 Two-dimensional elements

3.1.1 Orbit shape

From Kepler's first law, the shape of the orbit, or its path, is one of the four conic sections: circle, ellipse, parabola, or hyperbola [2]. This shape is defined by the orbit's eccentricity, e, which is the ratio of the difference and sum of the perigee and apogee distances.

$$e = \frac{R_{apogee} - R_{perigee}}{R_{apogee} + R_{perigee}} \tag{7}$$

For satellites orbiting the earth, only the circular and elliptical orbits are of interest. From the above formula, the circle is seen as a special case of the ellipse where the distances to perigee and apogee are equal. Keep in mind the radius at perigee must be greater than 6378 km (earth's radius) + 160 km (altitude above significant atmospheric effects). Since the satellite stays in elliptical orbit as long as $e < 1$, apogee could be much, much greater than the radius at perigee. However, at some point, gravitational forces from the sun and moon will come into play and invalidate the radius solution from the two-body problem presented earlier. An eccentricity of 0.7 is considered highly elliptical, and beyond that is seldom used.

3.1.2 Orbit size

The size of the orbit represents how much energy is in the orbit, and is calculated as the semi-major axis of the orbit. The larger the semi-major axis, the more energy is present. In fact, the specific mechanical energy of the orbit,, is defined as

$$\varepsilon = -G\frac{m_{earth}}{2a} = -\frac{\mu_{earth}}{2a} \tag{8}$$

Through calculations not shown here, the velocity in the satellite's direction of motion can be computed using:

$$V = \sqrt{2\left(\frac{\mu}{R} + \varepsilon\right)} \tag{9}$$

where R is the distance to the center of the earth. To give a sense of the speeds involved, a satellite in LEO will be traveling about 7.5 km/s, or 4000 mph. It is obvious from this equation that a circular orbit has a constant velocity. Less obvious, but naturally following, is the velocity of the satellite in an elliptical orbit is always changing. As the spacecraft passes perigee at maximum speed, it is slowing down as its kinetic energy is transferred into potential energy. When the satellite reaches apogee, it has reached its maximum potential energy. From there it gains speed as it falls towards perigee again, in a never-ending transfer of energy.

From Kepler's third law, the amount of time required for the satellite to make one orbit is known as the period of the orbit and is also defined by the size of the orbit [2].

$$P = 2\pi\sqrt{\frac{a^3}{\mu}} \tag{10}$$

3.1.3 True anomaly

The eccentricity and semi-major axis have defined the shape and size of the orbit, but where is the satellite along that path in space? The satellite's position on the path is measured counter-clockwise from perigee and is called the true anomaly, ν.

Much can be known about a satellite's path through space knowing size, shape, and true anomaly. To get the complete picture, we need to discuss the remaining three classical orbital elements. But before we do so, we must make the jump from the two-dimensional discussion so far into three-dimensional space. The orbital path described in this section can be rotated in three-dimensional space, and those rotations make up the remaining three classical orbital elements.

3.2 Frame of reference

To describe a location in space, we define a frame of reference that is nonrotating relative to the stars. The geocentric-equatorial coordinate system, with the origin at the earth's center and orthogonal vectors **I**, **J**, and **K** is one such example (**Figure 2**).

The fundamental plane is the **I**, **J** plane intersecting the equator, and **K** points to the north pole. To orient the fundamental plane, the principal direction, **I,** is defined to point towards the sun at vernal equinox, when the earth passes above the celestial equator of the sun on the first day of spring (**Figure 3**).

We can then describe the location of a satellite using a position vector, **R**, and a velocity vector, **V**, in the geocentric-equatorial coordinate system. Of note, each vector has three components, so together there are six components—the same number as the number of classical orbital elements. With six numbers, the satellite's path can be uniquely determined.

Since it is difficult to visualize the motion of a spacecraft in orbit using **R** and **V** vectors ([4234,2342,3] km and [7.5, 1, 1] km/s anyone?), another frame of reference

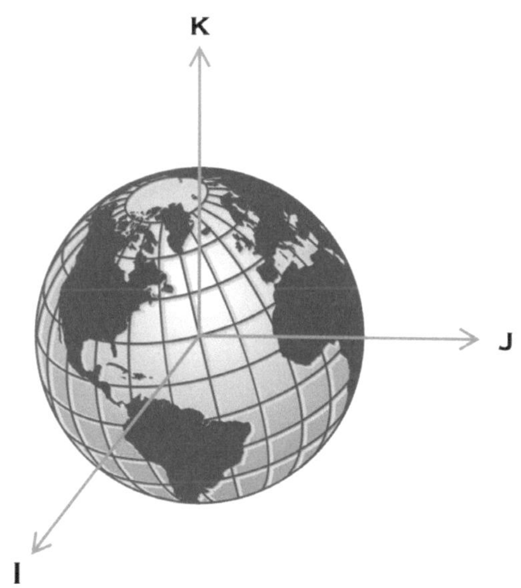

Figure 2.
Geocentric-equatorial coordinate system [3].

is used to describe locations of spacecraft in orbit around the earth. The Perifocal coordinate system uses the orthogonal unit vectors **P**, **Q**, and **W** to describe position. The satellite's orbital plane discussed in the previous section makes up the fundamental plane, with **P** pointing to periapsis and **Q** rotated 90° in the direction of satellite motion. **W** is then perpendicular to the orbital plane.

3.3 Three-dimensional elements

3.3.1 Inclination

Starting with an equatorial orbit, the orbital plane can be tilted up. The angle it is tilted up from the equator is referred to as the inclination angle, i. Since the center of the earth (the source of gravitational pull) must always be in the orbital plane, the point in the orbit where the satellite passes the equator on its way up is referred to as the ascending node, and the point where the satellite passes the equator on the way down is unsurprisingly referred to as the descending node. Drawing a line through these two points on the equator is what defines the line of nodes (**Figure 4**).

Inclining the orbital plane can be visualized as pivoting the plane about the line of nodes. Tilting the orbit 90° creates a polar orbit. A prograde orbit has an inclination between 0 and 90°. A retrograde orbit has an inclination between 90 and 180°. Tilting the orbital plane an extra 180° ($180 < i < 360$) results in the same plane in three-dimensional space.

3.3.2 Right ascension of the ascending node

The inclined orbit can be swiveled about the north pole by rotating the line of nodes counter-clockwise away from the direction of vernal equinox. The rotation of

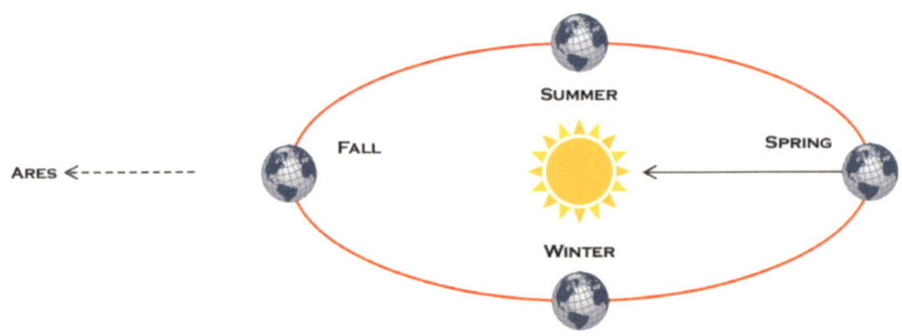

Figure 3.
Vernal equinox.

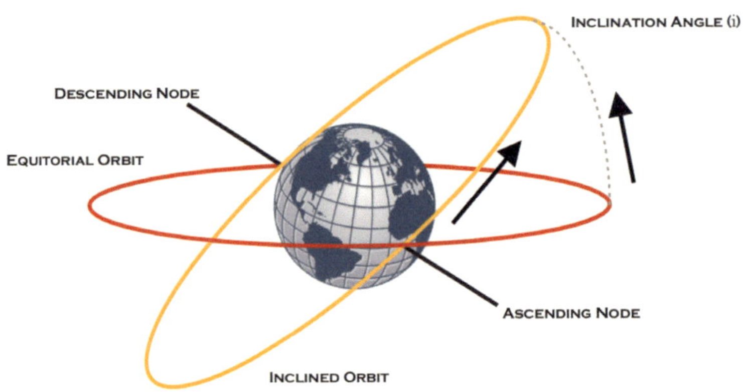

Figure 4.
Inclined orbit.

the right ascension of the ascending node (RAAN), Ω, can be any number between 0 and 360°. In the special case of an equatorial orbit, there is no ascending or descending node, therefore there is no line of nodes and omega is not defined (**Figure 5**).

3.3.3 Argument of perigee

The final rotation is not a rotation of the orbital plane, but the orientation of the orbit within the orbital plane. Rotating the orbit 90° counter-clockwise inside the orbital plane would put perigee at the ascending node. Rotating the orbit 270° would put perigee at the descending node. Perigee can be rotated 360°, so $0 < \omega < 360$. For the special case of a circular orbit, any rotation would result in the same orbit, so perigee is defined to be in the vernal equinox direction (**Figure 6**).

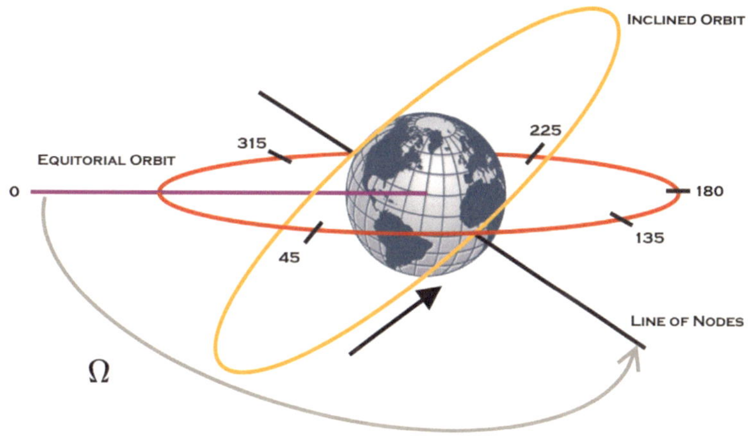

Figure 5.
Swiveled orbit.

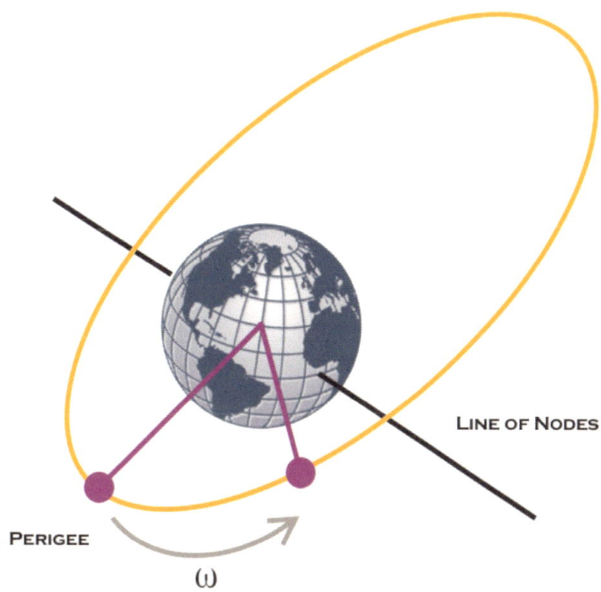

Figure 6.
Oriented orbit.

In summary, the six classical orbital elements are:

a, semi-major axis
e, eccentricity
i, inclination
Ω, right ascension of the ascending node (RAAN)
ω, argument of perigee
ν, true anomaly

The spacecraft's location in orbit can be visualized easily with these six values. Recall the **R** and **V** vectors also make up six components (each are three dimensional vectors), but do not paint the same intuitive mental picture as using the COEs.

3.4 Orbital perturbations

There are several disturbances that act on the satellite and can cause the orbital plane to shift over time. Most are related to factors we excluded to simply the math to a two-body problem.

For example, atmospheric drag is a force that takes energy out of the orbit with a perigee less than 700 km. Each pass at perigee happens with less speed, which reduces apogee. The effect is to circularize the orbit, and then spiral into the earth. The lower the perigee, the greater the effects of atmospheric drag.

Another example is the oblate earth. The earth is not a perfect sphere, which means its mass is not evenly distributed radially in all directions. Instead the mass of the earth is "squashed" like a pumpkin and produces what is referred to as the J2 effect. Since the mass of the earth is constant and gravity is conservative, the size and shape of the orbit is not impacted. Due to the symmetry above and below the equator, the inclination also does not change. However, the pull on the satellite from this "bulge" of the earth's mass will cause the line of nodes to rotate. Also due to symmetry, a polar orbit will not experience a J2 effect. Of note, the J2 can be used to create an orbital plane that always points to the sun. Satellites in LEO with $i = 98°$will rotate at the same 1°/day as the sun, creating a sun synchronous orbit.

Other disturbances include longitudinal drift, perigee rotation, 3rd body effects (gravitational pull from the sun, moon, jupiter, etc.), and solar radiation pressure. Of course, impacts with micro-meteoroids or other space debris can also impact the satellite's motion.

While many of the perturbations discussed in this section can be modeled with modern software, the simplified two-body problem is generally sufficient for a basic understanding of the satellite's location in space.

4. Useful orbits

Each of the orbits described below have been used successfully for space missions. They each have their own unique advantages and disadvantages, so selecting one for a mission hinges on which advantages are critical and which disadvantages can be lived with.

4.1 Low earth orbit (LEO)

LEOs can be circular or elliptical orbits with an altitude between 160 and 6000 km. Their relative closeness to the surface of the earth make them useful for several reasons. Since LEOs have the smallest size, they require the least amount of energy (smaller rocket) to get the satellite into orbit. Additionally, sensors collect more energy (such as light for photographs) the closer they are to the emitting source. Energy dissipates as the square of the distance, and these R2 losses only get worse the higher the satellite goes.

However, there are compromises to staying in LEO. As mentioned earlier, the atmosphere will impart drag on the satellites below about 700 km, robbing the orbit of energy and reducing the semi-major axis. This is a vicious circle, pun intended, and ultimately results in the satellite burning up during the transfer of kinetic energy into thermal energy.

Being that close to the earth's surface also means the satellite's sensors cannot "see" as much as if it were higher. The area on the earth that has line of sight to the spacecraft is called the footprint, and is larger the higher the satellite is. Consider what can be seen looking down from a tower compared to a helicopter, compared to an airplane. Staying too low comes with a limited view. Using multiple satellites in slightly offset orbits at the same altitude can overcome this limitation, but at significant expense.

4.2 Medium earth orbit (MEO)

At MEO, with an altitude of around 20,000 km, the orbits tend to be circular, though they do not need to be. This altitude equates to a period of 12 hours, which makes revisit times of the satellite consistent and predictable over a given area of the earth. Sending satellites higher also gives them a larger footprint across the surface of the earth, which means fewer satellites are necessary to cover the entirety of the earth's surface. Instead of 60+ satellites at LEO, 24 satellites can provide continuous coverage, as is done with GPS satellites.

Being above the atmosphere does have its drawbacks, as charged particles collected in regions called the Van Allen belts can adversely impact the electronics onboard the satellite. And even at this high altitude, the satellite is still not high enough to see the entire disk of the earth.

4.3 Geosynchronous earth orbit (GEO)

As the name implies, satellites at GEO complete one orbit in the same amount of time it takes the earth to make one rotation on its axis. Eq. (10) can be used to determine the semi-major axis to be 35,780 km. Like LEOs and MEOs, GEOs can also be circular or elliptical with any inclination. The high altitude means one satellite can see the entire disk of the earth, so only four satellites are required to provide continuous coverage over the entire surface of the earth.

The primary disadvantage at GEO are the R^2 losses of signals as they cross the vast distance to the surface. Of course, getting to GEO is no mean feat, and very large rockets are required to carry the energy to get there.

Geostationary orbits are special case of GEO, and deserve special mention here. They are defined to have a semi-major axis of 35,780 km, eccentricity of 0, an inclination near 0. These circular orbits allow the satellite to always be over the same point of the equator. This in turn means the ground station in communication with the satellite is always in view, which is especially useful for TV and radio satellites.

4.4 Highly elliptical orbit

The eccentricity of an elliptical orbit can be anywhere from 0 to less than 1. As the name suggests, the eccentricity of HEOs is far from 0. In fact, an eccentricity around 0.7 is common. One consequence is a very low perigee, which provides all the benefits found at LEO. And the orbit has a very high apogee, giving all the benefits of GEO. While it appears to be the best of both worlds, it also shares the disadvantages of each.

A special HEO is the Molniya orbit. This orbit has an inclination of 63 or 117° to keep the argument of perigee constant. It also has a 500 km perigee and a 40,000 km apogee to produce a semi-major axis corresponding to a 12-hour period.

5. Attitude

With the spacecraft's location described, the next step in establishing its pose is to define the orientation of the spacecraft. In later chapters we will be able to discuss equipment used for changing attitude and the control algorithms used to drive the satellite to the desired attitude.

5.1 Orientation

The spacecraft has its own three-dimensional orientation, known as body coordinates. The three orthogonal, right-handed unit vectors, **u**, **v**, and **w** are typically selected in some meaningful way, such as along the edges of the satellite body. This body coordinate system then needs to be compared to some inertial reference system so the changes can be measured.

5.1.1 Direction cosine matrix

Comparing the body coordinate system to the geocentric-equatorial coordinate system shows that each axis of one system can be represented as a vector sum of the three components of the other system.

In other words, **u**, **v**, and **w** (x, y, and z in **Figure** 7) each project some amount onto the **I** unit vector (e_1 of **Figure** 7) of the geocentric-equatorial coordinate system. The same is true for the **J** and **K** unit vectors. In matrix form,

$$A = \begin{bmatrix} u_1 & u_2 & u_3 \\ v_1 & v_2 & v_3 \\ w_1 & \mathrm{w}_2 & w_3 \end{bmatrix} \tag{11}$$

Thus, the direction cosine matrix specifies the orientation of the spacecraft relative to the inertial reference frame, and can be used to map a vector in one coordinate system to another. For example, to transform a vector in geocentric-equatorial coordinates to a vector in the body coordinate system through matrix multiplication:

$$\boldsymbol{a_b} = A\boldsymbol{a_{ge}} = \begin{bmatrix} u_1 & u_2 & u_3 \\ v_1 & v_2 & v_3 \\ w_1 & w_2 & w_3 \end{bmatrix} \begin{bmatrix} a_I \\ a_J \\ a_K \end{bmatrix} = \begin{bmatrix} a_u \\ a_v \\ a_w \end{bmatrix} \tag{12}$$

This can be seen as a single a rotation about some axis, which is called the Eigen axis. The eigenvector, **e**, is the unit vector in the direction the rotation is about. However, it is difficult to visualize this singular axis, so instead three rotations are made about the three principal orthogonal axes by an, **Φ**, to arrive at the same result as rotation about the single eigenvector.

$$\mathbf{a}_b = A_2 A_1 A_3 \mathbf{a}_{\mathbf{ge}} = \begin{bmatrix} a_u \\ a_v \\ a_w \end{bmatrix} \tag{13}$$

where the direction cosine matrix, A, in Eq. (11) is the product of:

$$\mathrm{A}_3 = \begin{bmatrix} \cos\Phi & \sin\Phi & 0 \\ -\sin\Phi & \cos\Phi & 0 \\ 0 & 0 & 1 \end{bmatrix} \tag{14}$$

$$\mathrm{A}_1 = \begin{bmatrix} 1 & 0 & 0 \\ 0 & \cos\Phi & \sin\Phi \\ 0 & -\sin\Phi & \cos\Phi \end{bmatrix} \tag{15}$$

$$\mathrm{A}_2 = \begin{bmatrix} \cos\Phi & 0 & -\sin\Phi \\ 0 & 1 & 0 \\ \sin\Phi & 0 & \cos\Phi \end{bmatrix} \tag{16}$$

Recall that matrix multiplication is not commutative, so the order of rotations definitely matters. The order is right to left, so Eq. (12) is a 3-1-2 rotation. Because transformations of an orthogonal matrix preserve the length of vectors, the transformation performed by the direction cosine matrix is seen to be a rotation of the original vector.

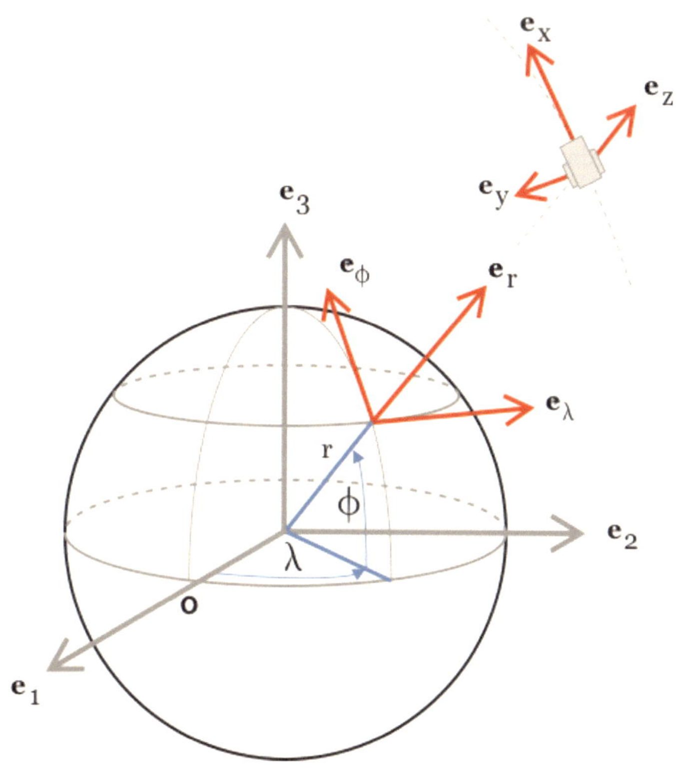

Figure 7.
Different coordinate systems. [4].

5.1.2 Quaternions

Parameterizing the direction cosine matrix with quaternions will help in future calculations, so it is presented here. Using the previously described terms for Eigen axis and angle of rotation, **e** and Φ, quaternions are defined as:

$$\boldsymbol{q} = \begin{bmatrix} q_1 \\ q_2 \\ q_3 \\ q_4 \end{bmatrix} = \begin{bmatrix} e_1 \sin\frac{\Phi}{2} \\ e_2 \sin\frac{\Phi}{2} \\ e_3 \sin\frac{\Phi}{2} \\ \cos\frac{\Phi}{2} \end{bmatrix} \tag{17}$$

with the constraint equation,

$$q_1^2 + q_2^2 + q_3^2 + q_4^2 = 1 \tag{18}$$

In terms of quaternions, the direction cosine matrix, A, can be written as:

$$A = \begin{bmatrix} q_1^2 - q_2^2 - q_3^2 + q_4^2 & 2(q_1q_2 + q_3q_4) & 2(q_1q_3 - q_2q_4) \\ 2(q_1q_2 - q_3q_4) & -q_1^2 + q_2^2 - q_3^2 + q_4^2 & 2(q_2q_3 + q_1q_4) \\ 2(q_1q_3 + q_2q_4) & 2(q_2q_3 - q_1q_4) & -q_1^2 - q_2^2 + q_3^2 + q_4^2 \end{bmatrix} \tag{19}$$

5.1.3 Euler angles

As seen in Eq. (18), three independent parameters are needed to describe the spacecraft's orientation. Another way to represent the spacecraft with only three parameters is through the use of Euler Angles. Instead of a single angle of rotation about the eigen axis, three rotations are made about the principle axes. The rotations can be made in any order and with any axis (i.e., A_{313} or A_{312}), but of course resulting in a different direction cosine matrix.

As with aircraft, these rotations are often referred to in terms pitch, roll, and yaw. By way of an example, let us consider a yaw, roll, pitch sequence. The first rotation around a principle axis is through an angle ϕ, then a second rotation around another principle axis by angle θ, a final rotation around the last principle axis by angle ψ. One example of this would be to rotate about the **K** axis, then the **I** axis and then the **J** axis. This is referred to as a 3-1-2 sequence and the direction cosine matrix can be written as:

$$\mathrm{A}_{312}(\phi, \theta, \psi) = \begin{bmatrix} \cos\psi\,\cos\phi - \sin\theta\,\sin\psi\,\sin\phi & \cos\psi\,\sin\phi + \sin\theta\,\sin\psi\,\cos\phi & -\cos\theta\,\sin\psi \\ -\cos\theta\,\sin\phi & \cos\theta\,\cos\phi & \sin\theta \\ \sin\psi\,\cos\phi + \sin\theta\,\cos\psi\,\sin\phi & \sin\psi\,\sin\phi - \sin\theta\,\cos\psi\,\cos\phi & \cos\theta\,\cos\psi \end{bmatrix} \tag{20}$$

Given the direction cosine matrix, the rotation angles for the 3-1-2 sequence can be calculated to be:

$$\begin{bmatrix} \theta \\ \phi \\ \psi \end{bmatrix} = \begin{bmatrix} arcsinA_{23} \\ -arctan\left(\dfrac{A_{21}}{A_{22}}\right) \\ -arctan\left(\dfrac{A_{13}}{A_{33}}\right) \end{bmatrix} \tag{21}$$

5.1.4 Small angle approximation

Sometimes the spacecraft will only rotate by a small amount. This is especially true as a move transient settles into the steady state attitude. Recall that for a small angle, θ,

$$cos\theta \cong 1 \tag{22}$$

$$sin\theta \cong \theta \tag{23}$$

$$sin\theta sin\theta = 0 \tag{24}$$

Using the small angle approximation reduces Eq. (20) to:

$$A_{312}(\phi, \theta, \psi) = \begin{bmatrix} 1 & \phi & -\psi \\ -\phi & 1 & \theta \\ \psi & -\theta & 1 \end{bmatrix} \tag{25}$$

For small angles, quaternions can be expressed in terms of the Euler angles by:

$$\boldsymbol{q} = \begin{bmatrix} q_1 \\ q_2 \\ q_3 \\ q_4 \end{bmatrix} \cong \begin{bmatrix} \frac{1}{2}\theta \\ \frac{1}{2}\psi \\ \frac{1}{2}\phi \\ 1 \end{bmatrix} \tag{26}$$

As with all algebraic calculations, care must be taken not to divide by zero. This is especially true when using trigonometric functions, but is also a concern when a quaternion is close to zero. Numerical techniques for handling these situations are beyond the scope of this introduction, but must be considered nonetheless.

5.2 Determination equipment

A variety of sensor are available to provide the information needed to determine the spacecraft's attitude. The constant position of the stars makes them especially useful for navigation. Star sensors take images of the star fields above them and compare the images to those in a catalog.

The magnetic field around the earth is not as constant, but has been modeled well enough to provide useful information for attitude determination. Magnetometers measure the strength of the electric field in a given axis. Using three orthogonal magnetometers can provide a good attitude estimate from the models.

The sun and moon locations at a given point in time are a known quantity and so sun sensors and moon sensors that detect them can be part of the solution. Similarly, horizon detectors can find the edge of the earth and use that information for reference.

Gyroscopes measure the rate of change and are especially useful in tracking changes to the satellite's attitude.

There are many methods to calculate the attitude of a spacecraft, including geometric, algebraic, covariance and q-method. However, the method used depends greatly on the spacecraft and its mission, so a discussion of determination methods is not presented here.

6. Control

Now that the attitude of the spacecraft can be defined and measured, the motion of the spacecraft can be controlled using control laws using any number of active and passive methods to achieve a set of desired input angles.

$$\begin{bmatrix} \theta_d \\ \phi_d \\ \psi_d \end{bmatrix} \tag{27}$$

We begin with a review of the kinematics involved [5].

6.1 Kinematics

The satellite is in a continuous state of free fall, and forces acting on the satellite will cause it to rotate about its center of mass. The resistance to that rotation is described by the moments of inertia and is represented by the inertia matrix

$$\mathrm{J} = \begin{bmatrix} J_{xx} & J_{xy} & J_{xz} \\ J_{yx} & J_{yy} & J_{yz} \\ J_{zx} & J_{zy} & J_{zz} \end{bmatrix} \tag{28}$$

The torque of the rotational motion in the inertial frame is calculated by

$$\mathrm{T} = \mathrm{J}\dot{\omega} \tag{29}$$

and from [5] the desired torques on the satellite in body coordinates becomes

$$\mathrm{T}_d = \mathrm{J}\dot{\omega}_d + \omega_d \times \mathrm{J}\omega_d \tag{30}$$

The required torques can calculated using the sinusoidal trajectory

$$\theta = \frac{1}{2}\left(\mathrm{A} + \mathrm{A}\sin\left(\omega_f \mathrm{t} + \varphi\right)\right) \tag{31}$$

The angular velocity of the body, ω_B, calculated from the torque

$$\mathrm{T} = \dot{\mathrm{H}}_i = \mathrm{J}\dot{\omega}_i + \omega_i \times \mathrm{J}\omega_i \tag{32}$$

The Euler angles are then found by integrating $\dot{\omega}_i$ to get

$$\omega^{NB} = \begin{bmatrix} \omega_x \\ \omega_y \\ \omega_z \end{bmatrix} \tag{33}$$

The satellite's pose in body coordinates relative to the orbital frame can then be found by taking the difference of the body and orbital frames relative to the inertial frame.

$$\omega^{OB} = \omega^{NB} - \omega^{NO} \tag{34}$$

6.2 Control

The motion of the satellite described by the kinetics of Section 6.1 must be controlled. There are two types of control systems, open loop and closed loop [6]. The open loop system has inputs that do not rely on the output. Examples would include systems where inputs are based on a clock, where the same input is put into the system at a given time. A closed loop system, on the other hand, closes the loop by sending the output information through a feedback mechanism to modify the input.

For satellite control, we are primarily concerned with comparing our observed state with some desired state using a closed loop control law. If the system is controllable, an unconstrained control vector, i.e., torque input, can take the system from an initial state to any other state in a finite interval of time [6].

One way to produce the required torque input is through the use of control moment gyros (CMGs). CMGs are momentum exchange devices, essentially rotating spinning discs which impart a rotation about the axis to which they are pointed. The larger the disc and the faster the spinning, the more torque is produced. A CMG inherently has a singular direction where no torque can be created [7]. To maintain controllability using critical components, a skewed pyramid orientation with advanced technique aim to avoid the impact of these singularities [8].

7. Conclusion

This chapter introduced the space environment and provided an overview of the forces acting on a satellite in orbit. The two-body problem was presented to show how the motion of satellites can be calculated and visualized in terms of the six COEs. The advantages and disadvantages of the most common orbits were discussed. Then the satellite's frame of reference was transformed to an inertial frame of reference with the direction cosine matrix or quaternions. Finally, basic spacecraft kinematics and dynamics were introduced along with the concept of controllability.

Author details

Henry Travis
Pebble Beach, California, United States of America

*Address all correspondence to: hdtravis@yahoo.com

IntechOpen

© 2020 The Author(s). Licensee IntechOpen. Distributed under the terms of the Creative Commons Attribution - NonCommercial 4.0 License (https://creativecommons.org/licenses/by-nc/4.0/), which permits use, distribution and reproduction for non-commercial purposes, provided the original is properly cited. BY-NC

References

[1] Wertz J, editor. Spacecraft Attitude Determination and Control. Ridel; 1978. p. 38. DOI: 3260.C65/9027709599. ISBN-13: 978-9027712042

[2] Bate R, Mueller D, White J. Fundamentals of Astrodynamics. Dover; 1971. p. 55. DOI: 73.157430/0486600610. ISBN-13: 978-0486600611

[3] Vallado D. Fundamentals of Astrodynamics and Applications. McGraw Hill; 1997. p. 113. DOI: 97.70117/0070668345. ISBN-13: 978-1881883128

[4] Roth. Reference Frames [Internet]. 1999. Available from: https://www.researchgate.net/figure/Reference-frames-Earth-Fixed-Reference-Frame-e-1-e-2-e-3-Local-North-Oriented_fig6_279402294 [Accessed: May 30, 2019]

[5] Smeresky B, Rizzo A, Sands T. Kinematics in the Information Age: Mathematics. Vol. 62018. p. 148

[6] Sands TA, Kim JJ, Agrawal BN. Nonredundant single-gimbaled control moment gyroscopes. Journal of Guidance, Control, and Dynamics. 2012; **35**(2):578-587

[7] Sands T. Control moment gyroscope singularity reduction via decoupled control. In: Proceedings of the IEEE SEC. 2009

[8] Ogata K. Modern Control Engineering. Prentice Hall; 2010. p. 7. ISBN-13: 978-0136156734

Chapter 2

Kinematics: On Direction Cosine Matrices

Brendon Smeresky and Alex Rizzo

Abstract

Motion mechanics (dynamics) comprises kinetics to describe the implications of applied forces and torques; and also kinematics (phoronomics). Developed in the 1700s, kinematics describes mathematical translations from one basis of measurement to another using common kinematic measurement variables like quaternions, Euler angles, and direction cosine matrices. Two ubiquitous rotation sequences are unquestionably adopted for developing modern direction cosine matrices from among the 12 potential options, stemming from applicability to aerospace systems, accuracy, and computation burden. This chapter provides a comprehensive reevaluation of all 12 options yielding a menu of options for accuracy and computational burdens, with the results illustrated compared to the ubiquitous two modernly adopted choices, broken into two rotational groups: symmetric rotations and nonsymmetric rotations. Validation will be provided by critical analysis of integration using step size to illustrate correlated minimal accuracy. No single rotational sequence is universally superior with respect to all figures of merit, enabling trade-space analysis between rotational sequences. One interesting revelation of one of the two ubiquitous sequences (the 3-1-3 symmetric sequence) is illustrated to have relatively less accuracy but lower computational burden than the other (the 3-2-1 nonsymmetric sequence). Meanwhile, a relatively unknown "2-3-1" rotational sequence is shown to have similar computational burden and accuracy.

Keywords: phoronomics, mechanics, kinetics, kinematics, direction cosines, Euler angles, space dynamics, digital computation, control systems, control engineering

1. Introduction

In 1775, Leonhard Euler developed motion phoronomics [1] which immediately blossomed in the next two centuries [2–29]. The space race between the now-defunct Soviet Union and the United State of the last century gave substantial impetus to development and adoption of motion kinematics together with survival imperatives driven by the nuclear cold war. The resultant lineage of literature contains seemingly countless technical and non-technical [30–62]. With this heritage the two most common rotational sequences used to calculate direction cosine matrices are referred to as "aerospace" sequences for nonsymmetric sequences (where the resulting angles are referred to as Tait-Bryan angles), while the symmetric sequences are oft referred to as "orbital" sequences (where the resulting angles are called proper "Euler Angles") [63].

In light of continued improvement in computational capabilities, the focus of this research is to evaluate all 12 rotation sequences comparing by mean and standard deviation of accuracy reflecting roll, pitch, and yaw angles; and also comparing by computational burden embodies by time required to perform the calculations. The chapter questions whether the 3-2-1 rotational sequence truly the best with respect to either of these figures of merit (statistical accuracy and computational burden). The results illustrate the two standard sequences are indeed good (with relative weaknesses). In particular the standard asymmetric sequence is more accurate, but slower than the standard symmetric sequence. On average, despite fewer mathematical steps, the symmetric rotations are on average slower to calculate. The 3-2-1 sequence is quickest to calculate amongst the asymmetric rotations, meanwhile the 2-3-2 and 1-2-1 are the quickest amongst the symmetric rotations.

Artificial intelligence and machine learning has evidenced the need for rapid calculations, so as motion mechanics incorporate adopt these new learning algorithms, the impact of this chapter become increasingly relevant in that options revealed in here illustrate simultaneous accuracy and favorable rapidity of calculation [62]. This chapter also complements other algorithmic advances [37–45] like system identification [55–59] including nonlinear adaptive forms and also control [46–54] for space guidance, navigation, and control (GNC) missions [35, 36, 60–65] in a time when the United States is developing and relying upon more advanced Machine Learning and AI products than ever before.

2. Materials and methods

One application of motion mechanics is the control of the attitude of spacecraft rotational maneuvers or even maintenance of a specified attitude. The key reminder is that Euler's moment equations governing rotational movement apply in a non-moving reference frame referred to as "inertial," which is a reference frame that has no meaning as a basis for measurement (i.e., it is not possible to identify a truly non-moving reference frame that can be used for measurement of angular position). This section of the chapter illustrates the method to numerically evaluate the options for kinematic expressions of rotations between chosen frames of reference (e.g., the body frame) and the inertial frame. MATLAB/SIMULINK depicted in **Figure 1** is used to create a simulation of rotation of spacecraft and necessary components of the simulation to make it relatively high-fidelity include aerodynamic and gravity gradient disturbances; kinematic expressions including quaternions, direction cosine matrices, and Euler angles; and even incorporation of the

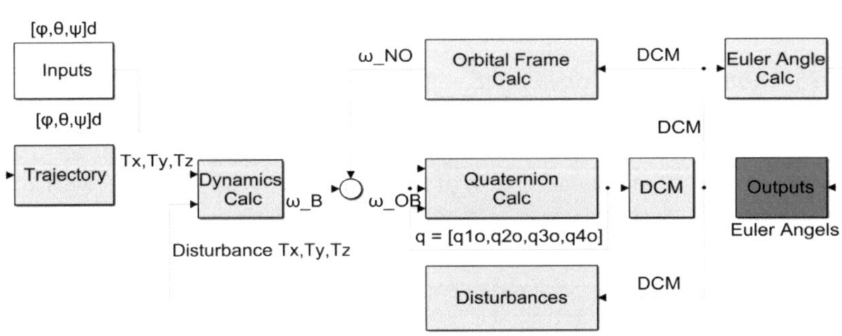

Figure 1.
MATLAB/SIMULINK simulation where controllers (not articulated) are fed a full-state trajectory autonomously generated, while the resulting motion is expressed in various forms of kinematics establishing the attitude that results in a specific calculations of disturbance torques.

motion of an object in a specified orbit. The simulation is elaborated in Section 2, while Section 3 will describe the experiments, with concluding results in Section 4.

2.1 Theory of dynamics

Mechanics and dynamics are synonyms. Interestingly, kinematics (which is also currently called phoronomics [13]) was referred to as "statics" in the era of Newton [2] indicating the lack of motion which is included in kinetics. This will be elaborated further in Section 2.1.2. Michael Chasle's theorems permit us to simply "invoke" Euler's moment equation to describe three-degrees of rotation and Newton's law to describe three-degrees of translation; together comprising a full mathematical description of so-called 6DOF motion, or motion in six-degrees. Euler describes rotational motion expressed in a moving body frame as $T = J\dot{\omega} + \omega \times J\omega$ [6], where [J] is a matrix of mass moments of inertia explained by Kane [23]. Measurements of rotational maneuvers are expressed in inertial coordinates by establishing an arbitrarily placed inertial reference frame [X_I, Y_I, Z_I], while kinematics relate the inertial coordinates to those expressed in the body reference frame [X_B, Y_B, Z_B]. References in the literature use the nomenclature "direction cosine matrix" [18], since the matrix is composed of projection components, where the dot-product projection operation is defined by the cosine of the angle between the two reference frames [17, 25, 26]. Individual vector components elaborate the orientation angle between reference frames [28].

2.1.1 Kinetics

Kinetics, or Dynamics, is the process of describing the motion of objects with focus on the forces involved. In the inertial frame, Newton's F = ma is applied but becomes Euler's $T = J\dot{\omega}$ when rotation is added, where $T = J\dot{\omega}$ is expressed in the inertial reference frame's coordinates, while $T = J\dot{\omega} + \omega \times J\omega$ from above is still measured in the inertial frame, but expressed in body coordinates.

Combining the Euler and Newton equations, we can account for all six degrees of freedom. In application, when an input angle [φ_d, θ_d, ψ_d] is commanded, the feedforward control uses Eq. (1) as the ideal controller with Eq. (2) as the sinusoidal trajectory to calculate the required torque [T_x, T_y, T_z] necessary to achieve the desired input angle. The Dynamics calculator then uses Eq. (3) to convert the torques into ω_B values, where ω_B is defined as the angular velocity of the body. In order to calculate this, the non-diagonal terms in Eq. (4) are neglected, removing coupled motion and leaving only the principle moments of inertia. Then, the inertia matrix J is removed from $\dot{\omega}$, and the remaining $\dot{\omega}$ is integrated into [ω_x, ω_y, ω_z], which is fed into the Kinematics block of the model to finally determine the outputted Euler Angles.

$$T_d = J\dot{\omega}_d + \omega_d \times J\omega_d \tag{1}$$

$$\theta = \frac{1}{2}\left(A + A\sin\left(\omega_f t + \varphi\right)\right) \tag{2}$$

$$T = \dot{H}_i = J\dot{\omega}_i + \omega_i \times J\omega_i \tag{3}$$

2.1.2 Kinematics, phoronomics, or "The Laws of Going"

Formulation of spacecraft attitude dynamics and control problems involves considerations of kinematics, especially as it pertains to the orientation of a rigid body that is in rotational motion. The subject of kinematics is mathematical in nature, because it does not involve any forces associated with motion. The kinematic representation of the orientation of one reference frame relative to another reference can also be expressed by introducing the time-dependence of Euler Angles. The so-called body-axis rotations involve successive rotations three times about the axes of the rotated body-fixed reference frame resulting in 12 possible sets of Euler angles. The so-called space-axis rotations instead involve three successive rotations using axes fixed in the inertial frame of reference, again producing 12 possible sets of Euler angles. Because the body-axis and space-axis rotations are intimately related, only 12 Euler angle possibilities need be investigated; and the 12 sets from the body-axis sequence are typically used [26]. Consider a rigid body fixed at a stationary point whose inertia ellipsoid at the origin is an ellipsoid of revolution whose center of gravity lies on the axis of symmetry. Rotation around the axis of symmetry does not change the Lagrangian function, so there must-exist a first integral which is a projection of an angular momentum vector onto the axis of symmetry. Three coordinates in the configuration space special orthogonal group (3) may be used to form a local coordinate system, and these coordinates are called the Euler angles.

Key tools of kinematics from which the Euler angles may be derived include direction cosines which describe orientation of the body set of axes relative to an external set of axes. Euler's angles may be defined by the following set of rotations: "rotation about x axis by angle and θ, rotation about z' axis by an angle ψ, then rotation about the original z-axis by angle φ". Eulerian angles have several "conventions: Goldstein uses [22] the "x-convention": z-rotation followed by x' rotation, followed by z' rotation (essentially a 3-1-3 sequence). Quantum mechanics, nuclear physics, and particle physics the "y-convention" is used: essentially a 3-2-3 rotation. Both of these have drawbacks, that the primed coordinate system is only slightly different than the unprimed system, such that, φ and ψ become indistinguishable, since their respective axes of rotation (z and z') are nearly coincident. The so-called Tait-Bryan convention in **Figure 2** therefore gets around this problem by making each of the three rotations about different axes: (essentially a 3-2-1 sequence) [22].

Kinematics is the process of describing the motion of objects without focus on the forces involved. The $[\omega_x, \omega_y, \omega_z]$ values from the Dynamics are fed into the Quaternion Calculator where Eqs. (5) and (6) yield q, the Quaternion vector. The Quaternions define the Euler axis in three dimensional space using $[q_1, q_2, q_3]$. About this axis, a single angle of rotation $[q_4]$ can resolve an object aligned in reference frame A into reference frame B. The Direction Cosine Matrix (DCM) then relates the input ω values to the Euler Angles using one of 12 permutations of possible rotation sequences, where multiple rotations can be made in sequence. Therefore, the rows of

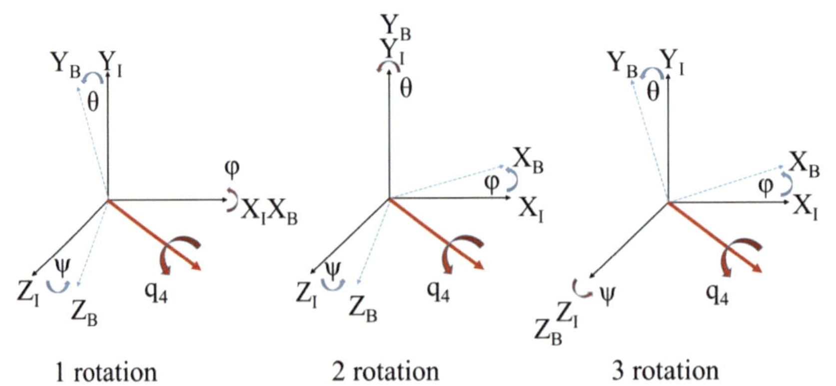

Figure 2.
Execution of a 3-2-1 rotation from CA to CB (left to right); blue-dotted arrows denote angle rotations. A direct rotation from CA to CB can be made about the Euler Axis, q_4 in red. The set of three rotations may be depicted as four rectangular parallelepipeds, where each contains the unit vectors of the corresponding reference frame [29].

the DCM show the axes of Frame A represented in Frame B, the columns show the axes of Frame B represented in Frame A, and φ, θ, and ψ are the angles of rotation that must occur in each axis sequentially to rotate from orientation A to orientation B, turning C^A to C^B. **Figure 2** depicts a 3-2-1 sequence to rotate from C^A to C^B, where the Euler Axis is annotated by the thickest line.

$$\begin{bmatrix} J_{xx}\dot{\omega}_x + J_{xy}\dot{\omega}_y + J_{xz}\dot{\omega}_z - J_{xy}\omega_x\omega_z - J_{yy}\omega_y\omega_z - J_{yz}\omega_z^2 + J_{xz}\omega_x\omega_y + J_{zz}\omega_z\omega_y + J_{yz}\omega_y^2 \\ J_{yx}\dot{\omega}_x + J_{yy}\dot{\omega}_y + J_{yz}\dot{\omega}_z - J_{yz}\omega_x\omega_y - J_{zz}\omega_x\omega_z - J_{xz}\omega_x^2 + J_{xx}\omega_x\omega_z + J_{xy}\omega_z\omega_y + J_{xz}\omega_z^2 \\ J_{zx}\dot{\omega}_x + J_{zy}\dot{\omega}_y + J_{zz}\dot{\omega}_z - J_{xx}\omega_x\omega_y - J_{xz}\omega_y\omega_z - J_{xy}\omega_y^2 + J_{yy}\omega_x\omega_y + J_{yz}\omega_z\omega_x + J_{xy}\omega_x^2 \end{bmatrix} = \begin{bmatrix} T_x \\ T_y \\ T_z \end{bmatrix} \quad (4)$$

$$\begin{bmatrix} \dot{q}_1 \\ \dot{q}_2 \\ \dot{q}_3 \\ \dot{q}_4 \end{bmatrix} = \frac{1}{2}\begin{bmatrix} 0 & \omega_3 & -\omega_2 & \omega_1 \\ -\omega_3 & 0 & \omega_1 & \omega_2 \\ \omega_2 & -\omega_1 & 0 & \omega_3 \\ -\omega_1 & -\omega_2 & -\omega_3 & 0 \end{bmatrix}\begin{bmatrix} q_1 \\ q_2 \\ q_3 \\ q_4 \end{bmatrix} = \frac{1}{2}\begin{bmatrix} q_4 & -q_3 & q_2 & q_1 \\ q_3 & q_4 & -q_1 & q_2 \\ -q_2 & q_1 & q_4 & q_3 \\ -q_1 & -q_2 & -q_3 & q_4 \end{bmatrix}\begin{bmatrix} \omega_1 \\ \omega_2 \\ \omega_3 \\ 0 \end{bmatrix} \quad (5)$$

$$\begin{bmatrix} 1-2(q_2^2+q_3^2) & 2(q_1q_2+q_3q_4) & 2(q_1q_3-q_2q_4) \\ 2(q_2q_1-q_3q_4) & 1-2(q_1^2+q_3^2) & 2(q_2q_3+q_1q_4) \\ 2(q_3q_1+q_2q_4) & 2(q_3q_2-q_1q_4) & 1-2(q_1^2+q_2^2) \end{bmatrix} = \begin{bmatrix} C_2C_3 & C_2S_3 & -S_2 \\ S_1S_2C_3-C_1S_3 & S_1S_2S_3+C_1C_3 & S_1C_2 \\ C_1S_2C_3-S_1S_3 & C_1S_2S_3-S_1C_3 & C_1C_2 \end{bmatrix} \quad (6)$$

2.1.3 The orbital frame

In order to more completely represent a maneuvering spacecraft, orbital motion must be included with the Kinematics. This relationship is represented in **Figure 1**, where the output of the DCM is fed into the Orbital Frame Calculator, and the second column of the DCM is multiplied against the orbital velocity of the spacecraft. The second column of the DCM represents the Y axis of Frame B projected in the X, Y, and Z axes of Frame A. This yields ω^{NO}, the orbital velocity relative to the Inertial Frame. Using Eq. (7), this velocity is removed from the velocity of the body relative to the Inertial Frame, leaving only the velocity of the body relative to the Orbital Frame for further calculations.

$$\omega^{OB} = \omega^{NB} - \omega^{NO} \quad (7)$$

2.1.4 Disturbances

Multiple disturbance torques exist that effect the motion of a spacecraft in orbit, two of which are addressed in this paper. The first is the disturbance due to gravity acting upon an object in orbit, where the force due to gravity decreases as the distance between objects increases. The force is applied as a scaling factor to the mass distribution around the Z axis of a spacecraft. This force applied to a mass offset from the center of gravity is calculated through the cross product found in Eq. (8) and yields an output torque about the Z axis.

The second disturbance is an aerodynamic torque due to the force of the atmosphere acting upon a spacecraft, which also decreases as the altitude increases. In Eq. (9), the force due to air resistance is calculated by scaling the direction of orbital velocity by the atmospheric density, drag coefficient, and magnitude of orbital velocity. This force then acts upon the center of pressure, which is offset from the center of gravity, and yields a torque about the Z axis, due to the cross product in Eq. (9).

The disturbances are additive and act upon the dynamics in **Figure 1**. Because the ideal feedforward controller is the dynamics, an offsetting component equal to the negative anticipated disturbances can be used to negate the disturbance torque. This results in nullifying the disturbances when the two are summed to produce ω^{OB}, the velocity of the body relative to the Inertial Frame.

$$T_g = 3\frac{\mu}{R^3}\hat{z} \times J\hat{z} \tag{8}$$

$$T_a = C_p \times f_a = C_p \times \left[\left(\rho_a V_R{}^2 A_p\right)\hat{V}_R\right] \tag{9}$$

2.2 Experimental setup

A model of the 12 DCM to Euler Angle rotations was implemented in Matlab and Simulink for this experiment. A [30, 0, 0] maneuver was commanded in the [φ, θ, ψ] channels, respectively. The expected runtime of each scenario was 15 s, comprised of a 5 s quiescent period, a 5 s maneuver time, and a 5 s post maneuver period for observations. The maneuver was initiated using a sinusoidal trajectory, calculated with $\omega_f = \pi/2$ and $\varphi = \pi/2$.

The simulated spacecraft had an inertia matrix of J = [2, 0.1, 0.1; 0.1, 2, 0.1; 0.1, 0.1, 2], the torque was initialized as T = [0, 0, 0], and the quaternion was initialized as q = [0, 0, 0, 1]. The spacecraft was simulated to fly at an altitude of 150 km, and received a drag coefficient of 2.5. For this experiment, both orbital motion and torque disturbances were turned off in order to simply the simulation.

The Matlab and Simulink models utilized the Runge-Kutta solver, with an ode4 back-end. Multiple step sizes were tested to determine accuracy variations for each of the rotations: 0.1, 0.001, and 0.0001 s. The trigonometric function used to mathematically solve for the Euler Angles was the atan2 function in Matlab.

Three figures of Merit were used to assess performance. The first two were the mean and standard deviation between the Euler Angles and Body Angles. The third was the calculation time for each rotation as a measure of complexity.

3. Experimental results and analysis

3.1 Euler angle calculations and post-processing

A relationship like Eq. (6) was created mathematically to relate the DCM and rotation matrices for each of the 12 rotation sequences. Then, φ, θ, and ψ were solved for, resulting in a mathematical process to determine the Euler Angles. This process was then coded in Matlab and Simulink, but the process was not perfect. Trigonometric quadrant errors caused the appearance of discontinuities from a commanded [30, 0, 0] maneuver. This artifact was resolved using post processing and further refinement of the DCM to rotation matrix derivations that correlated the 12 rotations found in **Figure 3** into two groups of six rotations: symmetric and non-symmetric rotations. To further define the groups, symmetric rotations would be 1-2-1 or 2-3-2, while non-symmetric rotations would include 1-3-2 or 3-1-2 rotations.

3.2 Euler angle to body angle accuracy

Accuracy was measured in the experiment by measuring the difference between the Body Angles and output Euler Angles. The expectation was that a perfectly accurate system would have a difference of zero. **Figure 4** depicts the deviation over time and **Table 1** provides the associated mean values and standard deviations for each of the rotations.

The six non-symmetric rotations show consistent error in φ, and only begin to deviate beyond the fifth decimal place in both mean error and standard deviation.

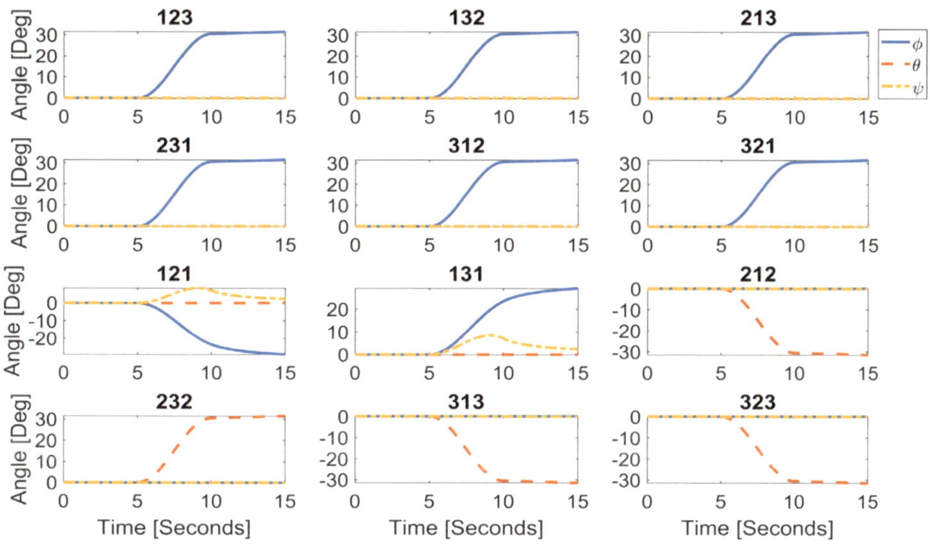

Figure 3.
Corrected Euler angles vs. time for all 12 DCM rotations.

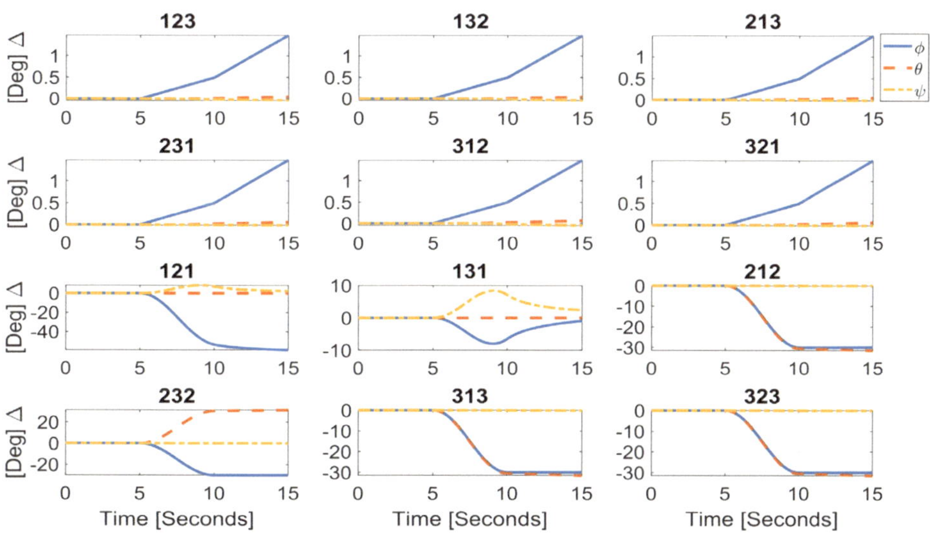

Figure 4.
Euler and body angle deviation, using a 0.1 step size.

	Mean			Standard deviation		
DCM	φ	θ	ψ	φ	θ	ψ
1-2-3	0.413	0.011	0.011	0.462	0.015	0.014
1-3-2	0.413	0.010	0.013	0.462	0.013	0.016
2-1-3	0.413	0.011	0.005	0.462	0.015	0.006
2-3-1	0.413	0.014	0.005	0.462	0.018	0.005
3-1-2	0.413	0.016	0.013	0.462	0.021	0.016
3-2-1	0.413	0.014	0.005	0.462	0.018	0.005
1-2-1	27.544	0.015	2.869	25.804	0.019	2.823
1-3-1	2.456	0.015	2.869	2.680	0.019	2.823
2-1-2	14.977	15.413	0.010	13.725	14.150	0.010
2-3-2	15.010	15.413	0.010	13.757	14.150	0.010
3-1-3	14.980	15.413	0.028	13.728	14.150	0.034
3-2-3	14.977	15.413	0.010	13.725	14.150	0.010

Table 1.
Mean and standard deviation for all 12 rotations, using a 0.1 step size.

While φ is commanded to change to 30°, θ, and ψ are expected to remain at zero, but show non-zero values due to error incurred by step size.

The six symmetric rotations are substantially harder to draw conclusions from because of the uncorrelated rotations. The mean error and standard deviation values are drastically different from each other in **Table 1** and visibly deviate in **Figure 4**. Therefore, further correlation is required to analyze accuracy. **Table 1** values were calculated over the 15 s simulation time, noting that some sequences had not reached steady-state values making their error values even larger compared to others in **Table 1** if the simulations had been run until steady state was reached.

3.3 Step size versus accuracy

This experiment implemented a variable step size to determine the accuracy delta resultant from the different step sizes. **Figure 5** depicts analysis using a step sizes of 0.001 s, which can be compared against **Figure 4**, which used a 0.1 s step size. The primary difference between **Figures 4** and **5** is the 2-order of magnitude increase of accuracy accompanying the two order of magnitude reduction in step size. A further reduction to a time step of 0.0001 s was made, with an additional order of magnitude increase in accuracy. Further reductions below this required more time than was feasible, but the trend holds that decreasing the step size increases accuracy. Furthermore, the relative accuracies between rotations held when the step sizes decreases, meaning the 1-3-2 and 3-1-2 rotations remained the most accurate.

3.4 DCM to Euler angle timing

All 12 rotation scenarios executed a maneuver within 5 s, with a standard pre and post maneuver observation period. However, actual runtimes sometimes exceeded this 15 s period; this is attributed to the complexity of the calculations and additional processes that were running at the time of the simulation. The results of each of the 12 rotations for each of three time steps are shown in **Table 2**. The simulation timing is effected by step size; therefore, the results can only be compared between

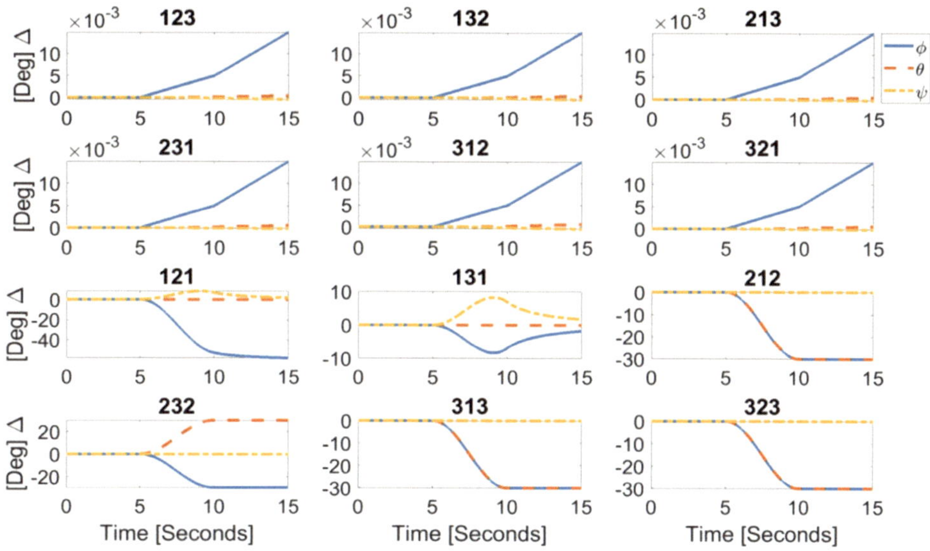

Figure 5.
Euler and body angle deviation, using a 0.001 step size.

	Execution time (s)		
DCM	0.1 step size	0.001 step size	0.0001 step size
1-2-3	8.408	11.836	28.433
1-3-2	1.533	6.789	22.187
2-1-3	1.419	6.978	22.102
2-3-1	1.188	4.436	23.259
3-1-2	1.549	4.302	20.971
3-2-1	1.018	3.475	21.420
1-2-1	0.952	3.715	20.505
1-3-1	1.190	4.082	23.331
2-1-2	1.015	3.860	21.005
2-3-2	0.931	3.710	21.410
3-1-3	0.939	3.789	20.908
3-2-3	1.091	3.955	22.044

Table 2.
Simulation run times for all 12 direction cosine matrices (DCM) rotations for a 30° roll maneuver, using 0.1, 0.001, and 0.0001 step sizes.

different rotations (vertically in the **Table 2**) and not between step sizes (horizontally in **Table 2**); however, relative comparisons between step sizes are valid.

Three observations can be made from the results in **Table 2**. The first is that the slowest rotation is the 1-2-3 rotation, by a significant amount depending upon the step size. The second is that on average, non-symmetric rotations were faster than symmetric rotations. This result is unique because the same algorithm with the same number of mathematical steps yielded different execution times. Lastly, the fastest overall rotation was the 2-3-2 rotation, with 3-2-1 as the fastest non-symmetric rotation.

4. Conclusions

This chapter on modern kinematics or motion phoronomics elaborated all 12 possible instantiations of direction cosine matrices with comparisons of numerical accuracy representing how accurately the chosen Euler angle represents the roll, pitch, and yaw expressions of rotations about x, y, z axes respectively. Additionally, comparison is made by using the figure of merit of computation burden expressed in time-necessary to perform calculations using each respective kinematic instantiation. The results were listed in a large table of options available for trade-offs, where symmetric sequences proved more difficult to compare and correlate, meanwhile the non-symmetric rotational sequences proved easier to correlate to roll, pitch, and yaw due to the ease of allocating independent angles.

The "trade-space" of options is a key elaboration, since none of the options were unanimously best using more than one figure of merit. If accuracy measured by mean error is most relatively important, 1-2-3, 1-3-2, 2-1-3, 2-3-1, 3-1-2, or 3-2-1 rotational sequences best represent roll, while the ubiquitous 3-2-1 sequence cannot best to represent pitch, where the 1-2-3 sequence is superior; while 2-1-3, 2-3-1, and 3-2-1 rotational sequences can most accurately reflect yaw. Instead if accuracy measured by standard deviation of errors was most important, the results were not identical. The most computationally efficient rotational sequence was the 2-3-2 rotation, while the 3-1-3 and 1-2-1 performed next in the list of best options. Oddly, the ubiquitous 3-2-1 sequence was merely the fifth fastest option.

The demonstration of relative inferiority of the standard ubiquitous options is a key novel development in the chapter, and the novelties were validated using a relatively high fidelity simulation of spacecraft attitude dynamics, but the novel development are valid for other forms of rotational motion mechanics like naval vessels, airplanes, and even robotics.

Future works will validate these results on laboratory spacecraft hardware simulators at the Naval Postgraduate School, and if successful flight in space is available on the international space station making the technology available to enhance the aforementioned applications of the technology [35–45].

Acknowledgements

We would like to thank our teacher, Dr. Timothy Sands for his guidance in developing this work, as well as our families for their support while we spent our time away from them developing this research, so thank you.

Conflict of interest

The authors declare no conflict of interest.

DOI: http://dx.doi.org/10.5772/intechopen.90461

Author details

Brendon Smeresky* and Alex Rizzo
Naval Postgraduate School Monterey, California, United States

*Address all correspondence to: bpsmeres@nps.edu

IntechOpen

© 2020 The Author(s). Licensee IntechOpen. Distributed under the terms of the Creative Commons Attribution - NonCommercial 4.0 License (https://creativecommons.org/licenses/by-nc/4.0/), which permits use, distribution and reproduction for non-commercial purposes, provided the original is properly cited.

References

[1] Euler L. (Euler) Formulae Generales pro Translatione Quacunque Corporum Rigidorum (General Formulas for the Translation of Arbitrary Rigid Bodies), Presented to the St. Petersburg Academy on 9 October 1775. and First Published in Novi Commentarii Academiae Scientiarum Petropolitanae 20, 1776, pp. 189–207 (E478) and Was Reprinted in Theoria motus corporum rigidorum, ed. nova, 1790, pp. 449-460 (E478a) and Later in His Collected Works Opera Omnia, Series 2, Volume 9, pp. 84-98. Available from: https://math.dartmouth.edu/~euler/docs/originals/E478.pdf [Accessed: August 22, 2018]

[2] Thompson W, Tait PG. Elements of Natural Philosophy. Cambridge, UK: Cambridge University Press; 1872

[3] Reuleaux F, Kennedy Alex BW. The Kinematics of Machinery: Outlines of a Theory of Machines. London, UK: Macmillan; 1876. Available from: https://archive.org/details/kinematicsofmach00reuluoft [Accessed: August 22, 2018]

[4] Wright TW. Elements of Mechanics Including Kinematics, Kinetics and Statics. New York, NY, USA/ Cambridge, MA, USA: D. Van Nostrand Company/Harvard University; 1896

[5] Merz JT. A History of European Thought in the Nineteenth Century. London, UK: Blackwood; 1903. p. 5

[6] Whittaker ETA. Treatise on the Analytical Dynamics of Particles and Rigid Bodies. Cambridge, UK: Cambridge University Press; 1904

[7] Whittaker ET. A Treatise on the Analytical Dynamics of Particles and Rigid Bodies. Cambridge, UK: Cambridge University Press; 1917

[8] Whittaker ET. A Treatise on the Analytical Dynamics of Particles and Rigid Bodies. Cambridge, UK: Cambridge University Press; 1927

[9] Whittaker ET. A Treatise on the Analytical Dynamics of Particles and Rigid Bodies. Cambridge, UK: Cambridge University Press; 1937

[10] Church IP. Mechanics of Engineering. New York, NY, USA: Wiley; 1908. p. 111

[11] Wright TW. Elements of Mechanics Including Kinematics, Kinetics, and Statics, with Applications. New York, NY, USA: Nostrand; 1909

[12] Study E, Delphenich DH. Foundations and goals of analytical kinematics. Sitzber. d. Berl. Math. Ges. 1913;**13**:36-60. Available from: http://neo-classical-physics.info/uploads/3/4/3/6/34363841/study-analytical_kinematics.pdf [Accessed: April 14, 2017]

[13] Gray A. A Treatise on Gyrostatics and Rotational Motion. London, UK: MacMillan; 1918. ISBN: 978-1-4212-5592-7 (Published 2007)

[14] Rose ME. Elementary Theory of Angular Momentum. New York, NY, USA: John Wiley & Sons; 1957. ISBN: 978-0-486-68480-2 (Published 1995)

[15] Kane TR. Analytical Elements of Mechanics. Vol. 1. New York, NY, USA/ London, UK: Academic Press; 1959

[16] Kane TR. Analytical Elements of Mechanics. Vol. 2: Dynamics. New York, NY, USA/London, UK: Academic Press; 1961

[17] Thompson W. Space Dynamics. New York, NY, USA: Wiley and Sons; 1961

[18] Greenwood D. Principles of Dynamics. Englewood Cliffs, NJ, USA: Prentice-Hall; 1965. ISBN: 9780137089741 (Reprinted in 1988 as 2nd ed.)

[19] Fang AC, Zimmerman BG. Digital Simulation of Rotational Kinematics; NASA Technical Report NASA TN D-5302. Washington, DC, USA: NASA; 1969. Available from: https://ntrs.nasa.gov/archive/nasa/casi.ntrs.nasa.gov/19690029793.pdf [Accessed: August 22, 2018]

[20] Henderson DM. Euler Angles, Quaternions, and Transformation Matrices—Working Relationships; As NASA Technical Report NASA-TM-74839. 1977. Available from: https://ntrs.nasa.gov/archive/nasa/casi.ntrs.nasa.gov/19770024290.pdf [Accessed: August 22, 2018]

[21] Henderson DM. Euler Angles, Quaternions, and Transformation Matrices for Space Shuttle Analysis; Houston Astronautics Division as NASA Design Note 1.4-8-020. 1977. Available from: https://ntrs.nasa.gov/archive/nasa/casi.ntrs.nasa.gov/19770019231.pdf [Accessed: August 22, 2018]

[22] Goldstein H. Classical Mechanics. 2nd ed. Boston, MA, USA: Addison-Wesley; 1981

[23] Kane T, David L. Dynamics: Theory and Application. New York, NY, USA: McGraw-Hill; 1985

[24] Huges P. Spacecraft Attitude Dynamics. New York, NY, USA: Wiley and Sons; 1986

[25] Wiesel W. Spaceflight Dynamics. 2nd ed. Boston, MA, USA: Irwin McGraw-Hill; 1989. p. 1997

[26] Wie B. Space Vehicle Dynamics and Control. Reston, VA, USA: AIAA; 1998

[27] Slabaugh GG. Computing Euler Angles from a Rotation Matrix. Vol. 6. 1999. pp. 39-63. Available from: http://www.close-range.com/docs/Computing_Euler_angles_from_a_rotation_matrix.pdf [Accessed: August 22, 2018]

[28] Vallado D. Fundamentals of Astrodynamics and Applications. 2nd ed. El Segundo, CA, USA: Microcosm Press; 2001

[29] Roithmayr CM, Hodges DH. Dynamics: Theory and Application of Kane's Method. New York, NY, USA: Cambridge University Press; 2016

[30] Sands T. Satellite electronic attack of enemy air defenses. In: Proceedings of the IEEE CDC; 2009. pp. 434-438. DOI: 10.1109/SECON.2009.5174119

[31] Sands T. Space mission analysis and design for electromagnetic suppression of radar. International Journal of Applied Electromagnetics. 2018;**8**:1-25. DOI: 10.5923/j.ijea.20180801.01

[32] Sands T, Lu D, Chu J, Cheng B. Developments in angular momentum exchange. International Journal of Aerospace Engineering. 2018;**6**:1-7. DOI: 10.5923/j.aerospace.20180601.01

[33] Sands TA, Kim JJ, Agrawal B. 2H singularity free momentum generation with non-redundant control moment gyroscopes. In: Proceedings of the IEEE CDC; 2006. pp. 1551-1556. DOI: 10.1109/CDC.2006.377310

[34] Sands T. Fine pointing of military spacecraft [PhD dissertation]. Monterey, CA, USA: Naval Postgraduate School; 2007

[35] Kim JJ, Sands T, Agrawal BN. Acquisition, tracking, and pointing technology development for bifocal relay mirror spacecraft. Proceedings of SPIE. 2007;**6569**:656907. DOI: 10.1117/12.720694

[36] Sands TA, Kim JJ, Agrawal B. Control moment gyroscope singularity reduction via decoupled control. In: Proceedings of the IEEE SEC; 2009. pp. 1551-1556. DOI: 10.1109/SECON.2009.5174111

[37] Sands T, Kim JJ, Agrawal BN. Nonredundant single-gimbaled control moment gyroscopes. Journal of Guidance, Control, and Dynamics. 2012; **35**:578-587. DOI: 10.2514/1.53538

[38] Sands T, Kim J, Agrawal B. Experiments in control of rotational mechanics. International Journal of Automation, Control and Intelligent Systems. 2016;**2**:9-22. ISSN: 2381-7534

[39] Agrawal BN, Kim JJ, Sands TA. Method and apparatus for singularity avoidance for control moment gyroscope (CMG) systems without using null motion. U.S. Patent 9567112 B1; 2017. Available from: https://calhoun.nps.edu/handle/10945/51921 [Accessed: August 22, 2018]

[40] Sands T, Kim JJ, Agrawal B. Singularity penetration with unit delay (SPUD). Mathematics. 2018;**6**:23. DOI: 10.3390/math6020023. Available from: http://www.mdpi.com/2227-7390/6/2/23/pdf [Accessed: August 22, 2018]

[41] Nakatani S, Sands T. Simulation of spacecraft damage tolerance and adaptive controls. In: Proceedings of the IEEE Aerospace Conference; 2014. pp. 1-16. DOI: 10.1109/AERO.2014.6836260

[42] Nakatani S, Sands T. Autonomous damage recovery in space. Int. J. Autom. Control Intell. Syst. 2016;**2**:22-36. ISSN: 2381-75

[43] Nakatani S, Sands T. Battle-damage tolerant automatic controls. Journal of Electrical and Electronic Engineering. 2018;**8**:10-23. DOI: 10.5923/j.eee.20180801.02

[44] Sands T, Mihalik R. Outcomes of the 2010 and 2015 nonproliferation treaty review conferences. World Journal of Social Science. 2016;**2**:46-51. DOI: 10.12691/wjssh-2-2-4. Available online at: http://pubs.sciepub.com/wjssh/2/2/4/index.html [Accessed: August 22, 2018]

[45] Sands T. Strategies for combating Islamic state. Social Science. 2016;**5**: 39-39. DOI: 10.3390/socsci5030039. Available from: www.mdpi.com/2076-0760/5/3/39/pdf [Accessed: August 22, 2018]

[46] Mihalik R, Camacho H, Sands T. Continuum of learning: Combining education, training and experiences. Education. 2017;**8**:9-13. DOI: 10.5923/j.edu.20180801.03

[47] Sands T, Camacho H, Mihalik R. Education in nuclear deterrence and assurance. Journal of Defense Studies and Resource Management. 2017;**7**: 166-166. DOI: 10.4172/2167-0374.1000166

[48] Sands T, Mihalik R. Theoretical context of the nuclear posture review. Journal of Social Sciences. 2018;**14**: 124-128. DOI: 10.3844/jssp.2018.124.128. Available from: http://thescipub.com/pdf/10.3844/jssp.2018.124.128 [Accessed: August 22, 2018]

[49] Sands T, Lorenz R. Physics-based automated control of spacecraft. In: Proceedings of the AIAA Space 2009 Conference and Exposition; Pasadena, CA, USA; 14–17 September 2009

[50] Sands T. Physics-Based Control Methods. In Advances in Spacecraft Systems and Orbit Determination. London, UK: InTechOpen; 2012. DOI: 10.5772/2408. Available from: https://www.intechopen.com/books/advances-in-spacecraft-systems-and-orbit-determination/physics-based-control-methods [Accessed: August 22, 2018]

[51] Sands T. Improved magnetic levitation via online disturbance decoupling. Physik Journal. 2015;**1**: 272-280

[52] Heidlauf P, Cooper M. Nonlinear Lyapunov control improved by an extended least squares adaptive feed forward controller and enhanced Luenberger observer. In: Proceedings of the International Conference and Exhibition on Mechanical & Aerospace Engineering; Las Vegas, NV, USA; 2–4 October 2017

[53] Cooper M, Heidlauf P, Sands T. Controlling chaos—Forced van der pol equation. Mathematics. 2017;**5**:70. DOI: 10.3390/math5040070. Available from: http://www.mdpi.com/2227-7390/5/4/70/pdf [Accessed: August 22, 2018]

[54] Sands T. Phase lag elimination at all frequencies for full state estimation of spacecraft attitude. Physik Journal. 2017;**3**:1-12

[55] Sands T. Nonlinear-adaptive mathematical system identification. Computation. 2017;**5**:47-59. DOI: 10.3390/computation5040047. Available online at http://www.mdpi.com/2079-3197/5/4/47/pdf [Accessed: August 22, 2018]

[56] Sands T, Kenny T. Experimental piezoelectric system identification. Journal of Mechanical Engineering and Automation. 2017;**7**:179-195. DOI: 10.5923/j.jmea.20170706.01

[57] Sands T. Space systems identification algorithms. Journal of Space Exploration. 2017;**6**:138-149. ISSN: 2319-9822

[58] Sands T. Experimental sensor characterization. Journal of Space Exploration. 2018;**7**:140. Available from: https://www.tsijournals.com/articles/experimental-sensor-characterization-13620.html. [Accessed: August 22, 2018]

[59] Sands T, Armani C. Analysis, correlation, and estimation for control of material properties. Journal of Mechanical Engineering and Automation. 2018;**8**:7-31. DOI: 10.5923/j.jmea.20180801.02. Available from: http://www.sapub.org/global/showpaperpdf.aspx?doi=10.5923/j.jmea.20180801.02 [Accessed: August 22, 2018]

[60] Sands T. Satellite electronic attack of enemy air defenses. In: Proceedings of the IEEE SEC; Atlanta, GA, USA; 5–8 March 2009. pp. 434-438

[61] Remarks by President Trump at a Meeting with the National Space Council and Signing of Space Policy Directive-3. Available from: https://www.whitehouse.gov/briefings-statements/remarks-president-trump-meeting-national-space-council-signing-space-policy-directive-3/ [Accessed: June 20, 2018]

[62] Sands T, Bollino K, Kaminer I, Healey A. Autonomous minimum safe distance maintenance from submersed obstacles in ocean currents. Journal of Marine Science and Engineering. 2018;**6**:98. Available from: http://www.mdpi.com/2077-1312/6/3/98/pdf [Accessed: August 22, 2018]

[63] Kuipers JB. Proceedings of the Quaternions and Rotation Sequences, Geometry, Integrability, and Quantization, Varna, Bulgaria, 1–10 September 1999. Sofia, Bulgaria: Coral Press; 2000

[64] Sands T. Optimization provenance of whiplash compensation for flexible space robotics. Aerospace. 2019;**6**:93

[65] Sands T. Comparison and interpretation methods for predictive control of mechanics. Algorithms. 2019;**12**:232

Chapter 3

Quaternion-Based Spacecraft Dynamic Modeling and Reorientation Control Using the Dynamically Equivalent Manipulator Approach

Elżbieta Jarzębowska and Marcin Kłak

Abstract

The chapter presents modification of a dynamically equivalent manipulator (DEM) method, which enables dynamic modeling of space robots and manipulators, including free-floating maneuvers, via their suitable substitution by ground-fixed manipulator models. DEM provides an attractive modeling and control design tool, since it enables conducting tests and experiments for space vehicles in earth laboratories. The modification of DEM method relies upon an introduction of quaternion-based modeling of manipulator attitude. Originally, DEM method was developed in Euler angles. The basic motivation for the presented DEM modification is to make dynamic and kinematic models more suitable for description of space robots and manipulator motions and their missions like debris removal, spacecraft servicing, space mining, and on-orbit docking and assemblies. It may also support control designs. The theoretical development is illustrated by an example of generation of spacecraft quaternion-based dynamics and simulation studies of its reorientation maneuvers.

Keywords: dynamically equivalent manipulator, quaternion-based dynamics, space robot attitude, free-floating maneuvers

1. Introduction

Possibilities of employing space robots and manipulators for variety of rescue, servicing, and reconnaissance missions have been addressed since the early 1980s (see, e.g., [1] and references there). A lot of research and theoretical studies address dynamic control and space missions for various space vehicles, but just a few experiments have been conducted on the orbit. Examples, from a very incomplete list, can be maintenance missions for the Hubble Space Telescope and the retrieval of the Space Flyer Unit as described in [2]. In these missions, however, the space crews manually operated manipulator arms. Autonomous target capture by an unmanned space robot can be another example of a challenging operation, first addressed theoretically through modeling and simulation studies by space robotic researchers. A couple of milestones marked the human's way to autonomous space robot operations toward service and exploration of the universe. The examples are the Robot Technology Experiment (ROTEX) developed by the German Aerospace Center [3]. A multisensory robot was flown on space shuttle COLUMBIA (STS-55) in 1993. Although the robot worked inside a work cell on the shuttle, several key technologies such as a multisensory gripper, tele-operation from the ground, shared autonomy, and time delay compensation by a predictive graphic display were successfully tested. One more example is the Engineering Test Satellite VII, in the area of satellite servicing, which is an unmanned spacecraft developed and launched by the National Space Development Agency of Japan [4]. In 2005, Demonstration of Autonomous Rendezvous Technology (DART) experimented with rendezvous and docking to another satellite. The mission failed due to defective autonomous navigation system, but lessons were learnt, and the next mission Orbital Express in 2007 was a success and demonstrated free-flying capture and refueling by an autonomous servicing satellite (ASTRO) [5, 6]. Another example of a promising servicing mission is SMART-OLEV (orbital life extension vehicle) implemented for life extension of GEO communication satellites [7]. The space debris removal problem is another instance of a complex task that requires multistage space manipulator (SM) missions including tracking, capturing, and debris safe removal. The strong need of working out effective methods of debris removal from the space opened new research areas and mission planning for space manipulator-based missions that were traditionally focused on on-orbit servicing of satellites. An intensive review of space debris removal problems including dynamic modeling and control can be found in [8] and references there. One more emerging research and future mission area is space mining, which needs to get frames of the scientific and future mission visions.

Before any experiment can be carried out on the orbit and before any spacecraft is launched to the orbit, intensive research; theoretical tests in fields of dynamic modeling, motion control, navigation, sensors, and vision; and related field theories have to be completed and verified.

This chapter focuses on dynamic modeling and reorientation of free-floating space manipulators dedicated to servicing tasks. The free-floating operation assumption requires the spacecraft thrusters to be turned off and the system linear and angular momentum to be conserved. This means that the spacecraft model is subjected to two constraints. One, the linear momentum conservation, generates the holonomic constraint equation, and the second, the angular momentum conservation, the nonholonomic constraints. Additionally, the uncontrolled space robot base makes the system underactuated, which means that there are less control inputs than degrees of freedom. Therefore, the free-floating space vehicles are classified as underactuated nonholonomic dynamical systems. The development of free-floating space manipulator mechanical models is then a complex task due to dynamic couplings, dynamic singularities, and nonholonomic constraints inherent to the system. There are many modeling methods, applied to a single spacecraft as well as to their formations that come from ground robotics and take advantage of the Lagrange approach and its modifications. However, space vehicles require more sophisticated modeling methods due to their specific properties and operation regimes. One of the recent modeling concepts was proposed by Liang et al. [9]. They proposed a new concept of a dynamically equivalent manipulator (DEM). In this formulation a free-floating space manipulator is substituted by a ground-fixed manipulator, whose first link is constrained by a spherical bearing. A proper scaling of physical parameters allows preserving kinematic and dynamic properties of the original space manipulator. In this original development, the attitude of the first DEM link is described by the Euler angles. Although this description is intuitive and well known in aviation, it is not suitable for dynamic modeling and control of space systems. Unlike quaternions, Euler angles are subject of gimbal lock and ambiguity. Considering free space manipulator rotation in space, quaternions are the more suitable parameters for attitude description. Not only they do not share Euler angles' drawbacks, but they are also computationally more efficient. However, implementation of quaternions reveals other challenges due to complex relations with respect to space manipulator angular velocities and the constraint equation they have to satisfy as parameters. Introduction of quaternion parameterization to the Lagrange-based dynamics modeling can be found in Nikravesh et al. [10]. There, the derivation procedure was developed for ground manipulators subjected to position constraints only.

This chapter contributes to the modification of DEM method to enable space manipulators and other spacecraft kinematic and dynamic presentations in quaternions. The modification of DEM enables dynamic modeling of space robots and manipulators, including free-floating maneuvers, via their suitable substitution by ground-fixed manipulator models. The modified DEM may provide then an attractive modeling and control design tool, since it enables conducting tests and experiments for space vehicles in earth laboratories. This may contribute to mission failure reduction and mission cost reductions. The modification of DEM method relies upon an introduction of quaternion-based modeling of manipulator dynamics and attitude. The basic motivation for the DEM modification is to make dynamic and kinematic models more suitable for description of space robots and manipulator motions and their missions like debris removal, spacecraft servicing, space mining, and on-orbit docking and assemblies. It may also support control designs. The novelty of this modeling is in the modification of DEM to enable spacecraft kinematic and dynamic presentation in quaternions. The chapter also provides a short overview of the frequently used spacecraft dynamic modeling methods, advantages, and shortcomings of the resulted models with respect to their applications to descriptions of new mission scenarios and control demands. The theoretical development of the quaternion-based DEM method is illustrated by simulation studies of an example of space manipulator attitude dynamics. The study, presented for the first time, is designed as a comparative one with respect to other modeling methods and provides a confirmation of the right approach from the modeling and simulation point of view. Also, the modeling approach proved its effectiveness when the space manipulator is added additional links.

2. Dynamic modeling of spacecraft: the existing approaches, modeling using quaternions, and advantages of the quaternion description

Majority of space robot dynamic models uses the Lagrange approach and its modifications with the generalized coordinates, joint coordinates, Denavit-Hartenberg parameters, or others. For example, following the popular derivation of a space robot dynamics, presented in, e.g., [4, 11, 12] and references there, for a simple free-floating model that consists of a base and a serial manipulator, the Lagrange-based dynamics can be presented in the form

$$\begin{bmatrix} H_b & H_{bm} \\ H_{bm}^T & H_m \end{bmatrix} \begin{bmatrix} \ddot{x}_b \\ \ddot{\Theta} \end{bmatrix} + \begin{bmatrix} c_b \\ c_m \end{bmatrix} = \begin{bmatrix} F_b \\ \tau_m \end{bmatrix} \tag{1}$$

where:
$H_m \in R^{n\times n}$ is the inertia matrix of the manipulator arms.
$H_b \in R^{6\times 6}$ is the inertia matrix of the base.

$H_{bm} \in R^{6\times n}$ is the coupling inertia matrix.
$c_b \in R^6$ is the velocity-dependent term for the base.
$c_m \in R^n$ is the velocity-dependent term for the arms.
$F_b \in R^6$ is the force and torque vector on the base mass center.
$\tau_m \in R^n$ is the torque on the manipulator joints.

The linear and angular velocities of the base are equal to $\dot{x}_b = \begin{bmatrix} v_b^T & \omega_b^T \end{bmatrix}^T$, the velocities of the end effector are equal to $\dot{x}_e = \begin{bmatrix} v_e^T & \omega_e^T \end{bmatrix}^T$, and the joint vector is $\Theta = \begin{bmatrix} \theta_1 & . & . & . & \theta_n \end{bmatrix}$. Then, the kinematics of the space robot can be presented as

$$\begin{bmatrix} v_e \\ \omega_e \end{bmatrix} = J_b \begin{bmatrix} v_b \\ \omega_b \end{bmatrix} + J_m \dot{\Theta} \tag{2}$$

where J_b and J_m are the Jacobian matrices that depend upon the base and the manipulator arms, respectively.

If there are no external forces and torques acting on a free-floating space system, the linear and angular momenta are conserved. If to assume that both of their initial values are equal to zeros, the momentum conservation equation yields

$$H_b \begin{bmatrix} v_b \\ \omega_b \end{bmatrix} + H_{bm}\dot{\Theta} = 0 \tag{3}$$

H_b is always non-singular, so (3) can be solved for the base velocities as

$$\begin{bmatrix} v_b \\ \omega_b \end{bmatrix} = -H_b^{-1}H_{bm}\dot{\Theta} = J_{bm}\dot{\Theta} \tag{4}$$

Inserting then Eq. (4) to Eq. (2), one can get the so-called generalized Jacobian matrix (GJM) of the form $\dot{x}_e = \left(J_m - J_b H_b^{-1} H_{bm}\right)\dot{\Theta} = J^*\dot{\Theta}$. It can be used to control the manipulator end effector by the resolved motion rate in the inertial space. Notice that the relation (3) is generally non-integrable and its structure is the same as for the nonholonomic kinematic constraint that comes from the no-slip wheel condition for wheeled vehicle dynamics and control. This is the reason for which Eq. (3) is called the nonholonomic constraint equation and the space robots are sometimes considered nonholonomic control systems and control designs for them follow nonlinear control technique methods. More details toward control of the space manipulator can be found in, e.g., [4, 13, 14].

Another modeling method, using the generalized coordinates is adopted in [13, 14]. It is based upon the generalized programmed motion equations (GPME) that enable incorporation of holonomic or nonholonomic constraints to the system dynamics. The GPME yield the smallest system of equations, i.e., the constraint reaction forces are eliminated during derivation. The GPME enable adding a desired trajectory for the end effector, written as a position constraint, and get the so-called reference dynamics that serves as a motion planner for a dynamic control model, which is also developed using the GPME. For more details about the use of the GPME for space manipulator dynamics and control, see [15].

Consider, as the GPME application illustrating example, a two-arm plane model of a space robot, as presented in **Figure 1**. The robot orientation is denoted by an angle θ and joint angles by a vector $q = \begin{bmatrix} q_1 & q_2 \end{bmatrix}^T$. The joint angles are not independent in the sense that they are counted from one link to another, but they do not add any position constraint equations.

A free-floating operation assumption requires the spacecraft thrusters to be turned off and the system linear and angular momentum to be zero. The condition

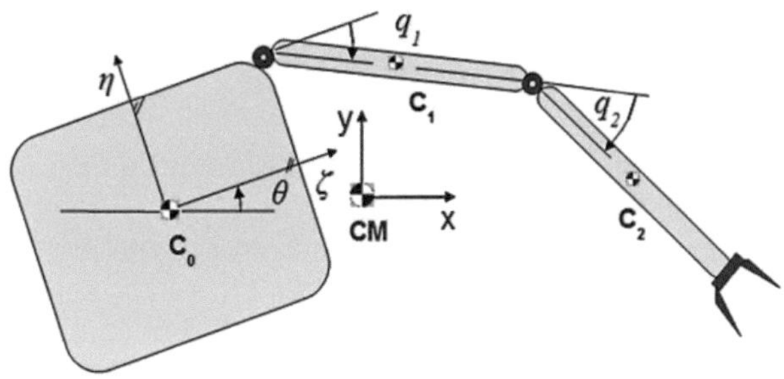

Figure 1.
Free-floating space manipulator [13].

of zero linear momentum is a holonomic constraint that can be integrated, and it is satisfied in this example by pinning the spacecraft at its mass center Co. The space manipulator angular momentum is equal to.

$$K = M(x_0\dot{y}_0 - \dot{x}_0 y_0) + I_0\dot{\theta} + m_1(x_1\dot{y}_1 - \dot{x}_1 y_1) + I_1(\dot{\theta} + \dot{q}_1) + m_2(x_2\dot{y}_2 - \dot{x}_2 y_2) + I_2(\dot{\theta} + \dot{q}_1 + \dot{q}_2). \quad (5)$$

This equation is the constraint equation on the space robot motion, and it is nonholonomic. The second constraint equation may come from a desired trajectory for the end effector E, e.g.,

$$X_d = \begin{bmatrix} x_E(t) \\ y_E(t) \end{bmatrix} = \begin{bmatrix} 1.11 - 0.3\cos(2\pi t/15) \\ 0.032 + 0.3\sin(2\pi t/15) \end{bmatrix} \quad (6)$$

as it is presented in [14]. The angular velocity of the base $\dot{\theta}$ can be determined from Eq. (5) for control purposes. The GPME enable deriving the constrained dynamics for a space robot subjected to the constraint Eqs. (5) and (6), and the Lagrange multipliers are eliminated at the equation derivation level. This constrained dynamics is referred to as the reference dynamics, and it serves as motion planner for a controller design.

Based upon the GPME, the robot dynamics described in the joint space is of the form

$$\begin{aligned} M(q,\theta)\ddot{q} + C(q,\theta,\dot{q}) &= \tau \\ \dot{\theta} &= D(q,\theta)\dot{q} \end{aligned} \quad (7)$$

where the second equation is the transformed angular momentum conservation Eq. (5). Notice, that the constraint reaction forces, i.e., Lagrange multipliers in the classical approach, are eliminated from Eq. (7). Eq. (7), being a dynamic control model, can be applied to design a tracking controller for the space robot, e.g., to track desired motion by the end effector Eq. (6).

Attitude dynamics of a space robot is of a special interest due to its reorientation maneuvers inherent to most of its operations. Attitude can be described in various ways. The most popular representations are rotation matrices, Euler angles, and quaternions. The quaternion originates in Euler's rotation theorem, and it describes attitude as a single rotation about a vector in 3D space.

A unit quaternion consists of four elements constrained by its norm. Thus, a quaternion has 3 degrees of freedom, and it is not the minimum representation, as, for instance, in the case of Euler angles. Quaternions come in different conventions, and in this chapter the Hamilton convention is adopted (see [16] for details). Specifically, the quaternion is represented as

$$\boldsymbol{q} = \begin{bmatrix} q_0 \\ \boldsymbol{q}_v \end{bmatrix} = \begin{bmatrix} q_0 & q_1 & q_2 & q_3 \end{bmatrix}^T \tag{8}$$

The scalar part of the quaternion is a function of rotation magnitude only. The latter elements describe direction of the rotation axis, preserving the unit norm. Describing the rotation magnitude as Θ and the vector of the rotation axis as $\bar{e}$, the formula for the quaternion yields

$$\boldsymbol{q} = \begin{bmatrix} \cos\left(\frac{\theta}{2}\right) \\ \bar{e}\sin\left(\frac{\theta}{2}\right) \end{bmatrix} \tag{9}$$

Quaternions can be easily related to the more intuitive space robot angular velocity vector $\boldsymbol{\omega}$ expressed in its body frame (x,y,z). These relations yield

$$\dot{\boldsymbol{q}} = \frac{1}{2}\boldsymbol{q}\otimes\begin{bmatrix} 0 \\ \boldsymbol{\omega} \end{bmatrix} = \frac{1}{2}\begin{bmatrix} \mathbf{0} & -\omega_x & -\omega_y & -\omega_z \\ \omega_x & \mathbf{0} & \omega_z & -\omega_y \\ \omega_y & -\omega_z & \mathbf{0} & \omega_x \\ \omega_z & \omega_y & -\omega_x & \mathbf{0} \end{bmatrix}\boldsymbol{q} \tag{10}$$

Eq. (10) applies the quaternion product described with the operator $\otimes$, and zero is appended to the velocity vector to form the so-called pure quaternion making the multiplication possible. However, a matrix multiplication form is also applicable.

In comparison to other representations, quaternions possess a couple of advantages:

- They are intuitive, unlike Euler angles where the sequential nature is more difficult to comprehend than a single rotation.
- The representation is not susceptible to gimbal lock as for the Euler angles.
- Any rotation can be presented as a continuous trajectory of quaternions.
- Quaternion algebra does not use trigonometric functions, just basic operations on numbers, and thus is usually more computationally efficient than Euler angles.
- Any rotation represented by quaternions can be linearly interpolated by efficient methods [16].
- Four elements construct a more compact representation than the nine-element rotation matrix.

There are also some disadvantages of adopting quaternion description, e.g.,

- The attitude is not represented uniquely; in particular $\boldsymbol{q}$ and $-\boldsymbol{q}$ describe the same rotation.
- Algebra behind quaternions requires some preprocessing work to start with this representation.

3. Spacecraft dynamic modeling using the dynamically equivalent manipulator approach modified for quaternion description application

The concept of mapping a free-floating space manipulator into equivalent fixed-base manipulator has been introduced by Liang, Xu, and Bergerman [9]. The dynamically equivalent manipulator preserves both kinematic and dynamic properties of a space manipulator. The method enjoys a couple of advantages. It allows to model a free-floating space robot with the use of classical modeling methods. Since reconstruction of space environment is complicated, DEM is also more suitable for experimental facilitating validation of guidance and control algorithms.

To map a free-floating space manipulator into a fixed-base robotic one, the base is replaced by another link. To reproduce the underactuated base, the link is fixed with a passive spherical joint. The latter joints are actuated according to the original space manipulator design. In **Figure 2** modeling structures of the (a) space manipulator (SM) and the (b) dynamically equivalent manipulator are shown. ϕ, θ, and ψ are Euler's angles, θ_i are joint angles, $\boldsymbol{u}_i$ is a vector of a rotation axis, $\boldsymbol{L}_i$ is a vector connecting joint $\mathrm{J_i}$ to the center of its mass C_i, $\mathbf{R}_i$ connects C_i to joint $\mathrm{J_{i+1}}$, and $\mathbf{W}_i$ is a vector from $\mathrm{J_i}$ to $\mathrm{J_{i+1}}$. All variables with a superscript "prime" refer to DEM.

Mass, inertia, and centers of masses of the DEM structure are scaled by transformations provided in [9]. Specifically

$$\begin{gathered} m_1' = m_1 \\ m_i' = \frac{M_t^2 m_i}{\sum_{k=1}^{i-1} m_k \sum_{k=1}^{i} m_k} \quad i = 2, \dots, n+1 \\ \mathrm{I'_i} = \mathrm{I_i} \ \mathrm{i} = 1, \dots, \mathrm{n}+1 \\ \boldsymbol{W}_1 = \boldsymbol{r}_1 \\ \mathrm{W_i} = \mathrm{r_i} + \mathrm{l_i} \ \mathrm{i} = 2, \dots, \mathrm{n}+1 \\ \boldsymbol{l}_{c1} = 0 \\ \mathrm{l_{ci}} = \frac{\sum_{\mathrm{k}=1}^{\mathrm{i}-1} \mathrm{m_k}}{\mathrm{M_t}} \mathrm{L_i} \ \mathrm{i} = 2, \dots, \mathrm{n}+1 \end{gathered} \tag{11}$$

In Eq. (11) $M_t = \sum_i m_i$ is a total mass of the space robot.

Equations of motion for a space robot as derived in [9] use Euler's angles for attitude representation. Due to the reasons emphasized in prior section, this description is not the most suitable for a space robot. This is why authors have introduced the quaternion representation to the DEM approach. Two concepts have

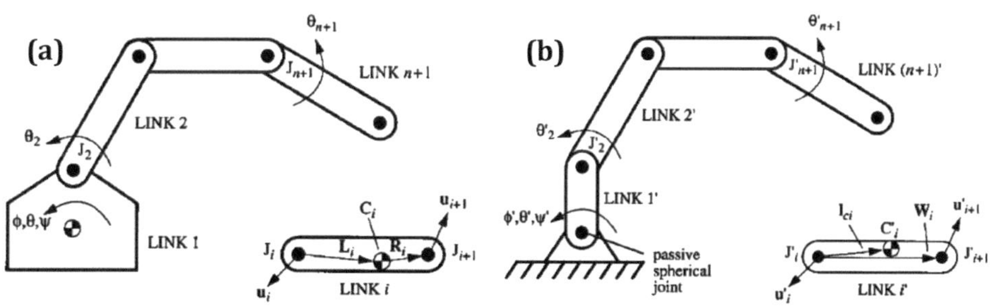

Figure 2.
Model structures of (a) space manipulator and (b) dynamically equivalent manipulator [9].

been researched. The first attempt was to develop the Lagrange equations using quaternions and then derive space robot equations of motion. This approach, however, occurred to be inefficient for an increasing number of manipulator links. Due to poor scalability, authors decided to model the space robot as a set of links, which for modeling purposes, are considered separate bodies subjected to position constraints. In this formulation each link has 6 degrees of freedom, and its state is described by the following 13-element state vector (time dependency is omitted for clarity):

$$\boldsymbol{x}_i = \left[\boldsymbol{r}_i^T \; \boldsymbol{v}_i^T \; \boldsymbol{q}_i^T \; \boldsymbol{\omega}_i^T \right]^T \tag{12}$$

where:

$\boldsymbol{r}_\mathrm{i}$ are global, translational coordinates of the center of mass of a body i.

$\boldsymbol{v}_\mathrm{i} = \dot{\boldsymbol{r}}_\mathrm{i}$ is global translational velocity.

$\boldsymbol{q}_\mathrm{i} = \boldsymbol{q}_\mathrm{iB}^\mathrm{I}$ is a quaternion rotating from the body to the inertial frame according to the formula:

$$\boldsymbol{p}^I = \boldsymbol{q}_\mathrm{B}^\mathrm{I} \otimes \boldsymbol{p}^B \otimes \boldsymbol{q}_\mathrm{B}^{\mathrm{I}\,*} \tag{13}$$

$\boldsymbol{\omega}_\mathrm{i}$ is the angular velocity determined in the body frame.

The Lagrange multipliers method is adopted due to the position constraints in the system. The equations governing DEM composed of b rigid bodies are of the following form:

$$\begin{bmatrix} \boldsymbol{M} & \boldsymbol{B}^T \\ \boldsymbol{B} & \boldsymbol{O} \end{bmatrix} \begin{bmatrix} \dot{\boldsymbol{x}} \\ \boldsymbol{\lambda} \end{bmatrix} = \begin{bmatrix} \boldsymbol{f} \\ \boldsymbol{\mu} \end{bmatrix} \tag{14}$$

In Eq. (14) $\boldsymbol{M} = \mathrm{diag}\left[m_1 \quad \boldsymbol{I}_1 \quad \ldots \quad m_b \quad \boldsymbol{I}_b \right]$ is a mass matrix, and $\boldsymbol{B}$ is a matrix satisfying the equation

$$\dot{\boldsymbol{\phi}} = \boldsymbol{B}\boldsymbol{w} + \boldsymbol{\phi}_t = \boldsymbol{O}, \tag{15}$$

where:

$\boldsymbol{\phi}$ represents the position constraint equation.

$$\boldsymbol{w} = \left[\boldsymbol{v}_\mathrm{i}^\mathrm{T} \; \boldsymbol{\omega}_\mathrm{i}^\mathrm{T} \right]^\mathrm{T}.$$

$\boldsymbol{\lambda}$ is a vector of Lagrange multipliers.

$\boldsymbol{f}$ is a vector of forces and torques.

$\boldsymbol{\mu}$ satisfies the equation.

$$\ddot{\boldsymbol{\phi}} = \boldsymbol{B}\dot{\boldsymbol{w}} - \boldsymbol{\mu} = \boldsymbol{O} \tag{16}$$

Further details related to the presented derivation can be found in [17].

The links of the space manipulator are connected by a pair of constraints that simulate a revolute joint. A position constraint of the form

$$\boldsymbol{\phi}_1 = \boldsymbol{r}_i + \boldsymbol{s}_{1i}^B - \boldsymbol{r}_j - \boldsymbol{s}_{1j}^B = \boldsymbol{O} \tag{17}$$

is needed to connect extremities of links i and j. $\boldsymbol{s}_1^B$ denotes a vector from the center of mass to the joint location, and it is expressed in the local coordinates.

Another equation is required to constrain the rotational motion to a single axis. This equation has the form

$$\boldsymbol{\phi}_2 = \boldsymbol{s}_{2i}^B \times \boldsymbol{s}_{2j}^B = \boldsymbol{O} \tag{18}$$

In Eq. (18) the vector $\boldsymbol{s}_{2k}^B$ governs the selected joint rotation vectors in their body frames. Eq. (18) preserves that those axes are parallel.

Constrained mechanical system models, when solved numerically, tend to exhibit unstable solutions, and instabilities increase with simulation time. To stabilize the solutions, a numerical stabilization method is welcome. One of them is the Baumgarte numerical stabilization method [18]. It requires that the differentiated constraint Eq. (16) is augmented as follows:

$$\ddot{\boldsymbol{\phi}} + 2\alpha\dot{\boldsymbol{\phi}} + \beta^2\boldsymbol{\phi} = \boldsymbol{O} \tag{19}$$

In Eq. (19) α and β are gains which have to be selected. The constraint Eq. (19) is numerically stable securing the constraint equation satisfaction during the system model motion. With the Baumgarte method introduced, Eq. (14) turns into

$$\begin{bmatrix} \boldsymbol{M} & \boldsymbol{B}^T \\ \boldsymbol{B} & \boldsymbol{O} \end{bmatrix} \begin{bmatrix} \dot{\boldsymbol{x}} \\ \boldsymbol{\lambda} \end{bmatrix} = \begin{bmatrix} \boldsymbol{f} \\ \boldsymbol{\mu} - 2\alpha\dot{\boldsymbol{\phi}} - \beta^2\boldsymbol{\phi} \end{bmatrix} \tag{20}$$

Eq. (20) is the final form of the space robot motion equations. They are to be solved in the numerical simulation study.

4. Example: spacecraft dynamic simulation studies

An experimental simulation study has been performed to verify, evaluate, and compare the correctness and possible applicability of the modified, quaternion-based DEM method. An example of a planar manipulator model using the original DEM method is presented in [9]. However, it is meaningless to verify a quaternion-based dynamic model on a plane. Thus, a spatial model is prepared for the simulation experiment. A space two-link manipulator model has been selected. Firstly, the properties of the space manipulator must be mapped to DEM. These are presented in **Tables 1** and 2.

The model of a space manipulator (SM) serving as a reference one has been developed in MATLAB Simscape. An open-loop torque applied to joints with the initial angular velocity vector in the direction perpendicular to joints' axes is supposed to reveal any potential inconsistency. The open-loop torque applied to the first joint J_2 is of the form of a sinusoidal signal of an amplitude of 0.5 Nm and period of 1 s, while the torque applied to the second joint J_3 is sinusoidal of an amplitude of 0.2 Nm and period of 1 s. The base of the space manipulator is not actuated and in the case of DEM joint J_1 remains passive. In the initial configuration,

Link number	$L_i[m]$	$R_i[m]$	$m_i[kg]$	$I_i\,[kg\ m^2]$
1	—	0.75	4	$1 * \mathbf{I}_3$
2	0.75	0.75	1	$0.2 * \mathbf{I}_3$
3	0.75	0.75	1	$0.2 * \mathbf{I}_3$

Table 1.
Space manipulator properties.

the manipulator arms are straightened, i.e., all angles as in **Figure 1** are equal to zeros. The angular velocity of 0.1 rad/s is applied around the initial links' axes. To verify the correctness of the quaternion-based DEM modified method, joint angles and the end effector position in the inertial reference frame are compared. The results are presented in **Figures 3** and **4**.

In **Figure 3(a)** the joint angles are compared. The obtained values of the angles overlap for the space manipulator (SM) and quaternion-based DEM. Both, space

Link number	$W_i[m]$	$l_{ci}[m]$	$m_i[kg]$	$I_i[kg\ m^2]$
1	0.5	0	4	$1*\mathbf{I}_3$
2	1.125	0.5	1.8	$0.2*\mathbf{I}_3$
3	1.375	0.625	1.2	$0.2*\mathbf{I}_3$

Table 2.
DEM properties.

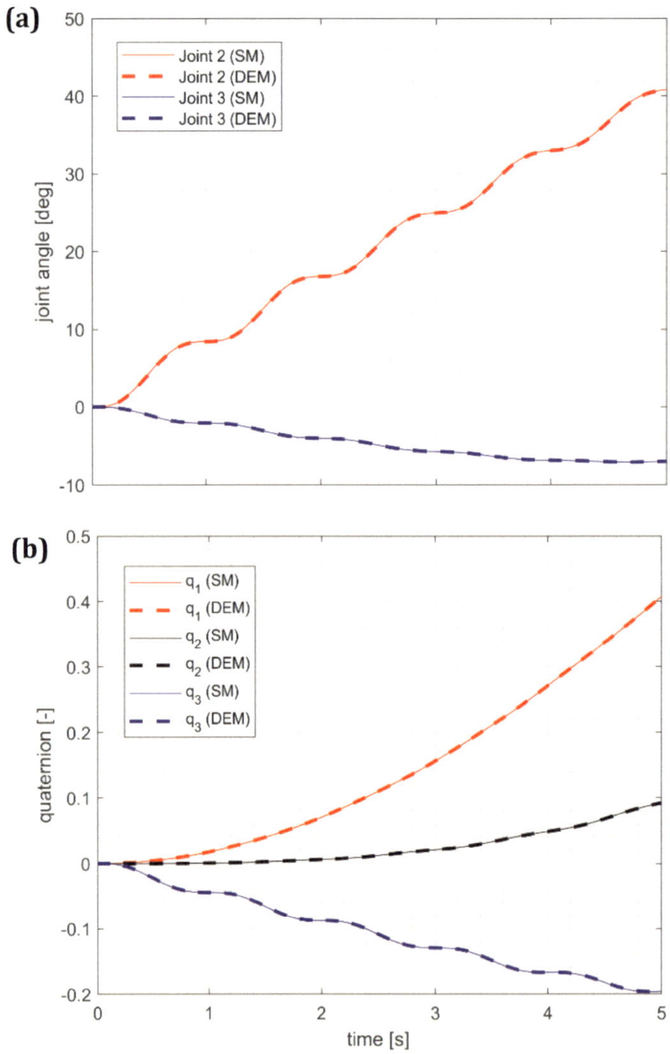

Figure 3.
(a) Comparison of joint angles for the space manipulator model and DEM quaternion-based model and (b) comparison of the quaternions of the space manipulator (SM) base and the equivalent first link of DEM.

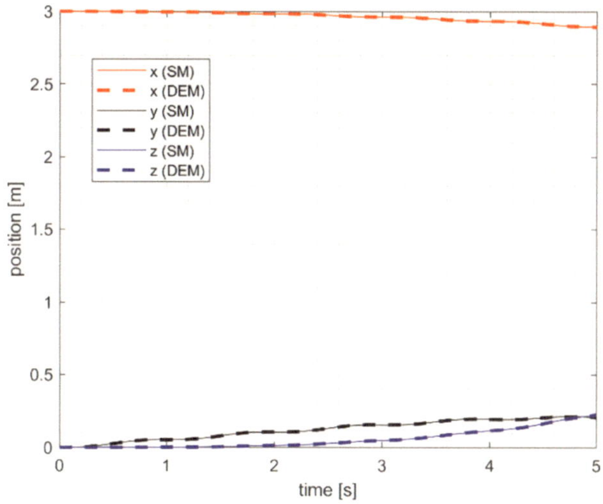

Figure 4.
Comparison of end effector positions for SM and DEM models.

manipulator's base and the corresponding first link of the DEM model have the same attitude through the entire simulation. Vector parts of quaternions representing their attitude are shown in **Figure 3(b)**. The positions in the inertial frame for both end effectors are presented in **Figure 4**. The quaternion-based DEM end effector achieves the same positions as the reference space manipulator one in the simulation run.

The numerical experiment demonstrated correctness of application of the modified quaternion-based dynamic DEM method. The base quaternion, joint angles, and the end effector positions are consistent between the reference (SM) and the developed (DEM) models. It may be concluded that the quaternion-based modified DEM is the good modeling tool, it is numerically efficient, and it is promising to be applied to study guidance algorithms, control systems, and design experimental setups for free-floating space manipulators.

5. Conclusions and future research prospects

The chapter presents a dynamics modeling method dedicated to free-floating spacecraft, specifically manipulators, based on a modified method of a dynamically equivalent manipulator. DEM enables dynamic modeling of space manipulators, e.g., free-floating maneuvers, via their suitable substitution by ground-fixed manipulator models. As a result, the space manipulator dynamics is equivalent to the ground one. This provides attractive modeling and control design tools, since it enables conducting tests and experiments for space manipulators in earth laboratories. The basic motivation for the DEM modification is to make dynamic and kinematic models suitable for description of arbitrary space manipulators maneuvers and their missions like debris removal, servicing, space mining, and on-orbit docking and assemblies. It may also support space manipulators attitude controller designs. The chapter contribution is the modification of DEM to enable space manipulator kinematic and dynamic representation in quaternions. The modified DEM method delivers a tool for conducting reliable simulation studies and tests for various maneuvers and mission scenarios for SM, and it offers a promising control design tool. The theoretical development of DEM method in quaternions is illustrated by a simulation study of a two-link space manipulator model. The space manipulator attitude dynamics has been compared to the results reported in the literature. The satisfactory results enhance the next studies to apply the quaternion-based DEM to design guidance algorithms and control systems for space manipulator missions.

Author details

Elżbieta Jarzębowska[1]* and Marcin Kłak[2]

1 Warsaw University of Technology, Power and Aeronautical Engineering, Institute of Aeronautics and Applied Mechanics, Warsaw, Poland

2 Warsaw University of Technology, Power and Aeronautical Engineering, Institute of Aeronautics and Applied Mechanics, GMV Innovating Solutions, Warsaw, Poland

*Address all correspondence to: elajarz@meil.pw.edu.pl

IntechOpen

© 2020 The Author(s). Licensee IntechOpen. Distributed under the terms of the Creative Commons Attribution - NonCommercial 4.0 License (https://creativecommons.org/licenses/by-nc/4.0/), which permits use, distribution and reproduction for non-commercial purposes, provided the original is properly cited. cc BY-NC

References

[1] Akin DL, Minsky ML, Thiel ED, Curtzman CR. Space applications of automation, robotics and machine intelligence systems (ARAMIS) phase II.NASA-CR-3734–3736. 1983

[2] Ohkami Y and Oda M. NASDA's activities in space robotics. In Proceedings of the 5th International Symposium on AI, Robotics and Automation in Space, iSAIRAS'99, June 1999, ESTEC, The Netherlands, 11–18. 1999

[3] Hirzinger G, Brunner B, Dietrich J, Heindl J. Sensor-based space robotic-ROTEX and its telerobotic features. IEEE Transactions on Robotics and Automation. 1993;**9**(5):649-663

[4] Yoshida K. Engineering test satellite VII flight experiments for space robot dynamics and control: Theories on laboratory test beds ten years ago, now in orbit. The International Journal of Robotics Research. 2003;**22**(5):321-335

[5] Ogilvie A, Allport J, Hannah M, Lymer J. Autonomous satellite servicing using the orbital express demonstration manipulator system. In: Proc. 9th International Symposium on Artificial Intelligence, Robotics and Automation in Space, Universal City, LA, USA, 26–29 February 2008. 2008

[6] Mulder T. Orbital express autonomous rendezvous and capture flight operations: Part 1 of 2—Mission description, AR&C exercises 1, 2 and 3. Advances in the Astronautical Sciences. 2008;**130**:649-651

[7] Kaiser C, Sjöberg F, Delcura JM, Eilertsen B. SMART-OLEV—An orbital life extension vehicle for servicing commercial spacecrafts in GEO. Acta Astronautica. 2008;**63**:400-410

[8] Ellery A. Tutorial review on space manipulators for space debris mitigation. Robotics. 2019;**8**:34. DOI: 10.3390/robotics8020034

[9] Liang B, Xu Y, Bergerman M. Mapping a space manipulator to a dynamically equivalent manipulator. Journal of Dynamic Systems, Measurement, and Control. 1998; **120**(1):1-7

[10] Nikravesh PE, Wehage RA, Kwon OK. Euler parameters in computational kinematics and dynamics. Part 1. Journal of Mechanisms, Transmissions, and Automation in Design. 1985;**107**(3): 358-365

[11] Xu YS, Kanade T. Space Robotics: Dynamics and Control. Boston, MA: Kluwer; 1992

[12] Papadopoulous E, Moosavian SAA. Dynamics and control of multi-arm space robots during chase and capture operations. In: Proc. IEEE International Conference on Intelligent Robots and Systems. 1994. pp. 1554-1561

[13] Jarzębowska E, Pietrak K. Constrained mechanical systems modeling and control: A free-floating space manipulator case as a multi-constrained system. Robotics and Autonomous Systems. 2014;**62**: 1353-1360

[14] Jarzębowska E, Pilarczyk B. Design of tracking controller for object interception in space. Discontinuity, Nonlinearity, and Complexity. 2017; **6**(4):435-443

[15] Jarzębowska E. Performance control of a spacecraft-robotic arm system—Desired motion tracking. In: Sąsiadek J, editor. Aerospace Robotics III. GeoPlanet: Earth and Planetary Sciences. Switzerland: Springer; 2019. ISSN 2190-5193

[16] Sola J. Quaternion kinematics for the error-state Kalman filter. arXiv preprint arXiv:1711.02508

[17] Nikravesh PE, Kwon OK, Wehage RA. Euler parameters in computational kinematics and dynamics. Part 2. Journal of Mechanisms, Transmissions, and Automation in Design. 1985;**107**(3): 366-369

[18] Baumgarte J. Stabilization of constraints and integrals of motion in dynamical systems. Computer Methods in Applied Mechanics and Engineering. 1972;**1**(1):1-16

Section 2

Control Techniques

Chapter 4

An Overview of Evolutionary Algorithms toward Spacecraft Attitude Control

Matthew A. Cooper and Brendon Smeresky

Abstract

Evolutionary algorithms can be used to solve interesting problems for aeronautical and astronautical applications, and it is a must to review the fundamentals of the most common evolutionary algorithms being used for those applications. Genetic algorithms, particle swarm optimization, firefly algorithm, ant colony optimization, artificial bee colony optimization, and the cuckoo search algorithm are presented and discussed with an emphasis on astronautical applications. In summary, the genetic algorithm and its variants can be used for a large parameter space but is more efficient in global optimization using a smaller chromosome size such that the number of parameters being optimized simultaneously is less than 1000. It is found that PID controller parameters, nonlinear parameter identification, and trajectory optimization are applications ripe for the genetic algorithm. Ant colony optimization and artificial bee colony optimization are optimization routines more suited for combinatorics, such as with trajectory optimization, path planning, scheduling, and spacecraft load bearing. Particle swarm optimization, firefly algorithm, and cuckoo search algorithms are best suited for large parameter spaces due to the decrease in computation need and function calls when compared to the genetic algorithm family of optimizers. Key areas of investigation for these social evolution algorithms are in spacecraft trajectory planning and in parameter identification.

Keywords: trajectory optimization, spacecraft control, artificial intelligence, genetic algorithm, particle swarm optimization, ant colony, artificial bee colony, cuckoo, firefly, swarm intelligence, evolutionary optimization

1. Introduction

Evolutionary algorithm use has been steadily increasing in the number of published papers corresponding to an increasing number of applications over the past 20 years [1–8]. Originating as an alternative to traditional mathematical optimization techniques, the techniques now span across almost every discipline to include data compression, traveling salesmen, image processing, and more importantly for spacecraft: control theory [9–13], system identification [14–19], and trajectory optimization [20–25]. Randomly searching a solution space to perform a global optimization routine can be computationally expensive and time consuming. Traditional methods such as stochastic parallel gradient decent, newton's method,

IntechOpen

and quadratic programming [26] are mathematical methods which rely mainly on the "steepness" of the gradients, or of the corresponding derivatives to follow the solution set to a zero-crossing gradient value and a potential optimum value. These methods are not easily implemented in discontinuous, or highly non-linear systems with a time-dependence such as in trajectory generation, and system identification of complex systems. One can perform a systematic, or random, search across an entire solution space, but the complex nature of some applications can limit the optimization to solution sets of only those within a small parameter space. Performing an exhausting search across an entire solution space can be considered the "brute force" method where the routine tries every single possible solution until the optimization criteria are met. For systems with a large parameter space with many variables, the computational cost might be too burdensome to arrive at a solution in a timely manner, and is certainly not relevant for a real-time application outside of a few cherry-picked examples. However, if a problem can be defined in an approachable way, evolutionary algorithms can provide a simpler and quicker way to find a viable solution. Due to the nature of these derivative free metaheuristic random search algorithms, the global optima may not be found but a suitable local optima that meets the optimization criteria can.

The most prominent evolutionary algorithm is the genetic algorithm officially introduced by John Holland in his 1975 book titled "Adaptation in Natural and Artificial Systems" [27] and its primary variants involving the concepts of chromosomes, elitism, parallel populations [28–30], and adaptation [31–33] which are derived from the concept of Darwinian evolution of animal species across many generations, also known as natural selection. Genetic Algorithms will be discussed more thoroughly in Section 2. The sister approach to natural selection based evolutionary algorithms are social-evolutionary algorithms also known as swarm intelligence which will be discussed in Section 3. Swarm intelligence is also a derivative free metaheuristic random search algorithm but with a slight modification on both selection criteria and on the definitions that spawn the "evolution". Swarm intelligence optimization algorithms for astronautical applications can be sub-categorized into particle optimization and combinatorics. Particle optimization includes particle swarm optimization [34], firefly algorithm [35], and cuckoo search algorithms [36] which focus on a large parameter space with a correspondingly large solution space that may be impossible to evaluate with traditionally exhaustive optimization routines. The artificial bee colony optimization [37], and ant colony optimization [38] algorithms are designed for a smaller investigation swarm but can successfully navigate a problem defined as an infinite set of combinations such as the commonly referred to problem of a traveling salesman visiting a large number of cities.

2. Genetic algorithm

Genetic algorithms have been a staple of heuristic artificial intelligence approaches since its inception in the 1960s and later more formally introduced by John Holland in 1975 [27]. In the 1990s this kind of random search global optimization routine became more mainstream through the use of greatly increased processing speeds brought on by the personal computers any more importantly GPUs. Just like other evolutionary algorithms, genetic algorithms rely on a very specific parameter space defining the population characteristics, parent selection, and success criteria.

Biesbroek presents a parametric study on the fundamental parameters within a genetic algorithm application in [39] via three cases toward spacecraft trajectory optimization. The fundamental parameters include population size, mutation probability, and cross-over probability. **Figure 1** illustrates the baseline genetic algorithm structure.

The first major step is in seeding a population. Seeding a population is done by presenting the algorithm an initial starting point to grow a population from. An alternative here is to randomly generate a population. A Gaussian distribution is one common approach. The parent population size is generally dependent on how many parameters are to be optimized via GA.

Holland described the genetic algorithm as being comprised of building blocks [27], which was later rederived by Goldberg [40] who related the population size with the quality of decisions and predicted that for an initial population of n, n^3 building blocks are potentially available in the algorithm. A rule of thumb is for m number of parameters the expected population required for convergence scales with the square root of the problem. More specifically, Harik showed that using probabilities, a more optimal population size can be described by Eq. (1) [41].

$$n = -2^{k-1} ln(\alpha) \frac{\sigma_{bb}\sqrt{\pi m}}{d} \quad (1)$$

where k is the order of building blocks, α is the probability of GA failure, σ_{bb} is the average noise expected on the quality of building blocks, m is the number of the building blocks within the parameter minus one, and d is the difference between the best and the second best fitness values. In this definition, the noise is a description of how the building blocks create a member via the genetic evolution of the algorithm and how the resulting combinations may interfere in finding the optimal solution. In other words, the parent population can create noise hindering the convergence efficiency. With the expectation on the population size that it scales with $\sqrt{m}$, it can be seen how quickly the required population and therefore computational need increases with the number of parameters. A more recently derived rule of thumb is that for "small" parameter sets, the population size is effective if scaled with the number of parameters with 10 m, and for larger spaces the population size scales with $ln(m)$ [42], where the definition of large is different for each author. For a simple parameter set GA can be quite effective for optimization

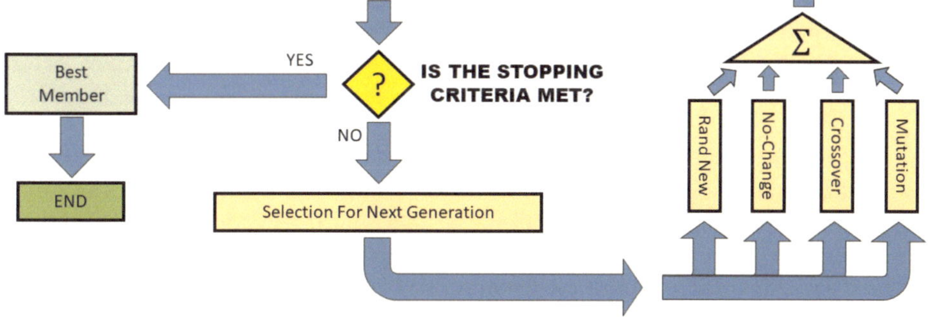

Figure 1.
Generic genetic algorithm flow diagram.

problems in many space based applications [23, 24, 28, 43–47] which stem from aeronautical control [20, 22], and ground-based robotic systems [48–50].

Looking again at **Figure 1**, the next step is to create children. Children are the result of statistical combinatorics of parents, and both of the mutation and cross-over of that parent population. The initial parent population will be evaluated and scored based on an objective function. The objective function is entirely user defined for the specific application. The objective function could be described with respect to minimizing the control effort needed to achieve a specified trajectory such as described in [51], or in minimizing the error between the actual and the desired trajectory in an attitude control scenario as presented in [52]. The objective function is the key component of any optimization and it is crucial to define it in such away as to minimize (or maximize) said function efficiently and precisely. This may seem obvious but the optimization routine will focus on the weighted parameter space defined by the objective function. If a control variable is not fully observable in the prescribed control law for a system, for example, the optimization may never converge to a viable solution set.

The objective function will be used to rate the performance of each member of the parent population. The population will then be ranked based on the threshold parameter specified in the algorithm. Different variations of GAs focus on certain threshold schemes to achieve faster convergence in various applications. In the basic GA scenario, those members who performed well enough to score below the threshold value (for a minimization problem) will form the parent population for the next generation. Additionally, a random subset of the original parent population will remain unmodified. Through mutation or cross-over based on mutation probability and cross-over probability. These parameters will define the statistical probability that a member of the population will be chosen for mutation or crossover. These probabilities typically start with a higher value and continually decrease on each subsequent generation to encourage the population to converge nicely to an optimal value. If a member is chosen for *mutation*, in this context, that will mean that a randomly chosen set of parameters within that member (if the parameter space is larger than one) will be adjusted via a Gaussian distribution function such that the amplitude of the specific parameter will have a statistical mean at the current value of the parameter. In short, a mutated member of the population will only have some parameters altered in value, not its entire chromosome, or set of parameters to optimize.

Cross-over is the next primary method by which the GA, alternatively described as a non-stochastic optimization approach, is taken. Non-stochastic optimization leads into the burgeoning field of deterministic artificial intelligence which entails more than there is space available in this chapter to discuss, and the reader is encouraged to review the following work compiled by Sands [53]. A cross-over is created by taking two parents, and through a predefined or randomly selected crossover point, they will be split and recombine as shown in **Figure 2** where Parent 1 and Parent 2 will now become Child 12 and Child 21 in the new population. After each member of the population is statistically chosen to be either modified or left alone it now is considered the children population. This new population is evaluated through the given objective function and the results become the segregated parent population for a new generation of possibilities. This cycle of parent-children function evaluations repeats until exit criteria are met. Common exit criteria may be to stop after a given number of generations have been evaluated, if the delta between the best performers across multiple generations shows no improvement, or ideally if the best performer of the current generation has met the performance objective desired. In this fashion, the population evolves over time via mutation and cross-over until an optimal solution is reached. In the best case, the optimal solution is

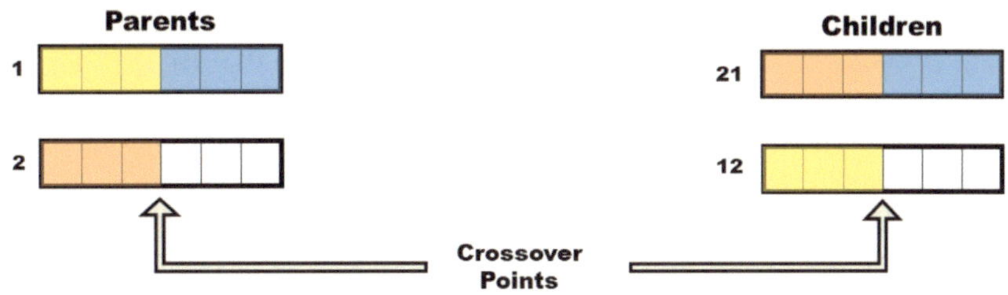

Figure 2.
Generic genetic algorithm crossover illustration.

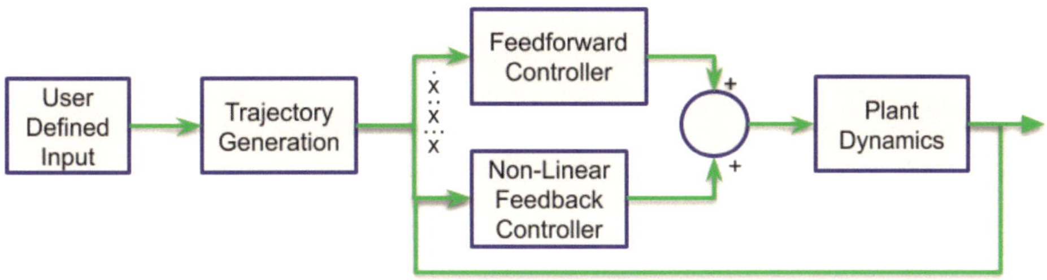

Figure 3.
Sample control flow diagram using the non-linear Van der pol equation as a system under test.

global, but is often local with the hyper-dimensional solution space too large to confirm one way or the other.

In the flow of the genetic algorithm framework, the algorithm is said to have "worked" when it reaches convergence. Convergence here is defined such that if the fitness of the entire population is decreasing (not counting stall generations where fitness is not improved) toward a global minimum, and that on the last generation the majority of the population has very similar fitness. An example is presented in **Figure 3** which illustrates a simple problem of tailoring the input of a trajectory system utilizing the dynamics based on the forced Van der Pol equation shown in Eq. (2). In this simple control example illustrated in blue and green, some arbitrary steady state value is the input to the system. It is then converted to a desired trajectory and sent to the feedforward controller. The feedforward controller builds a desired torque which is then sent as the control signal to the system dynamics, also know as the "plant". In an ideal situation, the equation used to determine the required control torque is perfectly understood such that the system is perfectly controllable. Here, the Van der Pol system converges to a desired state but only after through a large amount of transient states which can be detrimental to the physical stability of a mechanical system.

$$\frac{\partial^2 x}{\partial t^2} - \mu\left(1 - x^2\right)\frac{\partial x}{\partial t} + x = F(t) \tag{2}$$

$$K_i(x_m - x_d) + K_d(\ddot{x}_m - \ddot{x}_d) + \left(K_i(x_m - x_d)^2 - 1\right)K_p(\dot{x}_m - \dot{x}_d) \tag{3}$$

This control system breaks the desired input into three components, x, $\dot{x}$, and $\ddot{x}$ which represents angle, angular velocity, and angular acceleration, and are used to feed the feedforward controller of the system in order to prescribe the best torque command to the system dynamics described in Eq. (2). Using a non-linear PID feedback control scheme as seen in Eq. (3) where K_i, K_d, K_p are the integral, derivative, and proportional gains, and x_d, and x_m are the desired and measured output values respectively. Tuned by hand, one can achieve an RMS-error between the desired and measured circular trajectory in the phase-plane of this simulated system at $0.1767deg$. The x and $\dot{x}$ components are shown in **Figure 4a**. This type of highly non-linear feedforward – PID feedback system is challenging to classically tune therefore a GA was investigated. In this scenario the objective function is set to minimize the RMS-error between the desired x_d and $\dot{x}_d$ trajectory when compared to the actual trajectory x_m and $\dot{x}_m$. The objective function can be seen in Eq. (4). Using a GA with the identified parameters in **Table 1** an RMS-error of $0.1060deg$ is achieved representing a 40% reduction in error.

$$RMSe = \sqrt{\frac{1}{n}\sum_{i=1}^{n}\left(\begin{bmatrix} x_d \\ \dot{x}_d) \\ \ddot{x}_d \end{bmatrix} - \begin{bmatrix} x_m \\ \dot{x}_m) \\ \ddot{x}_m \end{bmatrix}\right)^2} \tag{4}$$

The reduction in RMS error is illustrated in **Figure 4b**. The dotted line is the desired trajectory whereas the solid line is the actual trajectory achieved within the

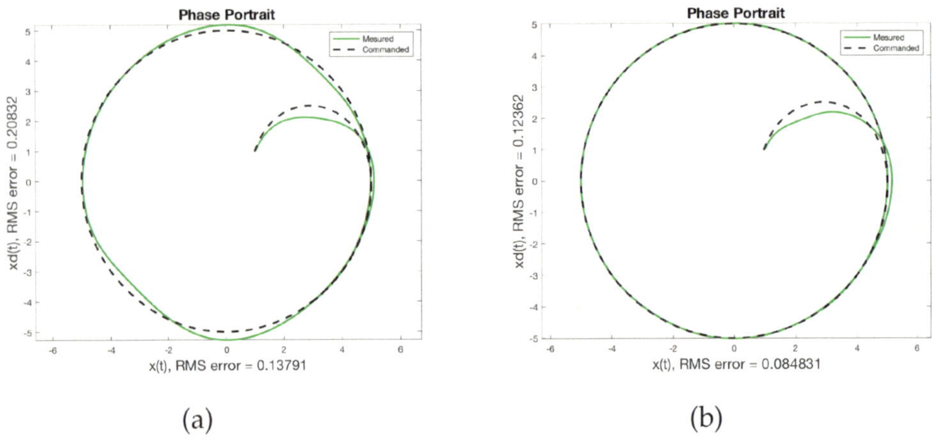

Figure 4.
Trajectory results using classically tuned PID feedback control on a highly non-linear system versus the results using a genetic algorithm tuned to minimize the RMS error. A 40% reduction is RMS-error is achieved. (a) Trajectory results using classically tuned PID controller, and (b) Trajectory Results using a GA tuned PID controller.

Parameter	Value	Result
Population size	200	Function call per generation
Population	3	Number of parameters to optimize
Mutation rate	10%	Probability to be selected for crossover
Crossover rate	80%	Probability to be selected for crossover
Lower bound a	[0,0,0]	Minimum values allowed in population
Upper bound a	[1000,1000,1000]	Maximum values allowed in population
Selection criteria	5% Elite	Choose the top 5% of population as parents
Max Stall generations	20	Exit criteria
Initial population range[a]	[0,20]	Initial population bounds

[a]*Constraints to the genetic algorithm.*

Table 1.
Genetic algorithm parameters for a highly nonlinear trajectory optimization of a Van der Pol system using PID feedback control.

phase plane of the system. The phase plane plots the angular position vs. the angular velocity. The phase plane plot can be used to monitor trajectory tracking when you are interested in more that one state in addition to highlighting any potential instability in the system.

With this example, the rate of convergence can be illustrated and examined. **Figure 5** highlights the fact that the GA implementation presented here required 8800 function calls, and 43 generations to converge within the exit criteria at 20 stall generations. **Figure 5** also implies that reasonable performance was achieved after only 10 generations.

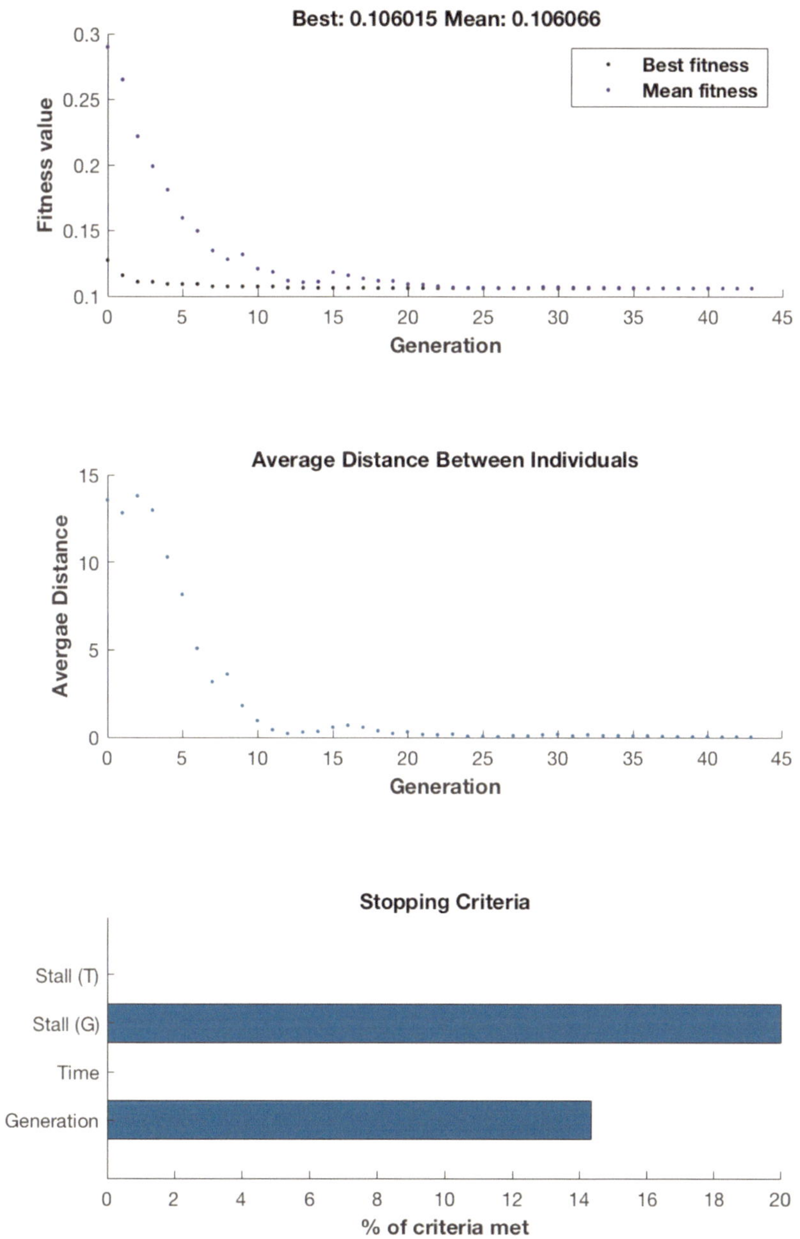

Figure 5.
Results of the genetic algorithm as it steps through each generation in optimizing a PID control variables.

Let us look again at the paper presented by Biesbroek [39], they look at a parameter space of only two in the application of optimizing the trajectory of a rocket such that a maximum horizontal distance, x, is achieved. The specific parameters are $V\left(\frac{m}{s}\right)$ & $m(kg)$, where V = velocity in meters per second, and m = mass in kilograms. The equation of motion for the dynamics of the rocket is defined in Eq. (5) along with the deferential equation to define the range x with respect to V, and m.

$$\dot{x} = V, \quad \dot{V} = \frac{T - D}{m} = \frac{V_E\beta - kV^2}{m}, \quad dx = -\frac{m}{kV}dV - \frac{V_E}{kV}dm \tag{5}$$

where T is thrust, D is the drag, V_E is the exhaust velocity, β is the mass ratio, and k is a constant. In this instance, the goal of the objective function is simply to find the values for V and m based on the equations of motion for the rocket and the differential equation defining the range. Using the equations defined in Eq. (5), it is then rearranged via Green's Theorem into the following objective function f in Eq. (6). Depending on the author, an objective function may also be referred to a fitness function when describing how well the converged solution "fits" the desired result via an error equation. A common approach is to use the root-mean-squared error (RMSe) between the resulting solution and the desired solution.

$$f = \sum_{k=1}^{n}\left[\frac{ln(V_k) - ln(V_{max,k})}{k} - \frac{V_E}{k}\right]\delta m + 7686.722 \tag{6}$$

Here, V_E is again the exhaust velocity, $V_{max,k}$ is the maximum velocity, and δm is the change in mass as the propellant is used. In this case δm and $V_{max,k}$ are the constraints that bound the solution space. Running through varying populations, cross-over and mutation populations lead to the quickest convergence with a population size of only 10,343 generations, a cross-over rate at 10% and mutation rate of only 1%. The optimal solution for two parameter problem presented can be calculated analytically to be equal to a distance of 9839 km. The most efficient GA solution, in terms of number of fitness function evaluations and therefore computational speed, required 3421 fitness function calculations and came to the correct solution such that the rocket should start with a full thrust until no more propellant is available, followed by a coasting arc to achieve maximum horizontal distance.

Another area of interest is in path planning. Jia presents a parallel evolutionary algorithm solution for real-time path planning for unmanned aerial vehicles (UAVs) in [29]. The Path planning problem for UAVs start with an initial point x_0, then the UAV needs to visit a series of waypoints, or stops, along its route, avoids potential dangers, or no-fly zones, before traveling back to x_0 or to another final landing zone as illustrated in **Figure 6**. In this scenario, a cost function might be described in such a fashion as to relate the fuel cost, a penalty of losing the UAV to a no-fly zone or a crash, and a reward for finishing the mission. The concept here takes a traditional genetic algorithm approach but modifies it through the use of competitive parallel generations. Each generation is evaluated through the objective function, but only the population with the best fitness value lives on whereas the other populations are re-initialized for the next generation. In the case of populations with similar peak fitness values, the population with the best overall fitness is chosen. The non-deterministic nature of path-planning presents an interesting challenge. The variability can stem from weather conditions, probabilities of danger to the UAV, and probabilities of failure modes of the UAV itself. The goal of using parallel competing populations is to mitigate the possibility of premature

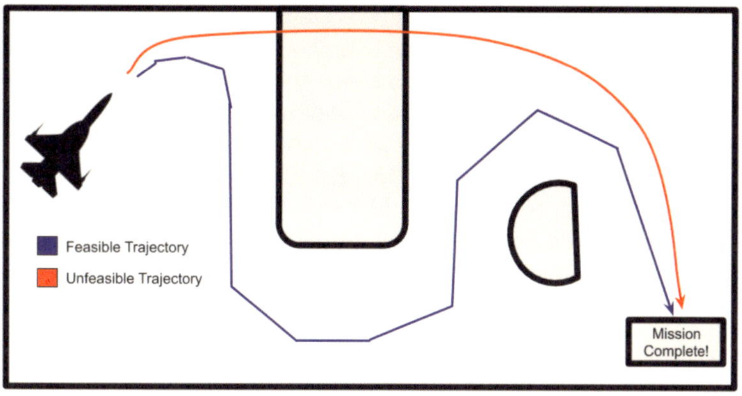

Figure 6.
Simple illustration of a UAV path plan with no-fly zones in gray.

convergence to a local minima. If the population with the best fitness stagnates in evolution to a local minima, the alternative parallel populations can potentially surpass the aforementioned population in fitness and therefore the stagnate population will become extinct and be re-initialized. This paper finalizes by highlighting the improvement of using two parallel evolutionary algorithm over a single in the ability to achieve a global minimum by 50% from 17%.

Taking the path planning application one step further leads us to the interplanetary path planning regime. Gage from Stanford University presented a novel paper in 1994 on interplanetary trajectory optimization using a genetic algorithm [43]. The preliminary design space for interplanetary spacecraft missions are highly complex, discontinuous, and anything beyond a two-body problem is primarily solved numerically. This initial investigation was on the viability of using a Genetic Algorithm as part of the team's suite of search methods toward finding viable interplanetary trajectories. It was shown that the computational need was reduced by almost four times from the more common grid-search method used up through the 1990s for designing interplanetary trajectories. The keys to the performance improvement were the constraints applied to objective functions which greatly penalized infeasible trajectories resultant from the parent-children, and to also artificially degrade the fitness of population members that were in close proximity within the GA's search space. This paper also noted that it can be helpful to restrict "mating" between parents with a separation $\leq \sigma$, where σ is a user-defined threshold distance between potential parents within the same search space. This can ensure that parents that are circling around two different local minimums (not yet decided if one is a global minimum) do not mate and produce offspring that result near neither of the two local minimums, and thus are unlikely to increase fitness. This restriction in mating can lead to multiple possible solutions sets evolving through the generations and ensure that a single strong local minimum does not dominate the evolution.

3. Swarm intelligence

The next major subset of evolutionary algorithms relies on the assumption of swarm intelligence and social evolution in time whereas the genetic algorithms previously discussed depends upon genetic evolution of the populations; which is metaheuristic in nature and is derived from the biological (and probabilistic) mechanisms describing the movement of swarms of birds, fish, and insects on their search for food or mates. Referring to this concept as metaheuristic means that these algorithms are high-level symbolic based approaches designed to utilize imperfect or incomplete data in order to identify or approximate a sufficient solution to the given optimization problems. Swarm intelligence algorithms are based on the simple individual actions of the swarm which can collectively be quite complex and result in self-organization, decentralization and cooperation utilizing what is also referred to as collective intelligence. The primary swarm intelligence algorithms to be discussed are particle swarm optimization (PSO), cuckoo search algorithm (CSA), firefly search algorithm (FA), ant colony optimization search algorithm (ACO), and artificial bee colony algorithm (ABC).

3.1 Particle swarm optimization

Particle swarm optimization (PSO) was first introduced by Kennedy and Eberhart in 1995 [34] as a data clustering algorithm [1], and follows a population-based evolutionary social algorithm [3] along side of what can be considered an individualized random search algorithm [54]. **Figure** 7 outlines the overarching procedure. Initially, the algorithm is seeded with a uniformly random distribution, which is called particles. These particles will result in many different values within the search space of the system of possibilities, and again, the individual particle performance is evaluated though an objective function just like with other optimization algorithms. In the general form of PSO, the algorithm utilizes a global topology which defines how the swarm communicates with each other. Utilizing the above mentioned communication, the swarm is aware of all other particle actions, success, and current velocity. Each particle's position within the search space (searching for the optimum value) is calculated with a corresponding velocity as defined by Eq. (7).

$$v_n(t) = \omega \times v_n(t-1) + a_1 r_1 \left(p_{B_n}(t-1) - x_n(t-1)\right) + a_2 r_2 \left(g_{B_n}(t-1) - x_n(t-1)\right) \tag{7}$$

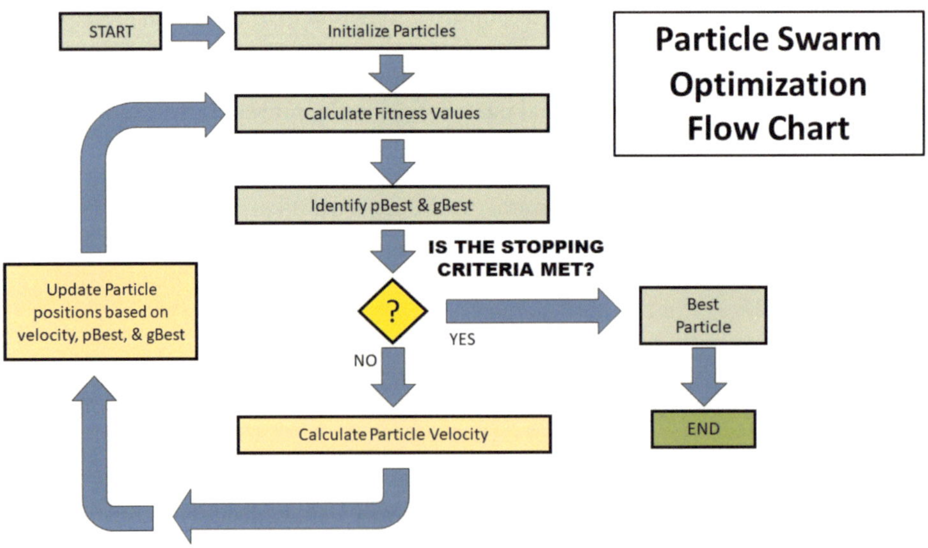

Figure 7.
Particle swarm optimization flow chart.

where v_n is the particle velocity in the next time step, a_1 and a_2 are constant weights on the effect due to social evolution, p_{B_n} is the individual particle's best result so far, g_{B_n} is the population's best result so far, ω is the particle's inertia, r_1 and r_2 are random numbers normally distributed from 0 to 1 to keep the random nature in the algorithm. Initially, the particles are generated with a small initial velocity and then grow as independent individuals within the swarm. PSO relies on the best particle solution position p_{B_n}, the best group solution position g_{B_n}, and the current particle velocity to calculate the particles next position in the solution space. Within that new particle position the objection function is then evaluated for fitness, along with all other particles in the population, and new p_{B_n} and g_{B_n} are calculated for the next iteration. The goal of particle swarm optimization algorithm is for all the particles to converge within the hyper-dimensional parameter space to the optimum value. With enough iterations, and a correspondingly well defined fitness function and PSO parameters, the algorithm will converge to global optima as illustrated in **Figure 8**. The algorithms parameters must be chosen such that the algorithm is balanced between two competing search notions: the particles exploring the unknown areas of the search space, and the particles exploiting the known areas of the search space in order to prevent premature convergence (if too heavily focused on exploiting) and non-convergence (if too heavily focused on exploring).

Alternatives and modifications to the PSO algorithm can include constrained velocities such as a minimum or maximum value (such that it will not grow unbounded), local biases defined by Euclidean distance between particles to define neighborhoods in order to prevent two sets of parents from different neighborhoods to attract to each other and reduce the overall fitness, and penalties for leaving the desired search space. Additionally, PSO can be hybridized with other approaches to utilize the lower computational cost of PSO but to decrease the randomness of the search if a general solution space is already known [55].

3.2 Firefly search algorithm

The firefly search algorithm (FA) is an interesting optimization technique based on the behavior of tropical fireflies who flash their abdomens with a bioluminescence chemical in order to both attract mates, and in some species to lure in prey such as the male of competing firefly species. With this behavior there are a few components that can lend itself toward the development of an interesting swarm optimization routine. Their light flashing pattern and intensity is also affected by their desire for mates or food, along with the distance another source is from the observing firefly. The definition on describing this pattern for what specifically

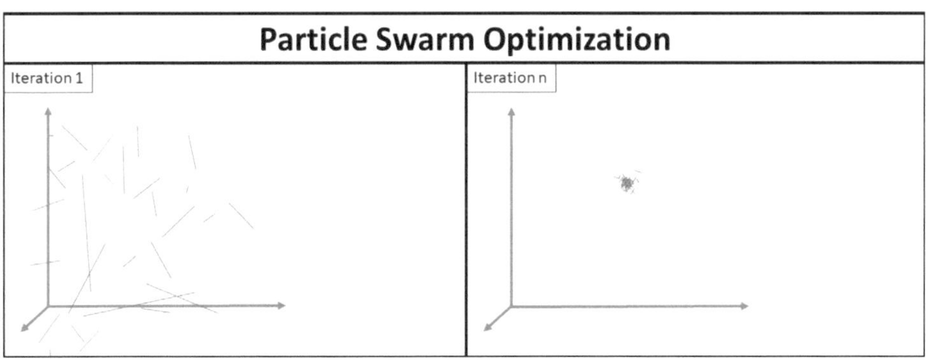

Figure 8.
Particle swarm optimization iteration illustration.

prompts the firefly to signal and how they signal there light is still unknown. But taking some of these concepts, an optimization routine can be developed. With this concept, the firefly algorithm was introduced by Yang [35]. The light intensity is a function of distance through the environment, which can be described by Eq. (8).

$$I_n = I_s e^{-\gamma r_{ns}^2}, \quad r_{ns} = \sqrt{(x_n - x_s)^2 + (y_n - y_s)^2}, \quad \beta \propto I_s e^{-\gamma r_{ns}^2} \tag{8}$$

where I_n is the intensity seen by the nth firefly which is considered to be the observer, I_s is the intensity of a nearby firefly also known as the source, and r_{ns} is the Euclidean distance between the nth firefly and the nearby firefly. With this equation an attractiveness calculation represented by β can be defined such that $I_n \propto \beta$ also shown in Eq. (8).

The FA routine is shown in **Figure 9**. This algorithm depends on an objective function just like the previously mentioned optimization algorithms. This objective function will evaluate the fitness of the fireflies, where the fitness will also determine the light intensity value at each firefly. An initial population of fireflies is generated and evaluated through the objective function. At this point the calculation for each firefly will be calculated to evaluate the attractiveness to its nearest neighbors. The attractiveness will affect the firefly's next step as shown in Eq. (9) [35].

$$x_{n_{t+1}} = x_{n_t} + \beta(x_n - x_s) + \varepsilon_n \tag{9}$$

where x_{n_t} is the current position of firefly n and $x_{n_{t+1}}$ will be the new position of firefly n. β is the attractiveness of firefly n to firefly s weighted by the distance $(x_n - x_s)$. ε_n is a uniform distribution random number to facilitate a random walk behavior toward the more attractive mates. Starting from a relatively uniform population density across the solution search space, the fireflies will iteratively converge to the nearest minima location as seen in **Figure 10**.

The firefly search algorithm and its variants have been applied to trajectory optimization [56–58], control parameter optimization [59, 60], and dynamics

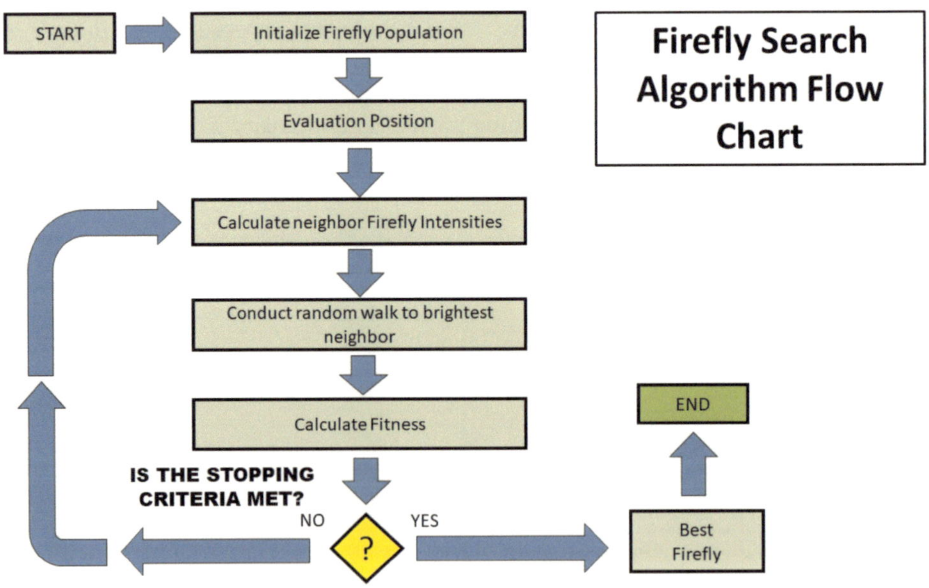

Figure 9.
Firefly search algorithm flow chart.

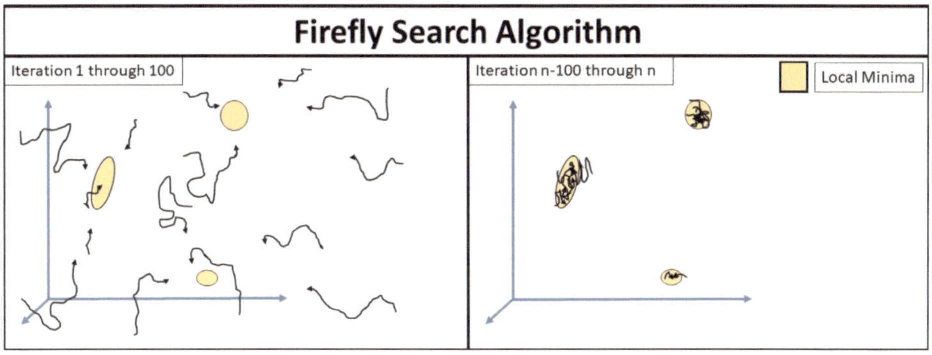

Figure 10.
Firefly search algorithm iteration illustration.

[61–63] in what can be considered as an introductory investigation by looking for initial successes in applying the FA toward these astronautical research areas.

3.3 Ant colony optimization search algorithm

The ant colony optimization (ACO) algorithm is another approach based off of the swarm behavior of insects, and was originally designed for combinatorial problems like the traveling salesman problem (TSP) where one tries to find the minimum path for an individual to traverse across n number of cities. In the instance of the traveling salesman, the salesman has a seemingly infinite number of combinations (but not really) to try but only wants to travel the shortest path, and to not repeat any stops along the way. The basic ant system, an earlier version of ACO, was presented by Dorigo in 1992 in his PhD thesis [38]. He presented a complex optimization algorithm based on the simultaneously simple and complex nature of foraging ants that would gain a large interest in the 1990s and later. Many variants and hybrids were presented by Dorigo and others. Most notably, offline pheromone updates, and pheromone evaporation was introduced which led to the more common ACO in 1999 [64, 65]. The concept of pheromones will be discussed shortly.

Figure 11 illustrates at a high-level the flow of the ant colony optimization routine. The algorithm is initiated with a given set of parameters and objective function. Next, an initial set of solutions is populated. This is the first round of traveling ants looking for an optimal solution. The given problem is defined and broken apart such that the optimization routine will look for the optimal set of these building blocks in order to minimize the objective functions, or distance in the realm of the traveling salesman. At this point the pheromone level will be calculated. The pheromone is laid out such that each ant lays the same amount of pheromone out on the path that it traverses. This pheromone makes the links between different combinatorial building blocks attractive. If more than one ant traverses a similar segment the pheromone level will increase on that path segment, with an end result of the most common path being the most attractive.

With the most traversed paths being the most attractive, one may notice that there could be a strong potential for the algorithm to get "stuck" in a local minima. The ACO algorithm includes what is know as a local pheromone update, which means that either only the last segment of a successful path will include the pheromone, or that the end segment will be delivered a heavier weight of pheromone by the one ant who achieved the best path in the current iteration. Alternatively, the best path so far (out of all the iterations) may receive that additional pheromone instead. At this point the solution space is evaluated to see if a viable global solution

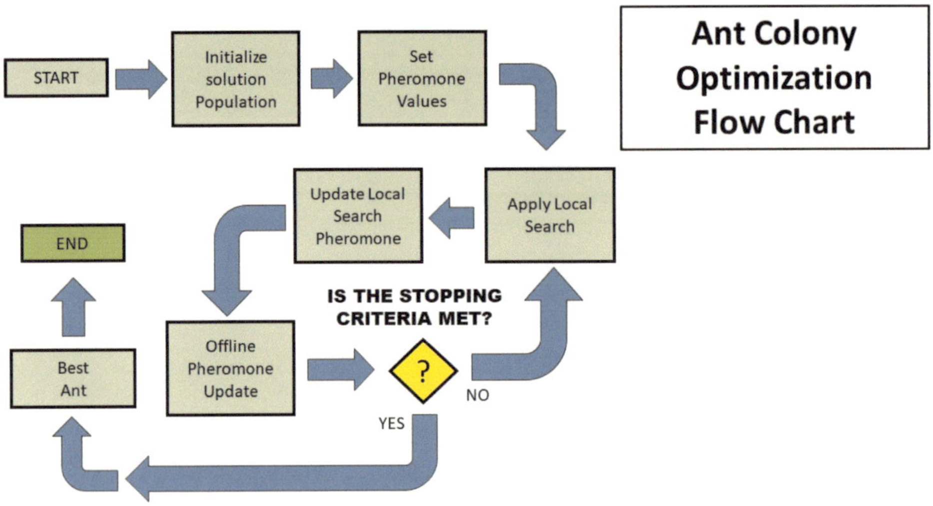

Figure 11.
Ant colony optimization search algorithm flow chart.

was found. If not, then the current ants will go through an exponential pheromone decrease before starting the next iteration in the optimization process. This pheromone decay reduces the impact of hysteresis on the solution space and helps prevent premature convergence.

On the following iteration the current ant population will then create a new set of paths to achieve a solution in the population. Each link will follow pseudorandom proportional rule, which uses a uniform distribution probability weighted by the segment pheromone values to decide on each link in the path. Once all ants create the new population of solutions another iteration of pheromone decisions will follow. This entire process will continue until convergence criteria are met.

Ant colony optimization has been successfully used for problem sets that can be discretized into a combinatory problem such as in path planning [66–68], trajectory optimization [21, 69], and even in spacecraft load bearing [70].

3.4 Artificial bee colony algorithm

Artificial bee colony (ABC) optimization is a derivative of both the bees system presented in 1997 by Sato and Hagiwara and the bee colony optimization by Teodorovic and Dell in 2005 [4], and was introduced later in 2005 by Karaboga [37]. The underlying algorithm flow chart is illustrated in **Figure 12** and involves three types of bees: onlookers, employed bees, and scouts. At the start of the algorithm, the routine parameters are initialized, and an initial population of food sources is generated via a uniform random distribution across the solution space. The population of food sources is discovered by employed bees and the quality of the food source is evaluated via an application specific fitness function. The employed bees then randomly search for a new food source, and if that food source is of better quality, then it becomes the primary food source. If not, then the new food source is abandoned. This is called a greedy search. Meanwhile, the onlooker bees observe the actions of the employed bees, and observe their communication dance through the lens of a randomly distributed variable. This represents the decision tree on which employed bee an onlooker will follow, or if it will create a new search. If the onlooker searches for a new food source, it will choose based on a random recombination of two solutions nearby. When a food source is not picked up by the onlookers when they transition to the employed bees, that food source is

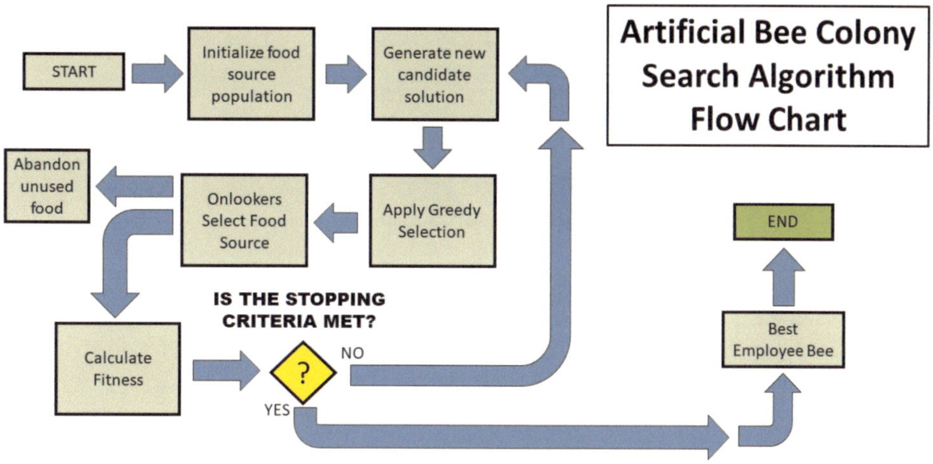

Figure 12.
Artificial bee colony algorithm flow chart.

now considered to be abandoned and will not be a part of the known current solution space. The next step is to repeat the search for new sources, a greedy selection, and the corresponding employed bee dance to randomly attract an onlooker to the known food source.

Similar to the ant colony optimization algorithm, the artificial bee colony algorithm is primarily designed for a combinatorics based problem where the potential solution can be discretized into an array of building blocks that can be rearranged and mutated to find an optimum solution. Additionally, the ABC algorithm has bee used for trajectory optimization [71], parameter optimization [72, 73], and remote sensing applications [74, 75].

3.5 Cuckoo-search algorithm

Another metaheuristic search algorithm is introduced with the cuckoo search algorithm (CSA). This approach mimics the parasitic brood behavior in certain species of the cuckoo bird. This type of bird has a fascinatingly aggressive reproduction strategy which is the heart of the CSA. Quite a few species of cuckoos participate in the obligate brood parasitism which means that they will lay their eggs in the nests of other birds [36]. The key to the success of the individual cuckoo is dependent on the cuckoo's ability to produce an egg that is able to mimic, or approximate, the host nest eggs such that the host nest mother bird can not distinguish the cuckoo egg from her own. The cuckoo egg will hatch before the host eggs, and ensure dominance in the nest and thus prolong their survival.

Taking this concept to an optimization algorithm it can be illustrated as seen in **Figure 13**. Step one is to initialize the algorithm and generate a population of host nests. Each host nest has a single potential solution to be compared against. The next step is to evaluate the initial population of nests with the defined fitness function. Once the fitness function is evaluated for each cuckoo egg, the next step is for the cuckoo to take flight via a classical Lvy flight path [36]. This is a type of step pattern is a heavy-tailed random walk similar to that of a fruit fly, which are observed to jolt out in a straight direction then randomly turn a sharp turn at a random angle. The desired behavior is described in Eq. (10).

$$x_{t+1} = x_t + sE_t \quad (10)$$

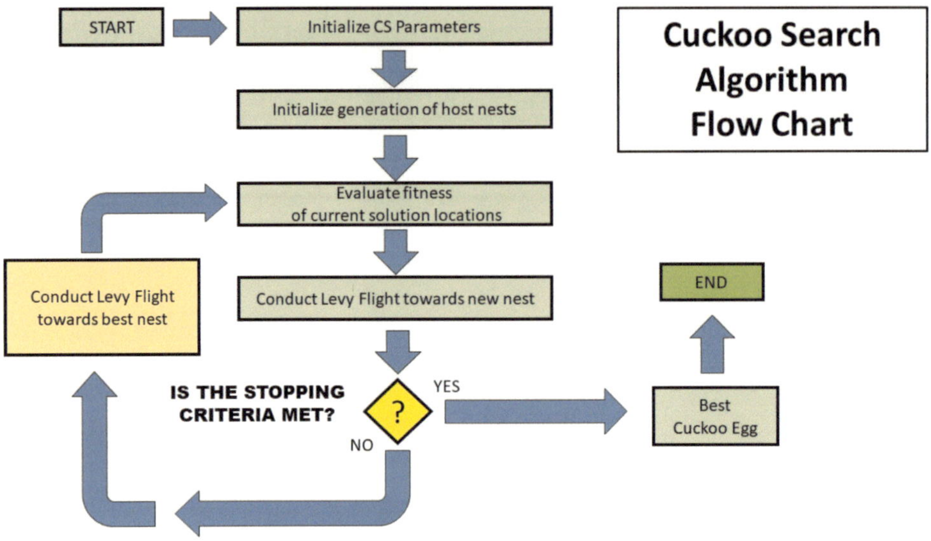

Figure 13.
Cuckoo-search algorithm flow chart.

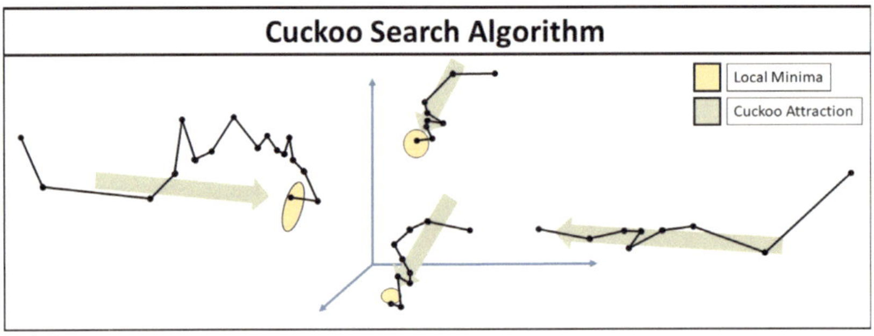

Figure 14.
Cuckoo-search algorithm iteration illustration.

where x_{t+1} is the position in the next iteration, s is the step size, and E_t is a zero-mean random Gaussian draw to mimic Lvy flights. Once each cuckoo take a flight to a "new" nest, the fitness function is then calculated again to evaluate the new population of potential solutions. If the solution criteria are not met, then this process of flying from nest to nest will continue until convergence as illustrated in **Figure 14**. Each individual will take a random walk biases by the direction of the current best solution within the group.

Yang's and Deb's seminal work on the CSA illustrated an enormous computational cost savings when compared to the genetic algorithm and the particle swarm optimization Algorithm as each algorithm was used to find the solution to a handful of standard challenging mathematical functions: Michalewiczs, Rosenbrocks, Schwefels, Rastrigins, with a 96, 89, 96, and 91% decrease in the number of required fitness function evaluations when compared to a GA solution respectively [36].

4. Conclusions

With a prevalence of evolutionary algorithms focused on solving trajectory generation, path planning, remote sensing, control theory, and parameter identification for aeronautical and astronautical applications it is a must to review the fundamentals of the most common evolutionary algorithms being used for those applications. Genetic algorithm, particle swarm optimization, firefly algorithm, ant colony optimization, artificial bee colony optimization, and the cuckoo search algorithm are presented and discussed with an emphasis on astronautical applications. In summary, the genetic algorithm and its variants can be used for a large parameter space but is more efficient in investigating a smaller parameter space, less than 1000 parameters in the chromosome. It is found that PID controller parameters, nonlinear parameter identification, and trajectory optimization are applications ripe for the genetic algorithm. Ant colony optimization, and artificial bee colony optimization are optimization routines more suited for combinatorics, such as with trajectory optimization, path planning, scheduling, and spacecraft load bearing. Particle swarm optimization, firefly algorithm, and cuckoo search algorithms are best suited for large parameter spaces due to the decrease in computation need and function calls when compared to the genetic algorithm family of optimizers. Key areas of investigation for these social evolution algorithms are in spacecraft trajectory planning, and in parameter identification.

Evolutionary algorithms have been shown to have a great potential to solve challenging problems that traditional optimization routines may not be able to tackle due to large computational need to support an exhaustive search of the solution space. The reader should now have the tools to take the foundational material presented here, to review the referenced sources, and conduct their own deep dive in an application area of interest.

Disclaimer

The views expressed in this paper are those of the author and do not reflect the official policy or position of the United States Air Force, Department of Defense, or U.S. Government.

Author details

Matthew A. Cooper[1*] and Brendon Smeresky[2]

1 Air Force Research Laboratory, Albuquerque, New Mexico, USA

2 Naval Postgraduate School, Monterey, California, USA

*Address all correspondence to: matthew.cooper.17@us.af.mil

IntechOpen

© 2020 The Author(s). Licensee IntechOpen. Distributed under the terms of the Creative Commons Attribution - NonCommercial 4.0 License (https://creativecommons.org/licenses/by-nc/4.0/), which permits use, distribution and reproduction for non-commercial purposes, provided the original is properly cited.

References

[1] Alam S, Dobbie G, Koh YS, Riddle P, Rehman SU. Research on particle swarm optimization based clustering: A systematic review of literature and techniques. Swarm and Evolutionary Computation. 2014;**17**:113

[2] Drugan MM. Reinforcement learning versus evolutionary computation: A survey on hybrid algorithms. Swarm and Evolutionary Computation. 2019; **44**:228-246

[3] Ekici B, Cubukcuoglu C, Turrin M, Sariyildiz IS. Performative computational architecture using swarm and evolutionary optimisation: A review. Building and Environment. 2019;**147**:356-371

[4] Rajasekhar A, Lynn N, Das S, Suganthan P. Computing with the collective intelligence of honey bees a survey. Swarm and Evolutionary Computation. 2017;**32**:2548

[5] Shehab M, Khader AT, Al-Betar MA. A survey on applications and variants of the cuckoo search algorithm. Applied Soft Computing. 2017;**61**: 1041-1059

[6] Betts JT. Survey of numerical methods for trajectory optimization. Journal of Guidance, Control, and Dynamics. Mar 1998;**21**(2):193-207

[7] Gotmare A, Bhattacharjee SS, Patidar R, George NV. Swarm and evolutionary computing algorithms for system identification and filter design: A comprehensive review. Swarm and Evolutionary Computation. 2017;**32**: 6884

[8] Ball JE, Anderson DT, Chan CS. Comprehensive survey of deep learning in remote sensing: Theories, tools, and challenges for the community. Journal of Applied Remote Sensing. 2017; **11**(04):1

[9] Ahmed R, Chaal H, Gu D-W. Spacecraft controller tuning using particle swarm optimization. Proceedings of 2009 ICCAS-SICE. 2009. DOI: 10.1016/j.apm.2017.07.024

[10] Laware A, Talange D, Bandal V. Evolutionary optimization of sliding mode controller for level control system. ISA Transactions. 2018;**83**:199213

[11] Ghiglino P, Forshaw JL, Lappas VJ. Online PID self-tuning using an evolutionary swarm algorithm with experimental quadrotor flight results. AIAA Guidance, Navigation, and Control (GNC) Conference. 2013

[12] Gazizov RR, Zabolotsky AM, Gazizov RR. Use of genetic algorithm and evolution strategy when revealing the worst case effects of crosstalk propagation in PCB bus of spacecraft autonomous navigation system. In: Proceedings of 19th International Conference of Young Specialists on Micro/Nanotechnologies and Electron Devices (EDM). 2018. DOI: 10.1109/ EDM.2018.8435093

[13] Ge X-S, Chen L-Q, Liu YZ. Attitude control of underactuated spacecraft through flywheels motion using genetic algorithm with wavelet approximation. In: Proceedings of Fifth World Congress on Intelligent Control and Automation. 2004. DOI: 10.1109/EDM.2018.8435093

[14] Iwasaki NMM. Evolutionary identification algorithm for unknown structured mechatronics systems using GA. In: Proceedings of 2000 26th Annual Conference of the IEEE Industrial Electronics Society. 2000. DOI: 10.1109/IECON.2000.972388

[15] Wakizono M, Hatanaka T, Uosaki K. Time varying system identification with immune based evolutionary computation. In: Proceedings of 2006 SICE-ICASE International Joint

Conference. 2006. DOI: 10.1109/SICE.2006.315098

[16] Kawada K, Yamamoto T. Evolutionary identification using closed-loop data for a mechanical system. Proceedings of SICE Annual Conference 2010. 2010

[17] Ghiglino P, Forshaw JL, Lappas VJ. Online evolutionary swarm algorithm for selftuning unmanned flight control laws. Journal of Guidance, Control, and Dynamics. 2015;**38**(4):772-782

[18] Whorton M. Closed loop system identification with genetic algorithms.

[19] Jiang Q, Zhang J. Fuzzy multi-objective evolutionary algorithm based structure identification of polynomial systems. In: Proceedings of the 33rd Chinese Control Conference. July 2014. pp. 6571-6576

[20] Mondoloni S. A genetic algorithm for determining optimal flight trajectories. In: Guidance, Navigation, and Control Conference and Exhibit. 1998

[21] Murrieta-Mendoza A, Hamy A, Botez RM. Four- and three-dimensional aircraft reference trajectory optimization inspired by ant colony optimization. Journal of Aerospace Information Systems. 2017;**14**(11):597-616

[22] Cheng Y-F, Balakrishnan S. Suboptimal atmospheric trajectory design using genetic algorithms with variable mutation. In: 33rd Aerospace Sciences Meeting and Exhibit. Jan 1995

[23] Subbarao K, Shippey BM. Hybrid genetic algorithm collocation method for trajectory optimization. Journal of Guidance, Control, and Dynamics. 2009;**32**(4):1396-1403

[24] Kumar GN, Ahmed MS, Sarkar A, Talole S. Reentry trajectory optimization using gradient free algorithms. IFAC-PapersOnLine. 2018;**51**(1):650-655

[25] Evans B, Walton S. Aerodynamic optimization of a hypersonic reentry vehicle based on solution of the Boltzmann BGK equation and evolutionary optimization. Applied Mathematical Modeling. 2017;**52**:215-240. DOI: 10.1016/j.apm.2017.07.024

[26] Du D-Z, Pardalos PM, Wu W. History of Optimization. Boston, MA: Springer US; 2009. pp. 1538-1542. DOI: 10.1007/978-0-387-74759-0 268

[27] Holland J. Adaption in Natural and Artificial Systems. Ann Arbor, MI, United States: The University of Michigan Press; 01 Dec 1975:206. ISBN 10: 0472084607. ISBN 13: 9780472084609

[28] Nyew HM, Abdelkhalik O, Onder N. Structured-chromosome evolutionary algorithms for variable-size autonomous interplanetary trajectory planning optimization. Journal of Aerospace Information Systems. 2015;**12**(3):314-328

[29] Jia D, Vagners J. Parallel evolutionary algorithms for uav path planning. In: AIAA 1st Intelligent Systems Technical Conference. 2004

[30] Qu X, Duan H. Three dimensional trajectory planning of unmanned aerial vehicles based on quantum differential search. In: Proceedings of the 33rd Chinese Control Conference. 2014

[31] Anh HPH, Son NN, Nam NT. Adaptive evolutionary neural control of perturbed nonlinear serial pam robot. Neurocomputing. 2017;**267**:525-544

[32] Kargupta H. Gene expression: The missing link in evolutionary computation. 1997;**9**. Available from: https://www.osti.gov/biblio/524858-gene-expression-missing-link-evolutionary-computation

[33] Goldberg DE, Korb B, Deb K. Messy genetic algorithms: Motivation, analysis, and first results. In: Complex Systems. Vol. 5. Complex Systems Publications, Inc; 1989

[34] Kennedy J, Eberhart R. Particle swarm optimization. In: Proceedings of ICNN'95 – International Conference on Neural Networks. Available from: https://ieeexplore.ieee.org/document/488968/authors#authors

[35] Yang X-S. Firefly algorithm. In: Nature-Inspired Metaheuristic Algorithms. 2nd ed. Luniver Press; 2010

[36] Yang X-S, Deb S. Cuckoo search via levy flights. In: Proceedings of 2009 World Congress on Nature & Biologically Inspired Computing. 2009

[37] Karaboga D. An idea based on honey bee swarm for numerical optimization. In: Technical report TR06. Kayseri/Trkiye: Erciyes University, Engineering Faculty Computer Engineering Department; Oct 2005

[38] Dorigo M. Optimization, learning and natural algorithms [PhD dissertation]. Italy: Politecnico di Milano; 1992

[39] Biesbroek R. Study of genetic algorithm settings for trajectory optimisation. In: 54th International Astronautical Congress of the International Astronautical Federation, the International Academy of Astronautics, and the International Institute of Space Law. 2003

[40] Goldberg DE. Sizing populations for serial and parallel genetic algorithms. In: Proceedings of the Third International Conference on Genetic Algorithms. Mar 1989. pp. 70-79, also TCGA Report 88004

[41] Harik G, Cant-Paz E, Goldberg DE, Miller BL. The gambler's ruin problem, genetic algorithms, and the sizing of populations. Evolutionary Computation. 1999;7(3):231-253

[42] Chen T, Tang K, Chen G, Yao X. A large population size can be unhelpful in evolutionary algorithms. Theoretical Computer Science. 2012;**436**:54-70

[43] Cage P, Kroo I, Braun R. Interplanetary trajectory optimization using a genetic algorithm. In: Astrodynamics Conference. 1994

[44] Wuerl A, Crain T, Braden E. Genetic algorithm and calculus of variations-based trajectory optimization technique. Journal of Spacecraft and Rockets

[45] Yokoyama N, Suzuki S. Modified genetic algorithm for constrained trajectory optimization. Journal of Guidance, Control, and Dynamics. 2005;**28**(1):139-144

[46] Yokoyama N, Suzuki S. Trajectory optimization via modified genetic algorithm. In: AIAA Guidance, Navigation, and Control Conference and Exhibit. 2003

[47] Silva PR, Abreu IS, Forte PA, Amaral HMCD. Genetic algorithms for satellite launcher attitude controller design. Inteligencia Artificial. 2019; **22**(63):150-161

[48] Chai R, Savvaris A, Tsourdos A, Xia Y, Chai S. Solving multiobjective constrained trajectory optimization problem by an extended evolutionary algorithm. IEEE Transactions on Cybernetics. 2018:114

[49] Tighzert L, Aguercif T, Fonlupt C, Mendil B. Intelligent trajectory planning and control of a humanoid robot using a new elitism based selfish gene algorithm. In: 2017 6th International Conference on Systems and Control (ICSC). May 2017

[50] Azarkaman M, Aghaabbasloo M, Salehi ME. Evaluating GA and PSO

evolutionary algorithms for humanoid walk pattern planning. In: 2014 22nd Iranian Conference on Electrical Engineering (ICEE). 2014

[51] Huang H, Zhou H, Cai Y. Study on multi-path planning and tracking control of the ucav based on evolutionary algorithm. In: 2015 15th International Conference on Control, Automation and Systems (ICCAS). Oct 2015

[52] Cooper MA, Heidlauf PT. Nonlinear feed forward control of a perturbed satellite using extended least squares adaptation and a luenberger observer. Journal of Aeronautics & Aerospace Engineering. Jan 2018;**07**(01)

[53] Sands T. Deterministic Artificial Intelligence. IntechOpen; 2019. ISBN: 978-1-78984-112-1

[54] Brownlee J. Swarm algorithms. Lulu.com, 2011. [Online]. Available from: http://www.CleverAlgorithms.com

[55] Pontani M, Conway BA. Particle swarm optimization applied to space trajectories. Journal of Guidance, Control, and Dynamics. Oct 2010;**33**(5)

[56] Akhmedova S, Stanovov V, Erokhin D, Semenkina O. Success-history based biology inspired algorithms for global trajectory optimization. IOP Conf. Series: Materials Science and Engineering. 2019;**537**. DOI: 10.1088/1757-899X/537/5/052008

[57] Liu C, Gao Z, Zhao W. A new path planning method based on firefly algorithm. In: 2012 Fifth International Joint Conference on Computational sciences and Optimization. 2012. DOI: 10.1109/CSO.2012.174

[58] Inghilterra G, Arrigoni S, Braghin F, Cheli F. Firefly algorithm-based nonlinear mpc trajectory planner for autonomous driving. In: 2018 International Conference of Electrical and Electronic Technologies for Automotive. 2018

[59] Raja NSM, Manic KS, Rajinikanth V. Firefly algorithm with various randomization parameters: An analysis. In: Proceedings of SEMCCO 2013, Part I, LNCS 829. Switzerland: Springer International Publishing; 2013. pp. 110-121

[60] Meena S, Chitra K. An approach of firefly algorithm with modified brightness for pid and i-pd controllers of siso systems. Journal of Ambient Intelligence and Humanized Computing. Berlin Heidelberg: Springer; Mar 2018. pp. 1–9. DOI: 10.1007/s12652-018-0747-x

[61] Mahapatra G, Mahapatra S, Banerjee S. A study of firefly algorithm and its application in non-linear dynamic systems. International Journal of Trend in Scientific Research and Development. 2018;**2**(2)

[62] Alao OJ, Metu MO, Amibor IP, Ibe CC, Oluyombo OW. Firefly optimization design and simulation of a single-axis helmholtz coils for spacecraft components testing. International Journal of Engineering and Applied Sciences. 2018;**5**(11)

[63] Shabbir F, Omenzetter P. Application of firefly algorithm to the dynamic model updating problem. 2015

[64] Corne D, Glover F, Dorigo M. New Ideas in Optimization, Advanced Topics in Computer Science Series. McGraw-Hill; 1999. Available from: https://books.google.com/books?id=-LYHngEACAAJ

[65] Dorigo M, Birattari M, Stutzle T. Ant colony optimization: Artificial ants as a computational intelligence technique. IEEE Computational Intelligence Magazine. Nov 2006; **1556-603X/06**

[66] Ma Y-N, Gong Y-J, Xiao C-F, Gao Y, Zhang J. Path planning for autonomous underwater vehicles: An ant colony algorithm incorporating alarm pheromone. IEEE Transactions on Vehicular Technology. 2019;**68**(1): 141-154

[67] Ntagiou E. Earth observation and data relay constellation missions' planning with ant colony optimization [PhD dissertation]. University of Surrey; 2019

[68] Zhang T-J, Shen H-X, Shang X, Wang J, Li H-N. Optimal mission planning of leo debris collecting. In: Proceedings of 2018 SpaceOps Conference. 2018. DOI: 10.2514/6.2018-2412

[69] Wase V. High-thrust interplanetary spacecraft trajectory optimization using cuda [PhD dissertation]. Institutionen fr informationsteknologi: Uppsala Universitet; 2018

[70] de Oliveira Alves GF, Becceneri JC, Sandri S. A balancing heuristic for spacecraft equipment layout optimization. In: 2015 IEEE International Conference on Fuzzy Systems (FUZZIEEE). Aug 2015. pp. 1-8

[71] Murrieta-Mendoza A, Botez RM, Bunel A. Four-dimensional aircraft en route optimization algorithm using the artificial bee colony. Journal of Aerospace Information Systems. 2018; **15**(6):307-334

[72] Zhong SA, Dong YF. Foundations and practical applications of cognitive systems and information processing. Advances in intelligent systems and computing. In: Artificial Bee Colony Algorithm for Parametric Optimization of Spacecraft Attitude Tracking Controller. Vol. 215. Berlin, Heidelberg: Springer; 2014

[73] Ding L, Wu H, Yao Y. Chaotic artificial bee colony algorithm for system identification of a small-scale unmanned helicopter. International Journal of Aerospace Engineering. 2015. DOI: 10.1155/2015/801874

[74] Banerjee S, Bharadwaj A, Gupta D, Panchal V. Remote sensing image classification using artificial bee colony algorithm. International Journal of Computer Science and Informatics. 2012;**2**(3). DOI: 10.1155/2015/801874

[75] Liu X, Li D, Dong N, Ip WH, Yung KL. Non-cooperative target detection of spacecraft objects based on artificial bee colony algorithm. IEEE Intelligent Systems. 2019:1-1

Chapter 5

Spacecraft Guidance Sensing at Relativistic Velocities

Emanuele Calabrò

Abstract

In this chapter a solution to the problem of planning an interstellar voyage at relativistic velocities by automatic control was proposed. To this aim, position and velocity of a relativistic interstellar spacecraft can be found by means of automatic measurements onboard of the aberrated angular distances between three quasars, at least. Indeed, this set can represent a reliable inertial reference frame due to the circumstance that quasars can be considered fixed in the space due to their large distances from Earth. To this aim, the so-called apical latitude and longitude of some quasars can be obtained from their right ascension α and declination δ in astronomical catalogues, using some mathematical expressions to provide the aberrated coordinates of a relativistic spacecraft during an interstellar space mission. The algorithm used in this study showed that the accuracy of determining the aberrated apical coordinates of a spacecraft increases significantly using quasars with aberrated apical latitude values lower than 45° in the direction of motion, suggesting that one or more normal-sized telescopes aboard the spacecraft can use quasars within a cone with angular aperture of about 45°, providing aberrated apical latitudes of the spacecraft with errors ranging from 10^{-7} to 10^{-9}.

Keywords: apical aberrated coordinates, quasars, relativistic velocities, interstellar navigation, spherical astronomy, spacecraft automatic control

1. Introduction

The history of human civilization is characterized by a natural tendency of extending the limits of human exploration. This is surely the most important reason of the exploration of interstellar spaces. Furthermore, the exponential increase in energy requirement by mankind may be considered another reason of the exploration of interstellar spaces. Nevertheless, also the research of extraterrestrial life can be considered an input to explore interstellar spaces.

In this regard, several projects to plan an automated spacecraft throwing toward the nearest interstellar systems have been proposed up to now. Project Orion proposed a mission toward the star closest to Earth, Alpha Centauri, using the nuclear pulse propulsion system, a mission which would take about 140 years [1, 2]. Project Daedalus followed the guidelines that the spacecraft could be designed to allow for a variety of target stars, reaching its destination within a human lifetime, using electron-driven D/He_3 fusion reactions, to accelerate the spaceship up to 12% of the velocity of light [3]. Project Icarus has been recently proposed to revise some aspects of the original Project Daedalus, as the choice of fuel to be used as a propellant [4]. Further possible techniques for propulsion of an interstellar spacecraft have been proposed up to now [5–7]. Nevertheless, an upper limit to the velocity of an interstellar spacecraft exists, because, hypothesizing a velocity comparable to the light velocity, the spaceship will get a weight more than 2000 ton. Hence, a reasonable value of velocity of an interstellar spaceship would not exceed 0.3 c for an interstellar voyage, and it is expected to last about 30 years, at least. Such long time forces us to plan navigation and guidance of the spacecraft by means of automated control on-board the spacecraft. Indeed, sending a signal from a spaceship to Earth at a distance of several light years would ask for an extremely long time, as the signal would travel at the velocity of light, making out of the question any Earth-side control of an interstellar mission.

Otherwise, mankind's exploration of space has been characterized by the extraordinary achievements of both robotic and manned space missions. The success of robotic space missions was due to the development of automated space navigation systems that have enabled the determination of the spacecraft's position and velocity, providing accuracies for traversing interplanetary distances and obtaining precise landings on the surface of the moon and of some planet of the solar system. The required position and velocity of a space mission to support trajectory corrections can be obtained by the current and predicted values of the spacecraft's position and velocity, provided by ground and on-board guidance and control systems.

In contrast, hypothesizing an interstellar voyage, no navigation and guidance control can be carried out by control systems on Earth because of the very long distances between Earth and stars. An interstellar spacecraft should check automatically its trajectory, calculating direction and modulus of its velocity by means of automatic measurements on-board. To this aim, a celestial reference frame is needed so that a fixed coordinate system can provide an instantaneous determination of the spacecraft's position with respect to the celestial reference frame. Hence, the spacecraft's trajectory can be compared with knowledge of the destination stellar object, and maneuver control can be applied, determining velocity changes to rectify the spaceship's trajectory.

Previous studies showed that a celestial reference frame constituted by three quasars, at least, can be successfully used to determine position and velocity of an interstellar spacecraft [8]. Indeed, quasi-stellar objects (quasars) can be considered a reliable inertial reference frame for an interstellar voyage because they are point-like stellar objects and their proper motion can be neglected due to their extremely long distance. Furthermore, the accuracy of determining the aberrated coordinates of an interstellar spacecraft can be improved using a set of quasars whose aberrated apical latitudes are within a cone with an angular aperture of 45° and the axis in the direction of motion of the spacecraft.

2. The celestial reference frame for an interstellar space mission

Previous space missions within the solar systems have been carried out up to now in a space reference frame associated with the planetary ephemeris represented by a solar system barycentric frame aligned with the planetary ephemeris. In previous space missions, space radio tracking has been performed by means of Doppler and range systems and very-long-baseline interferometry (VLBI), so that accurate information regarding corrections to be carried out to the spacecraft's trajectory were obtained [9].

Otherwise, automatic measurements on-board an interstellar spacecraft should be carried out to check the prefixed trajectory, comparing computed values of position and velocity with expected values of position and velocity so that the spaceship's trajectory can be automatically rectified toward its target. To this aim, the primary step to plan an interstellar space mission is the choice of a reliable inertial reference frame, using computing and motion sensors on-board for tracking spacecraft's position, orientation, and velocity to support trajectory corrections.

The discovery of radio pulsars led to the idea of using pulsar timing observations for interstellar navigation [10]. Indeed, pulsars are rotating neutron stars that emit beams of electromagnetic radiation, and they are bright enough to be used in a space mission. Nevertheless, some limitations reduce their effectiveness in navigation and guidance of interstellar space missions. Indeed, neighboring celestial objects are broadband radio sources that can obscure weak pulsar signals [11]. Furthermore, propagation of radio signals is in phase lags of variable and unpredictable duration so that they set the limitation on accuracy. The most relevant limitation is that at radio frequencies that pulsars emit, radio-based systems on-board would require too large antennas impracticable for a spacecraft. Furthermore, optical observations of pulsars during interstellar navigation would be impractical because of the small number of detectable optical pulsars [10–12].

X-ray pulsars were recently considered to overhead these limitations. Indeed, an X-ray telescope of normal-size dimension can be required to detect X-ray pulsars. The basic concept of interstellar space missions using X-ray pulsars was recently described [12–14]. Nevertheless, other limitations have to be considered. First, long-term observations of X-ray pulsars highlight irregularities in the pulse rate. Second, irregularities in the spacecraft's clock could cause an error in measurement of time of arrivals of pulsars' beam. Third, the pulse shape may differ between the X-ray and the radio wave bands producing an offset between the time of arrivals measured using the different bands. Finally, pulsar timing ephemeris obtained from long-term ground-based radio observations may be not reliable because pulsars' proper motion cannot be negligible and reducing uncertainties that arise from pulsar position errors is critical.

Otherwise, quasars' position can be considered stationary in the sky because of their large distance from the observer, deduced by very high redshift values. Indeed, spectra of the most numerous quasars can be explained only by a cosmological redshift due to the expansion of the universe. The accretion of material on a central, massive black hole can explain the observed high quasar energy fluxes. Hence, quasars can be considered a reliable inertial reference frame because their proper motion can be neglected due to their extreme distance and bright and point-like appearance. As regards this topic, the International Celestial Reference Frame (ICRF) was proposed, and it represents a catalogue of extragalactic radio sources observed with VLBI; the majority of them are quasars and are distributed around the sky [15]. The ICRF was successively developed using an extended list of sources that was adopted by the International Astronomical Union in 2009 for a second realization of a new catalogue named ICRF2, which provides absolute coordinates for 3414 sources with errors within 0.1 mAs (milliarcseconds) and the orientation of the axes that can be considered fixed within 0.01 mAs [16, 17].

Proper motions of sources could be taken into account to improve the reliability of ICRF2. The dominant proper motion of the major parts of sources is related to internal structural changes that can produce apparent motions several 100 μAs/yr (microarcseconds per year), that is, an order of magnitude larger than proper motions due to the secular aberration drift. However, proper motion due to internal structural changes was detected to be relevant only for unstable sources. A selection of stable sources could be made for a realization of a catalogue of quasars to be used for interstellar space missions. Instead, secular aberration drift is an apparent change in the velocity of distant objects caused by the acceleration of the solar system barycenter directed toward the Galactic Center. This effect may cause apparent proper motion of all quasars by an estimated average value 4–6 μAs/yr and direction toward the points with equatorial coordinates α = 266° and δ = −29° [18]. Also this error may be considered negligible, but it could be taken into account for accurate planning of interstellar space mission.

Hence, quasars can represent a reliable inertial reference frame to be used for interstellar space missions, because their proper motions can be neglected. In the optical domain, the Hipparcos catalog is currently used for optical astrometry, due to the launch in 1989 of the ESA space-astrometry satellite Hipparcos, which was aligned to the ICRF within 0.6 mAs for the orientation at 1991.25. Nevertheless, other ambitious space-astrometry projects will provide astrometry measurements in the optical domain. The ESA Gaia mission, which will survey about 109 stellar objects brighter than 20 magnitude (mag), with expected accuracies in the 7–25 μAs, ranges down to 15 mag and sub-mAs accuracies at the limit 20 mag. The observations of about 500,000 quasars will provide the Gaia extragalactic reference frame (GCRF), a kinematically nonrotating system close to 0.3 μAs/yr and a positional precision reaching 50 μAs. Of these, only the quasars with the most accurate positions with magnitude lesser than 18 will be used to define a new celestial reference frame in the optical domain, the Large Quasar Reference Frame (LQRF) [19]. The final catalogue is expected around 2021, but with intermediate data that are expected to be available by 2015. An accurate alignment between the two celestial reference frames, the LQRF and the ICRF, can be carried out using only 10% of the current ICRF sources for the alignment with the future Gaia frame, but further multistep VLBI observational projects have been planned to observe new VLBI sources suitable for the alignment with the future Gaia frame [19].

Another space-astrometry project is the Space Interferometry Mission PlanetQuest Light (SIM-Lite), which consists of an optical interferometer system with a baseline of 6 m and a 30-cm guide telescope that would search 65 nearby stars for planets of masses down to one Earth mass, achieving 8 μAs accuracy on the nineteenth magnitude objects and 4 μAs for objects up to 14 mag that would constitute a new astrometric grid.

3. The apical coordinate system

The inertial reference frame for interstellar missions can be represented by a spherical coordinate system where the spacecraft is at rest. This system is termed the “apical system” and is represented in **Figure 1**, where the origin of the system represents the position of the spaceship, OV is the direction of the motion, P is the North equatorial pole, and Q is a quasar. The *apical latitude* θ is measured from the direction of the spacecraft’s velocity OV to the quasar Q. The *apical longitude* l is measured from the plane that contains the direction of velocity and the direction of the equatorial pole to the plane containing the same vector of velocity and the quasar (see **Figure 1**).

The apical coordinates of a quasar can be related to its astronomical coordinates the *right ascension* α and the *declination* δ (that are known from the ICRF2 catalogue) by means of some equations derived from spherical astronomy [20, 21]:

$$\sin\theta = \sin\delta_v \sin\delta + \cos\delta_v \cos\delta \cos(\alpha_v - \alpha) \tag{1}$$

$$\cos\theta = \frac{\cos\delta \sin(\alpha_v - \alpha)}{\sin l} \tag{2}$$

$$\cos l = \frac{\sin\delta \cos\delta_v - \cos\delta \cos\delta_v \cos(\alpha_v - \alpha)}{\cos\theta} \tag{3}$$

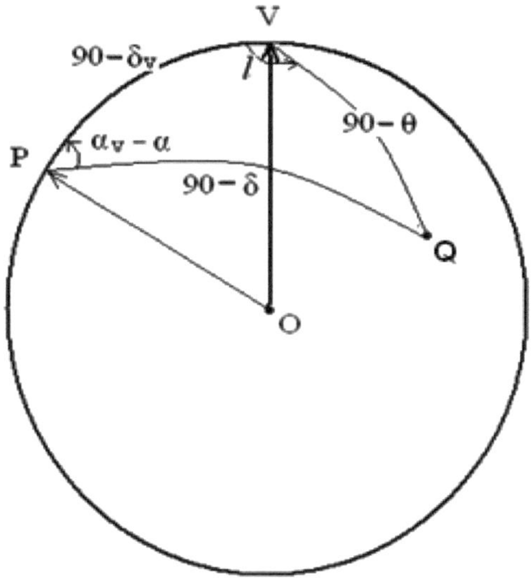

Figure 1.
The apical latitude θ and the apical longitude l of a quasar in the apical system.

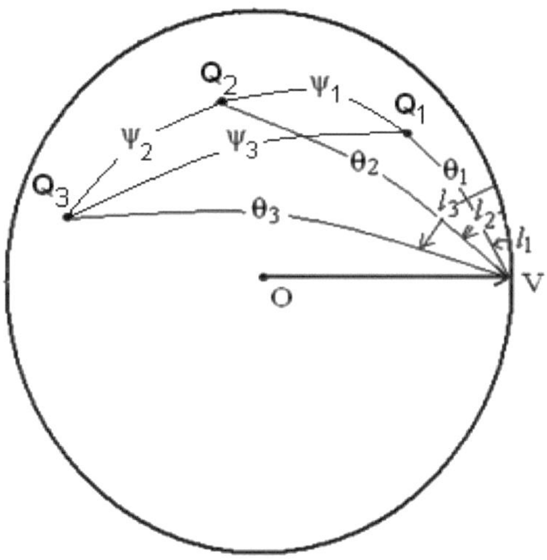

Figure 2.
The apical coordinates θ_i and l_i (i = 1, 2, 3) of three quasars in the apical system (where the spacecraft is at rest).

where α_v and δ_v are the stationary coordinates of the vector velocity of the spacecraft that have to be recomputed during the voyage.

The spaceship should determine its position by means of the apical coordinates of a number of quasars. It was shown that spacecraft's position and velocity can be determined using only three quasars by means of automatic measurements on-board the spacecraft of the angular distances ψ_i between the quasars [8].

Indeed, the angular distances ψ_1, ψ_2, and ψ_3 between three quasars named Q_1, Q_2, and Q_3, pointed out in **Figure 2**, can be related to their apical coordinates applying the II Gauss formula to the spherical triangles represented in **Figure 2**:

$$\cos\psi_1 = \cos\theta_1\cos\theta_2 + \sin\theta_1\sin\theta_2\cos(l_2 - l_1) \quad (4)$$

$$\cos\psi_2 = \cos\theta_3\cos\theta_2 + \sin\theta_3\sin\theta_2\cos(l_3 - l_2) \quad (5)$$

$$\cos\psi_3 = \cos\theta_3\cos\theta_1 + \sin\theta_3\sin\theta_1\cos(l_3 - l_1) \quad (6)$$

Measurements on-board the spaceship of the angular distances ψ_1, ψ_2, and ψ_3 could be used to obtain the apical coordinates, determining the position and velocity of a spacecraft.

4. The aberrated apical coordinates at relativistic velocities

The most relevant effect to be taken into account at relativistic velocities is represented by the change in direction of a stellar object because the point of view of an object from a moving observer depends on its velocity, and this change is not negligible if the velocity is comparable to the velocity of light. Indeed, according to the relativistic aberration, during the motion of an object, its apical coordinates θ_i change into θ_i' as follows [22]:

$$\cos\theta' = \frac{\cos\theta + \beta'}{1 + \beta'\cos\theta} \quad (7)$$

$$\sin\theta' = \frac{\sin\theta}{\gamma\left(1 + \beta'\cos\theta\right)} \quad (8)$$

where $\beta' = {}^{v}/_{c}$, $\gamma = {}^{1}\Big/\sqrt{1-\beta'^{2}}$
(v and c are the velocities of the spacecraft and of light, respectively).

As the velocity of a spacecraft during an interstellar space mission should range from 0.1 to 0.3 c, the apical coordinates θ_i and l_i of the spacecraft (i = 1, 2, 3) and the angular distances ψ_i between the quasars should change in the "aberrated coordinates" θ_i', l_i', and ψ_i', respectively. However, the aberrated coordinates of the spaceship can be related to its apical coordinates applying some relation analogue to Eqs. (4)–(6).

In particular, the spacecraft's velocity value V_{j+1} at the time (j + 1) and the related aberrated coordinates can be obtained from the spacecraft's velocity V_j and the aberrated coordinates at the time (j) by means of measurements of three quasars at least and some expressions derived from spherical astronomy [20, 21]. The apical and aberrated coordinates of the velocities V_1 (j = 1) and V_2 are pointed out in **Figure 3**. The vectors V_1 and V_2 may be represented in the apical system by their apical coordinates $[(\theta_{A,1}, l_{A,1}), (\theta_{B,1}, l_{B,1}), (\theta_{C,1}, l_{C,1})]$ and $[(\theta_{A,2}, l_{A,2}), (\theta_{B,2}, l_{B,2}), (\theta_{C,2}, l_{C,2})]$, respectively.

Nevertheless, only measurements of the aberrated angular distances ψ_i'' (i = 1, 2, 3) between the quasars can be carried out aboard the spacecraft. Applying the II Gauss' formulae to the spherical triangles ABV_2 and $A''B''V_2$, BCV_2 and $B''C''V_2$, and CAV_2 and $C''A''V_2$, the following relations can be obtained:

$$\cos\psi_1'' = \cos\theta_{A_2}''\cos\theta_{B_2}'' + \sin\theta_{A_2}''\sin\theta_{B_2}''\cos E_{2,1}'' \quad (9)$$

$$\cos\psi_2'' = \cos\theta_{B_2}''\cos\theta_{C_2}'' + \sin\theta_{B_2}''\sin\theta_{C_2}''\cos E_{2,2}'' \quad (10)$$

$$\cos\psi_3'' = \cos\theta_{A_2}''\cos\theta_{C_2}'' + \sin\theta_{A_2}''\sin\theta_{C_2}''\cos E_{2,3}'' \quad (11)$$

Assuming that the dihedral angles $E_{i,j}$ (i, j = 1, 2) between two quasars do not change because of their large distances from the observer (i.e., $\cos E_{2,i} = \cos E_{2,i}''$, i = 1, 2, 3), we can relate Eqs. (9)–(11) to analogue expressions where the not-aberrated coordinates appear:

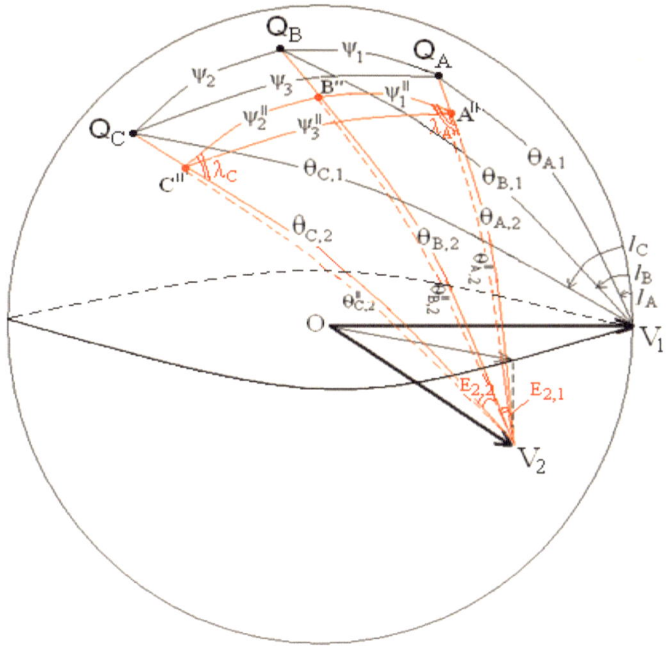

Figure 3.
The apical coordinates of a spacecraft at two velocities V_1 ($\theta_{i,1}$; i = A, B, C) and V_2 ($\theta_{i,2}$; i = A, B, C). The aberrated angular distances ψ_1, ψ_2, and ψ_3 between three quasars (Q_A, Q_B, Q_C) and the aberrated coordinates $\theta_{A,2}$, $\theta_{B,2}$, and $\theta_{C,2}$ referred to velocity V_2 are pointed out.

$$\frac{\cos\psi_1'' - \cos\theta_{A_2}''\cos\theta_{B_2}''}{\sin\theta_{A_2}''\sin\theta_{B_2}''} = \frac{\cos\psi_1 - \cos\theta_{A_2}\cos\theta_{B_2}}{\sin\theta_{A_2}\sin\theta_{B_2}} \tag{12}$$

$$\frac{\cos\psi_2'' - \cos\theta_{B_2}''\cos\theta_{C_2}''}{\sin\theta_{B_2}''\sin\theta_{C_2}''} = \frac{\cos\psi_2 - \cos\theta_{B_2}\cos\theta_{C_2}}{\sin\theta_{B_2}\sin\theta_{C_2}} \tag{13}$$

$$\frac{\cos\psi_3'' - \cos\theta_{A_2}''\cos\theta_{C_2}''}{\sin\theta_{A_2}''\sin\theta_{C_2}''} = \frac{\cos\psi_3 - \cos\theta_{C_2}\cos\theta_{A_2}}{\sin\theta_{C_2}\sin\theta_{A_2}} \tag{14}$$

Applying the formulation of the relativistic aberration to the aberrated coordinates θ_{A2}, θ_{B2}'', and θ_{C2}'', the apical coordinates θ_{A2}, θ_{B2}, and θ_{C2} can be obtained, respectively, by means of the following equations, derived from Eqs. (7) and (8):

$$\sin\theta_{A_2}'' = \frac{\sin\theta_{A_2}}{\gamma\left(1+\beta''\cos\theta_{A_2}\right)} \quad \cos\theta_{A_2}'' = \frac{\cos\theta_{A_2}+\beta''}{1+\beta''\cos\theta_{A_2}} \tag{15}$$

$$\sin\theta_{B_2}'' = \frac{\sin\theta_{B_2}}{\gamma\left(1+\beta''\cos\theta_{B_2}\right)} \quad \cos\theta_{B_2}'' = \frac{\cos\theta_{B_2}+\beta''}{1+\beta''\cos\theta_{B_2}} \tag{16}$$

$$\sin\theta_{C_2}'' = \frac{\sin\theta_{C_2}}{\gamma\left(1+\beta''\cos\theta_{C_2}\right)} \quad \cos\theta_{C_2}'' = \frac{\cos\theta_{C_2}+\beta''}{1+\beta''\cos\theta_{C_2}} \tag{17}$$

The set of Eqs. (15)–(17) allows to express the set of Eqs. (12)–(14) as a function of the apical coordinates of the spaceship θ_{A2}, θ_{B2}, and θ_{C2} and of the aberrated modulus of the velocity β'' (in unit of c). Applying again the II Gauss' equation and the sinus theorem to the spherical triangles $B''C''V_2$ and $A''B''V_2$, we obtain

$$\cos\theta''_{B_2} = \cos\psi''_2\cos\theta''_{C_2} + \sin\psi''_2\sin\theta''_{C_2}\cos\lambda_C \tag{18}$$

$$\cos\theta''_{B_2} = \cos\ \psi''_1\cos\theta''_{A_2} + \sin\psi''_1\sin\theta''_{A_2}\cos\lambda_A \tag{19}$$

$$\sin\lambda_A = \frac{\sin E_{2,1}\sin\theta''_{B_2}}{\sin\psi''_1}\quad \sin\lambda_C = \frac{\sin E_{2,2}\sin\theta''_{B_2}}{\sin\psi''_2} \tag{20}$$

The set of equations reported above allows to obtain the apical coordinates θ_{A2}, θ_{B2}, and θ_{C2} (at time $j = 2$) as a function of the apical coordinates θ_{A1}, θ_{B1}, and θ_{C1} (at time $j = 1$) and of the measured quasars' angular distances ψ_1'', ψ_2'', and ψ_3''. Hence, this algorithm can provide the direction and modulus of velocity that have to be used to rectify the trajectory of the spacecraft toward its target.

5. Applying the algorithm to determine the spacecraft's aberrated apical coordinates

The solution of the set of equations reported above can provide the exact values of the aberrated coordinates of a spaceship that are required for navigation and guidance during an interstellar space mission. Nevertheless, previous results showed that the accuracy of determining the aberrated coordinates depends on the apical coordinates of the quasars that are used. Indeed, it was shown that applying the algorithm to typical apical latitudes around 90° of three quasars, the best accuracy in the determination of apical coordinates can be obtained using quasars with apical longitudinal angular distances around 90° and 180° [8].

In this simulation study, instead of simulating a variation of quasars' apical longitudes, quasars' apical latitudes ranging from 5° to 120° were used, using also the value of the spaceship's aberrated velocity modulus $\beta = 0.1$ (in unit of c) and the values $\theta = l = 0.5°$ that represent a typical change in direction of the motion of the spacecraft. The results of this simulation study were reported in **Tables 1–12** (in the Appendix section) where quasars' aberrated angular distances, spacecraft's aberrated apical latitudes, and aberrated velocity modulus values were reported as a function of typical quasars' apical latitudes and longitudes. Looking at the results reported in **Tables 1–12**, it appears that the aberrated velocity modulus values obtained from the input value $\beta'' = 0.1$ are all very close to this value, confirming the reliability of the algorithm. In addition, the variations of the aberrated velocity modulus decreased with the decrease of quasars' apical latitude, getting the best accuracy using quasars' aberrated apical latitudes lower than 45°. The average of the values of spacecraft's aberrated velocity modulus obtained at typical apical longitudes was plotted as a function of quasars' aberrated apical latitude, and a sigmoidal fit was used with upper and lower asymptotes equal to 7.08×10^{-7} and 4.88×10^{-8}, respectively (see **Figure 4**). A statistical analysis was carried out applying Student's t-test for comparison between two groups: the group of quasars whose aberrated apical latitude are $\theta \leq 45°$ and the group of quasars with $\theta > 90°$, with $p < 0.05$ considered significant. The t-test provided the result that the group of quasars with $\theta \leq 45°$ is significantly different in comparison to the other group ($p < 0.01$), showing that the accuracy of determining the spacecraft's aberrated velocity β'' increases using quasars' aberrated apical latitude $\theta \leq 45°$. Hence, the result of this simulation study has confirmed that a celestial reference frame consisting of three quasars can be successfully used for interstellar navigation regardless of their apical coordinates, but the best accuracy in the determination of spacecraft's apical coordinates can be obtained using quasars whose aberrated apical latitudes are lower than 45°.

Apical latitudes (deg) $\theta_A = 5; \theta_B = 5; \theta_C = 5$				
Apical longitude values (deg) l_A l_B l_C	**Quasars' aberrated angular distances (rad)** Ψ_1'' Ψ_2'' Ψ_3''	**Aberrated apical latitudes (rad)** θ_{A2}'' θ_{B2}'' θ_{C2}''	**Aberrated velocity** $\beta'' = v''/c$	**Uncertainty of aberrated apical latitudes** $\Delta\theta''$ **(rad)**
0	6.036476×10^{-2}	7.104854×10^{-2}	0.1	1.58638×10^{-9}
45	6.036476×10^{-2}	7.352113×10^{-2}		
90	0.1115786	7.104854×10^{-2}		
0	0.1366982	7.104854×10^{-2}	0.1	9.794508×10^{-9}
120	0.1366982	8.311642×10^{-2}		
240	0.1366908	7.104854×10^{-2}		
0	0.1572874	7.104854×10^{-2}	0.1	7.521397×10^{-8}
170	0.1400199	8.672082×10^{-2}		
45	6.036476×10^{-2}	7.351951×10^{-2}		
45	0.157287	7.352113×10^{-2}	0.1	5.19094×10^{-7}
215	0.1400161	8.377577×10^{-2}		
90	6.036476×10^{-2}	7.104854×10^{-2}		
90	6.03667×10^{-2}	7.926836×10^{-2}	0.1	1.489584×10^{-9}
45	0.1115813	7.352033×10^{-2}		
135	6.03667×10^{-2}	7.362574×10^{-2}		
135	0.1115866	8.466642×10^{-2}	0.1	4.037413×10^{-9}
45	0.1458454	7.352113×10^{-2}		
180	0.0603697	7.940599×10^{-2}		
225	0.1458474	8.475729×10^{-2}	0.1	1.405048×10^{-8}
90	0.1115829	7.104854×10^{-2}		
0	0.1458528	7.927288×10^{-2}		
180	0.1458503	8.684048×10^{-2}	0.1	1.168942×10^{-8}
45	6.036476×10^{-2}	7.352113×10^{-2}		
90	0.1115866	7.104854×10^{-2}		

Table 1.
Quasars' aberrated angular distances, spacecraft's aberrated apical latitudes, uncertainties, and aberrated velocity as a function of quasars' apical latitudes $\theta_A = \theta_B = \theta_C = 5°$.

Apical latitudes (deg) $\theta_A = 10; \theta_B = 10; \theta_C = 10$				
Apical longitude values (deg) l_A l_B l_C	**Quasars' aberrated angular distances (rad)** Ψ_1'' Ψ_2'' Ψ_3''	**Aberrated apical latitudes (rad)** θ_{A2}'' θ_{B2}'' θ_{C2}''	**Aberrated velocity** $\beta'' = v''/c$	**Uncertainty of aberrated apical latitudes** $\Delta\theta''$ **(rad)**
0	0.1204418	0.1500399	0.1	3.873324×10^{-9}
45	0.1204418	0.1524055		
90	0.2228695	0.1500403		
0	0.2732715	0.1500399	0.1	1.859782×10^{-8}
120	0.2732715	0.1619809		
240	0.2732428	0.1500399		
0	0.3146759	0.1500399	0.1	1.405435×10^{-7}
170	0.2799516	0.1657227		
45	0.1204418	0.1524067		

Apical latitudes (deg) $\theta_A = 10; \theta_B = 10; \theta_C = 10$				
Apical longitude values (deg) l_A l_B l_C	Quasars' aberrated angular distances (rad) Ψ_1'' Ψ_2'' Ψ_3''	Aberrated apical latitudes (rad) θ_{A2}'' θ_{B2}'' θ_{C2}''	Aberrated velocity $\beta'' = v''/c$	Uncertainty of aberrated apical latitudes $\Delta\theta''$ (rad)
45	0.3146734	0.1524059	0.1	2.326587×10^{-7}
215	0.279938	0.1626599		
90	0.1204418	0.1500399		
90	0.1204502	0.1580701	0.1000007	3.756036×10^{-9}
45	0.2228789	0.1524059		
135	0.1204502	0.1525068		
135	0.2229005	0.1635804	0.1	7.86745×10^{-9}
45	0.2916495	0.1524059		
180	0.1204621	0.1582079		
225	0.2916580	0.1636748	0.1	2.711515×10^{-8}
90	0.2228851	0.1500399		
0	0.2916783	0.1580879		
180	0.2916696	0.1658484	0.1	2.471196×10^{-8}
45	0.1204418	0.1524059		
90	0.2229005	0.1500399		

Table 2.
Quasars' aberrated angular distances, spacecraft's aberrated apical latitudes, uncertainties, and aberrated velocity as a function of quasars' apical latitudes $\theta_A = \theta_B = \theta_C = 10°$.

Apical latitudes (deg) $\theta_A = 20; \theta_B = 20; \theta_C = 20$				
Apical longitude values (deg) l_A l_B l_C	Quasars' aberrated angular distances (rad) Ψ_1'' Ψ_2'' Ψ_3''	Aberrated apical latitudes (rad) θ_{A2}'' θ_{B2}'' θ_{C2}''	Aberrated velocity $\beta'' = v''/c$	Uncertainty of aberrated apical latitudes $\Delta\theta''$ (rad)
0	0.2385968	0.3083875	0.1	8.906437×10^{-9}
45	0.2385968	0.3107102		
90	0.4434312	0.3083873		
0	0.5454995	0.3083875	0.1	3.621693×10^{-8}
120	0.5454995	0.3203031		
240	0.5453854	0.3083873		
0	0.6301567	0.3083875	0.1	2.721964×10^{-7}
170	0.5591298	0.3241316		
45	0.2385968	0.3107106		
45	0.6301463	0.3107102	0.1	3.33884×10^{-7}
215	0.5590756	0.3209947		
90	0.2385968	0.3083875		
90	0.2386291	0.3163490	0.1000001	8.764387×10^{-9}
45	0.4434671	0.3107102		
135	0.2386296	0.3108094		
135	0.4435533	0.3219346	0.1	1.532988×10^{-8}
45	0.5829566	0.3107102		
180	0.2386760	0.3164873		

Apical latitudes (deg) $\theta_A = 20; \theta_B = 20; \theta_C = 20$				
Apical longitude values (deg) l_A l_B l_C	**Quasars' aberrated angular distances (rad)** Ψ_1'' Ψ_2'' Ψ_3''	**Aberrated apical latitudes (rad)** θ_{A2}'' θ_{B2}'' θ_{C2}''	**Aberrated velocity** $\beta'' = v''/c$	**Uncertainty of aberrated apical latitudes** $\Delta\theta''$ **(rad)**
225 90 0	0.5829908 0.4434926 0.5830727	0.3220308 0.3083875 0.3164183	0.1	5.315188×10^{-8}
180 45 90	0.5830376 0.2385965 0.4435540	0.3242611 0.3107100 0.3083875	0.1000005	5.07998×10^{-8}

Table 3.
Quasars' aberrated angular distances, spacecraft's aberrated apical latitudes, uncertainties, and aberrated velocity as a function of quasars' apical latitudes $\theta_A = \theta_B = \theta_C = 20°$.

Apical latitudes (deg) $\theta_A = 30; \theta_B = 30; \theta_C = 30$				
Apical longitude values (deg) l_A l_B l_C	**Quasars' aberrated angular distances (rad)** Ψ_1'' Ψ_2'' Ψ_3''	**Aberrated apical latitudes (rad)** θ_{A2}'' θ_{B2}'' θ_{C2}''	**Aberrated velocity** $\beta'' = v''/c$	**Uncertainty of aberrated apical latitudes** $\Delta\theta''$ **(rad)**
0 45 90	0.3521147 0.3521147 0.6590953	0.4675786 0.4698990 0.4675784	0.1000003	1.502422×10^{-8}
0 120 240	0.8153257 0.8153257 0.8150669	0.4675787 0.4795486 0.4675787	0.1	5.38223×10^{-8}
0 170 45	0.9472179 0.8364417 0.3521149	0.4675787 0.4834285 0.4699020	0.1	4.057065×10^{-7}
45 215 90	0.9471941 0.8363186 0.3521149	0.4698993 0.4802482 0.4675791	0.1	4.579924×10^{-7}
90 45 135	0.3521857 0.6591761 0.3521866	0.4755584 0.4698993 0.4699984	0.1	1.484936×10^{-8}
135 45 180	0.6593686 0.8733605 0.3522882	0.4811999 0.4698993 0.4756976	0.1	2.243538×10^{-8}
225 90 0	0.8734387 0.6592328 0.8736262	0.4812975 0.4675787 0.4756282	0.1	7.921381×10^{-8}
180 45 90	0.8735464 0.3521149 0.6593703	0.4835601 0.4698993 0.4675786	0.1	7.698898×10^{-8}

Table 4.
Quasars' aberrated angular distances, spacecraft's aberrated apical latitudes, uncertainties, and aberrated velocity as a function of quasars' apical latitudes $\theta_A = \theta_B = \theta_C = 30°$.

Apical latitudes (deg) $\theta_A = 40; \theta_B = 40; \theta_C = 40$				
Apical longitude values (deg) l_A l_B l_C	Quasars' aberrated angular distances (rad) Ψ_1'' Ψ_2'' Ψ_3''	Aberrated apical latitudes (rad) θ_{A2}'' θ_{B2}'' θ_{C2}''	Aberrated velocity $\beta'' = v''/c$	Uncertainty of aberrated apical latitudes $\Delta\theta''$ (rad)
0	0.4584757	0.6280316	0.1000001	2.306603×10^{-8}
45	0.4584757	0.6303636		
90	0.8665489	0.6280315		
0	1.080622	0.6280316	0.1000001	7.157303×10^{-8}
120	1.080622	0.6400977		
240	1.080156	0.6280323		
0	1.266572	0.6280316	0.1	5.437029×10^{-7}
170	1.110074	0.6440277		
45	0.4584759	0.6303676		
45	1.266527	0.6303637	0.1	5.889684×10^{-7}
215	1.109851	0.640806		
90	0.4584759	0.6280328		
90	0.4585961	0.6360657	0.1000002	2.284043×10^{-8}
45	0.8666899	0.6303634		
135	0.4585977	0.6304636		
135	0.867028	0.6417693	0.1	2.902025×10^{-8}
45	1.161710	0.6303636		
180	0.4587704	0.6362065		
225	1.161853	0.6418682	0.1	1.05652×10^{-7}
90	0.8667899	0.6280316		
0	1.162195	0.6361363		
180	1.162049	0.6441610	0.1000003	1.036194×10^{-7}
45	0.4584756	0.6303634		
90	0.8670309	0.6280313		

Table 5.
Quasars' aberrated angular distances, spacecraft's aberrated apical latitudes, uncertainties, and aberrated velocity as a function of quasars' apical latitudes $\theta_A = \theta_B = \theta_C = 40°$.

Apical latitudes (deg) $\theta_A = 45; \theta_B = 45; \theta_C = 45$				
Apical longitude values (deg) l_A l_B l_C	Quasars' aberrated angular distances (rad) Ψ_1'' Ψ_2'' Ψ_3''	Aberrated apical latitudes (rad) θ_{A2}'' θ_{B2}'' θ_{C2}''	Aberrated velocity $\beta'' = v''/c$	Uncertainty of aberrated apical latitudes $\Delta\theta''$ (rad)
0	0.5081233	0.7088562	0.1000001	2.821275×10^{-8}
45	0.5081233	0.7111977		
90	0.9658515	0.7088549		
0	1.210536	0.7088562	0.1	8.064029×10^{-8}
120	1.210536	0.7209837		
240	1.209941	0.7088549		
0	1.427298	0.7088562	0.1000001	6.160802×10^{-7}
170	1.244587	0.7249402		
45	0.5081232	0.7111935		

	Apical latitudes (deg) $\theta_A = 45; \theta_B = 45; \theta_C = 45$			
Apical longitude values (deg) l_A l_B l_C	**Quasars' aberrated angular distances (rad)** Ψ_1'' Ψ_2'' Ψ_3''	**Aberrated apical latitudes (rad)** θ_{A2}'' θ_{B2}'' θ_{C2}''	**Aberrated velocity** $\beta'' = v''/c$	**Uncertainty of aberrated apical latitudes** $\Delta\theta''$ **(rad)**
45 215 90	1.427240 1.244302 0.5081233	0.7111977 0.7216963 0.7088572	0.1	6.578899×10^{-7}
90 45 135	0.5082715 0.9660284 0.5082732	0.7169275 0.7111976 0.7112963	0.1000002	2.795166×10^{-8}
135 45 180	0.9664523 1.304455 0.5084859	0.7226662 0.7111977 0.7170680	0.1	3.206421×10^{-8}
225 90 0	1.304639 0.9661539 1.305079	0.7227656 0.7088564 0.7169986	0.1	1.192555×10^{-7}
180 45 90	1.304891 0.5081232 0.9664560	0.7250745 0.7111977 0.7088549	0.1000001	1.173458×10^{-7}

Table 6.
Quasars' aberrated angular distances, spacecraft's aberrated apical latitudes, uncertainties, and aberrated velocity as a function of quasars' apical latitudes $\theta_A = \theta_B = \theta_C = 45°$.

	Apical latitudes (deg) $\theta_A = 55; \theta_B = 55; \theta_C = 55$			
Apical longitude values (deg) l_A l_B l_C	**Quasars' aberrated angular distances (rad)** Ψ_1'' Ψ_2'' Ψ_3''	**Aberrated apical latitudes (rad)** θ_{A2}'' θ_{B2}'' θ_{C2}''	**Aberrated velocity** $\beta'' = v''/c$	**Uncertainty of aberrated apical latitudes** $\Delta\theta''$ **(rad)**
0 45 90	0.5985235 0.5985235 1.152048	0.8719386 0.8743041 0.8719379	0.1	4.217321×10^{-8}
0 120 240	1.461552 1.461552 1.460648	0.8719386 0.8842108 0.871938	0.1000001	9.980699×10^{-8}
0 170 45	1.751104 1.505919 0.5985236	0.8719386 0.8882255 0.8743089	0.1	7.747448×10^{-7}
45 215 90	1.751009 1.505481 0.5985233	0.8743038 0.8849332 0.8719363	0.1000006	8.090532×10^{-7}
90 45 135	0.5987303 1.152306 0.5987327	0.8801010 0.8743043 0.8744049	0.1	4.180844×10^{-8}
135 45 180	1.152921 1.584652 0.5990289	0.8859171 0.8743041 0.8802433	0.1000003	3.75483×10^{-8}

Apical latitudes (deg) $\theta_A = 55; \theta_B = 55; \theta_C = 55$				
Apical longitude values (deg) l_A l_B l_C	Quasars' aberrated angular distances (rad) Ψ_1'' Ψ_2'' Ψ_3''	Aberrated apical latitudes (rad) θ_{A2}'' θ_{B2}'' θ_{C2}''	Aberrated velocity $\beta'' = v''/c$	Uncertainty of aberrated apical latitudes $\Delta\theta''$ (rad)
225 90 0	1.584939 1.152488 1.585625	0.8860181 0.8719386 0.8801725	0.1000002	1.483053×10^{-7}
180 45 90	1.585333 0.5985235 1.152927	0.8883619 0.8743042 0.871938	0.1	1.467092×10^{-7}

Table 7.
Quasars' aberrated angular distances, spacecraft's aberrated apical latitudes, uncertainties, and aberrated velocity as a function of quasars' apical latitudes $\theta_A = \theta_B = \theta_C = 55°$.

Apical latitudes (deg) $\theta_A = 65; \theta_B = 65; \theta_C = 65$				
Apical longitude values (deg) l_A l_B l_C	Quasars' aberrated mangular distances (rad) Ψ_1'' Ψ_2'' Ψ_3''	Aberrated apical latitudes (rad) θ_{A2}'' θ_{B2}'' θ_{C2}''	Aberrated velocity $\beta'' = v''/c$	Uncertainty of aberrated apical latitudes $\Delta\theta''$ (rad)
0 45 90	0.6744785 0.6744785 1.31547	1.037221 1.039617 1.037223	0.1000003	6.49072×10^{-8}
0 120 240	1.693532 1.693532 1.692253	1.037222 1.049665 1.037221	0.1	1.224615×10^{-7}
0 170 45	2.078023 1.74993 0.674478	1.037221 1.053743 1.039607	0.1000006	9.741514×10^{-7}
45 215 90	2.077874 1.749306 0.6744784	1.039617 1.050399 1.037222	0.1000002	9.992526×10^{-7}
90 45 135	0.6747422 1.315813 0.6747456	1.045493 1.039617 1.039722	0.1000002	6.436005×10^{-8}
135 45 180	1.316635 1.851687 0.6751235	1.051398 1.039617 1.045640	0.1000002	4.208667×10^{-8}
225 90 0	1.852109 1.316056 1.853118	1.051501 1.037221 1.045566	0.1000004	1.831758×10^{-7}
180 45 90	1.852688 0.6744785 1.316643	1.053883 1.039617 1.037223	0.1000002	1.820385×10^{-7}

Table 8.
Quasars' aberrated angular distances, spacecraft's aberrated apical latitudes, uncertainties, and aberrated velocity as a function of quasars' apical latitudes $\theta_A = \theta_B = \theta_C = 65°$.

Apical latitudes (deg) $\theta_A = 75; \theta_B = 75; \theta_C = 75$				
Apical longitude values (deg) l_A l_B l_C	**Quasars' aberrated angular distances (rad)** Ψ_1'' Ψ_2'' Ψ_3''	**Aberrated apical latitudes (rad)** θ_{A2}'' θ_{B2}'' θ_{C2}''	**Aberrated velocity** $\beta'' = v''/c$	**Uncertainty of aberrated apical latitudes** $\Delta\theta''$ **(rad)**
0	0.7328074	1.205014	0.1	1.086075×10^{-7}
45	0.7328074	1.207440		
90	1.446703	1.205018		
0	1.892254	1.205014	0.1	1.559243×10^{-7}
120	1.892254	1.217651		
240	1.890555	1.205017		
0	2.406806	1.205014	0.1000002	1.289855×10^{-6}
170	1.962084	1.221801		
45	0.7328069	1.207434		
45	2.406558	1.207444	0.1000001	1.301131×10^{-6}
215	1.961239	1.218397		
90	0.7328073	1.205012		
90	0.7331199	1.213411	0.1000007	1.076865×10^{-7}
45	1.447127	1.207444		
135	0.7331235	1.207545		
135	1.448146	1.219414	0.1000007	4.551616×10^{-8}
45	2.091397	1.207444		
180	0.7335716	1.213555		
225	2.091990	1.219517	0.1000006	2.351534×10^{-7}
90	1.447428	1.205014		
0	2.093409	1.213484		
180	2.092804	1.221941	0.1000006	2.348527×10^{-7}
45	0.7328073	1.207444		
90	1.448155	1.205017		

Table 9.
Quasars' aberrated angular distances, spacecraft's aberrated apical latitudes, uncertainties, and aberrated velocity as a function of quasars' apical latitudes $\theta_A = \theta_B = \theta_C = 75°$.

Apical latitudes (deg) $\theta_A = 90; \theta_B = 90; \theta_C = 90$				
Apical longitude values (deg) l_A l_B l_C	**Quasars' aberrated angular distances (rad)** Ψ_1'' Ψ_2'' Ψ_3''	**Aberrated apical latitudes (rad)** θ_{A2}'' θ_{B2}'' θ_{C2}''	**Aberrated velocity** $\beta'' = v''/c$	**Uncertainty of aberrated apical latitudes** $\Delta\theta''$ **(rad)**
0	0.7806323	1.46195	0.1000006	4.35259×10^{-7}
45	0.7806323	1.464438		
90	1.559071	1.461949		
0	2.076409	1.461950	0.1000007	3.782526×10^{-7}
120	2.076409	1.474906		
240	2.074203	1.461950		
0	2.875982	1.461950	0.1000005	3.594844×10^{-6}
170	2.163023	1.479170		
45	0.7806325	1.464444		

Apical latitudes (deg) $\theta_A = 90; \theta_B = 90; \theta_C = 90$				
Apical longitude values (deg) l_A l_B l_C	Quasars' aberrated angular distances (rad) Ψ_1'' Ψ_2'' Ψ_3''	Aberrated apical latitudes (rad) θ_{A2}'' θ_{B2}'' θ_{C2}''	Aberrated velocity $\beta'' = v''/c$	Uncertainty of aberrated apical latitudes $\Delta\theta''$ (rad)
45 215 90	2.875224 2.161893 0.7806324	1.464438 1.475672 1.461953	0.1000002	3.465266×10^{-6}
90 45 135	0.7809882 1.559575 0.7809924	1.470553 1.464438 1.464544	0.1000007	4.313473×10^{-7}
135 45 180	1.560786 2.330901 0.7815026	1.476717 1.464438 1.470704	0.1000008	4.824766×10^{-8}
225 90 0	2.331752 1.559933 2.333790	1.476824 1.461950 1.470629	0.1000007	5.535144×10^{-7}
180 45 90	2.332921 0.78063222 1.560797	1.479315 1.464438 1.461949	0.1000007	5.63842×10^{-7}

Table 10.
Quasars' aberrated angular distances, spacecraft's aberrated apical latitudes, uncertainties, and aberrated velocity as a function of quasars' apical latitudes $\theta_A = \theta_B = \theta_C = 90°$.

Apical latitudes (deg) $\theta_A = 105; \theta_B = 105; \theta_C = 105$				
Apical longitude values (deg) l_A l_B l_C	Quasars' aberrated angular distances (rad) Ψ_1'' Ψ_2'' Ψ_3''	Aberrated apical latitudes (rad) θ_{A2}'' θ_{B2}'' θ_{C2}''	Aberrated velocity $\beta'' = v''/c$	Uncertainty of aberrated apical latitudes $\Delta\theta''$ (rad)
0 45 90	0.7737214 0.7737214 1.542527	−1.415936 −1.413387 −1.415934	0.1000007	3.270342×10^{-7}
0 120 240	2.048211 2.048211 2.046092	−1.415936 −1.402644 −1.415936	0.1000008	8.354338×10^{-8}
0 170 45	2.770738 2.131879 0.7737212	−1.415936 −1.398259 −1.413391	0.1000007	1.055704×10^{-6}
45 215 90	2.770202 2.130802 0.7737214	−1.413387 −1.401856 −1.415935	0.1000008	1.061392×10^{-6}
90 45 135	0.7740707 1.543019 0.7740752	−1.407115 −1.413387 −1.413275	0.1000007	3.237373×10^{-7}
135 45 180	1.544199 2.292566 0.7745759	−1.400782 −1.413387 −1.406957	0.1000006	4.780046×10^{-8}

DOI: http://dx.doi.org/10.5772/intechopen.87981

Apical latitudes (deg) $\theta_A = 105; \theta_B = 105; \theta_C = 105$				
Apical longitude values (deg) l_A l_B l_C	**Quasars' aberrated angular distances (rad)** Ψ_1'' Ψ_2'' Ψ_3''	**Aberrated apical latitudes (rad)** θ_{A2}'' θ_{B2}'' θ_{C2}''	**Aberrated velocity** $\beta'' = v''/c$	**Uncertainty of aberrated apical latitudes** $\Delta\theta''$ **(rad)**
225	2.293366	−1.400672	0.1000007	1.896711×10^{-7}
90	1.543368	−1.415936		
0	2.295284	−1.407037		
180	2.294466	−1.398109	0.1000007	1.861201×10^{-7}
45	0.7737215	−1.413387		
90	1.544210	−1.415934		

Table 11.
Quasars' aberrated angular distances, spacecraft's aberrated apical latitudes, uncertainties, and aberrated velocity as a function of quasars' apical latitudes $\theta_A = \theta_B = \theta_C = 105°$.

Apical latitudes (deg) $\theta_A = 120; \theta_B = 120; \theta_C = 120$				
Apical longitude values (deg) l_A l_B l_C	**Quasars' aberrated angular distances (rad)** Ψ_1'' Ψ_2'' Ψ_3''	**Aberrated apical latitudes (rad)** θ_{A2}'' θ_{B2}'' θ_{C2}''	**Aberrated velocity** $\beta'' = v''/c$	**Uncertainty of aberrated apical latitudes** $\Delta\theta''$ **(rad)**
0	0.7084652	−1.145326	0.1000007	1.25511×10^{-7}
45	0.7084652	−1.142719		
90	1.391237	−1.145329		
0	1.806544	−1.145326	0.1000007	1.622084×10^{-8}
120	1.806544	−1.131707		
240	1.805039	−1.145324		
0	2.256094	−1.145326	0.1000007	8.497113×10^{-8}
170	1.870123	−1.127201		
45	0.7084649	−1.142729		
45	2.255899	−1.142719	0.1000008	8.359305×10^{-8}
215	1.869381	−1.130897		
90	0.7084652	−1.145327		
90	0.7087568	−1.136293	0.1000006	1.239996×10^{-7}
45	1.391625	−1.142719		
135	0.7087601	−1.142611		
135	1.392558	−1.129794	0.1000007	4.385023×10^{-8}
45	1.986321	−1.142719		
180	0.7091784	−1.136131		
225	1.986831	−1.129681	0.1000007	9.03231×10^{-10}
90	1.391901	−1.145327		
0	1.988053	−1.136214		
180	1.987532	−1.127048	0.1000007	1.083175×10^{-9}
45	0.7084652	−1.142719		
90	1.392566	−1.145324		

Table 12.
Quasars' aberrated angular distances, spacecraft's aberrated apical latitudes, uncertainties, and aberrated velocity as a function of quasars' apical latitudes $\theta_A = \theta_B = \theta_C = 120°$.

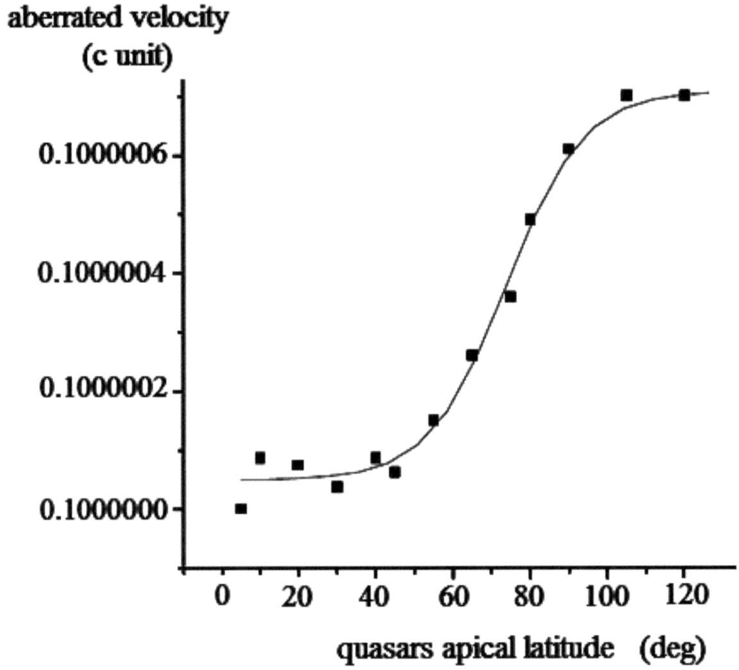

Figure 4.
The averaged aberrated velocity β (in c unit) of a relativistic interstellar spacecraft (with β = 0.1) as a function of quasars' aberrated apical latitude.

Furthermore, as the aberrated coordinates of an interstellar spaceship are related to its aberrated velocity by means of Eqs. (15)–(17) the minimization of the uncertainty in the determination of the spacecraft's aberrated velocity provides an increase in accuracy of determining the aberrated apical coordinates using quasars in that range of aberrated apical latitudes.

This result suggests that one or more normal-sized telescopes aboard the spacecraft can carry out feasible maneuvers along the direction of motion of the spaceship for the automatic measurements of quasars' angular distances because quasars to be used are within a cone with the axis in the direction of motion of the spaceship and an angular aperture of 45° (see **Figure 5**). The large number of quasars whose coordinates have been measured in radio and optical domains and quasars' uniform distribution over the sky [17, 19] can ensure the feasibility of this design.

Furthermore, the limit of accuracy of determining the aberrated coordinates and velocity of an interstellar spacecraft depends on the technique which can be used aboard the spaceship for measuring angular distances between quasars. As described in the previous sections, a positional precision close to 50 μAs for quasars with magnitude lesser than 18 should be reached by means of Gaia space mission which will define a new celestial reference frame in the optical domain, the LQRF. We have performed a simulation study assuming that angular measurements between quasars can be carried out on-board the spaceship with errors within 1 mAs. It may be considered a reasonable estimate of accuracy of automatic angular measurements aboard an interstellar spacecraft, because it represents a value conservatively much smaller than that will be reached in the future astrometry space missions mentioned above. In addition, coordinates' evolution of "stable" quasars is assumed to be around 0.2 mAs [19], so that this uncertainty cannot influence measurements of quasars' angular distance aboard the spacecraft within the assumed accuracy of 1 mAs.

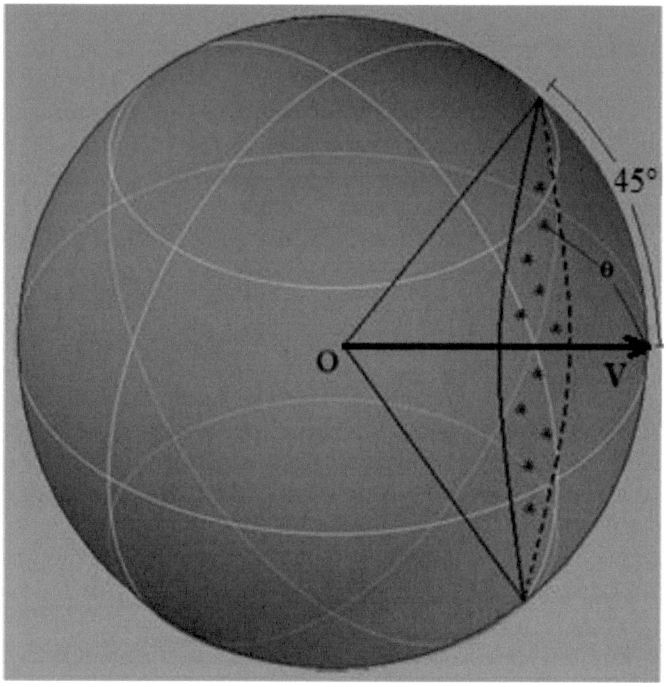

Figure 5.
The accuracy of determining the spacecraft's aberrated velocity and apical coordinates increases using quasars' aberrated apical latitude within a cone with the axis in the direction of motion of the spaceship and an angular aperture of 45°.

Hence, the uncertainty of spacecraft's aberrated apical latitude can be related to the errors of angular measurements carried out on-board the spaceship by means of the following equation derived from Eqs. (18)–(20):

$$\Delta\theta'' = \frac{(\cos\lambda \sin\psi\, \Delta\lambda + \sin\lambda \cos\psi\, \Delta\psi) \sin E - \cos E \sin\lambda \sin\psi\, \Delta E}{\cos\theta'' \sin E^2} \tag{21}$$

The uncertainty values $\Delta\theta''$ of the aberrated apical latitude were computed using the values obtained from this simulation study, assuming that a reasonable estimate of uncertainty of measurements on-board the spacecraft is $\Delta\psi = \Delta\lambda = \Delta E = 1$ mAs. The results of this computation were reported in the last columns of **Tables 1–12**.

Looking at the values $\Delta\theta''$ reported in these columns, it appears that the relative error of the aberrated apical latitude $\frac{\Delta\theta''}{\theta''}$ decreases with a decrease of the aberrated apical latitude, providing the lowest-order relative error values ranging from 10^{-7} to 10^{-9} using aberrated apical latitudes lesser than 45°. This result is in agreement with the previous result regarding the increase in accuracy of determining the spacecraft's aberrated velocity β'' which was obtained using quasars' aberrated apical latitude $\theta \leq 45°$.

6. Conclusions

In this chapter an inertial celestial reference frame represented by three quasars, at least, was described, which can be used for future interstellar space missions at relativistic velocities. The equations to determine the aberrated apical coordinates of a spacecraft as a function of the astronomical coordinates of a set of quasars were derived from spherical astronomy.

In particular, a simulation to increase the accuracy in the determination of the aberrated coordinates of a relativistic spacecraft during an interstellar space mission was carried out. The uncertainty of measurements in navigation control can be minimized selecting the set of quasars. The accuracy of determining the aberrated velocity and the aberrated apical coordinates of a spacecraft increased significantly ($p < 0.01$) using an inertial reference frame formed by quasars with aberrated apical latitudes lower than 45°. This result suggests feasible design techniques for measurements of quasars' aberrated angular distances aboard the spaceship within a cone with the axis in the direction of motion of the spaceship and an angular aperture of 45°.

Further simulation was performed assuming that measurements of quasars' angular distances can be carried out on-board the spacecraft with accuracy within 1 mAs. The uncertainty of the aberrated apical latitudes of a spacecraft was obtained in this simulation providing small errors ranging from 10^{-7} to 10^{-9} using quasars' apical latitudes less than 45°.

Finally, further corrections can be carried out taking into account corrections to aberrated coordinates due to Doppler shift, secular aberration drift, and the expansion of the universe.

A. Appendix

Aberrated velocity and apical latitude of a spaceship as a function of quasars' aberrated apical latitude

Author details

Emanuele Calabrò
CISFA "Interuniversity Consortium of Applied Physical Sciences" (Consorzio Interuniversitario di Scienze Fisiche Applicate), Messina, Italy

*Address all correspondence to: e.calabro@yahoo.com

IntechOpen

© 2020 The Author(s). Licensee IntechOpen. Distributed under the terms of the Creative Commons Attribution - NonCommercial 4.0 License (https://creativecommons.org/licenses/by-nc/4.0/), which permits use, distribution and reproduction for non-commercial purposes, provided the original is properly cited. (cc) BY-NC

References

[1] N.A.S.A. A report by the solar system exploration division. T.O.P.S: Towards other planetary systems. 1993

[2] Dyson F. Project Orion the Atomic Spaceship 1957–1965. USA: Penguin Books; 2002

[3] Bond A et al. Project Daedalus—Report on the BIS starship study. Journal of the British Interplanetary Society. 1978;**31**:S1-S192

[4] Long KF, Obousy RK, Tziolas AC, Mann A, Osborne R, Presby A, et al. Project Icarus: Son of daedalus—Flying closer to another star. Journal of the British Interplanetary Society Daedalus. 2009;**62**(11-12):403-414. arXiv: 1005.3833v1

[5] Frisbee RH. Advanced space propulsion for the 21st century. Journal of Propulsion and Power. 2003;**19**(6): 1129-1154

[6] Piefer G. Advanced fusion for space application. Prepared for Neep 602. 2000

[7] Andrews DG. Interstellar transportation using today's physics. In: Joint Propulsion Conference & Exhibit; Huntsville, Alabama; 2003

[8] Calabrò E. Relativistic aberrational interstellar navigation. Acta Astronautica. 2011;**69**:360-364

[9] Ma C, Feissel M. Definition and realization of the international celestial reference system by VLBI astrometry of extragalactic objects. IERS Technical Note No. 23; International Earth Rotation and Reference System Service (Paris: Observatoire de Paris); 1997

[10] Thornton CL, Border JS. "Radiometric Tracking Techniques for Deep-Space Navigation", Deep-Space Communications and Navigation Series. USA: Jet Propulsion Laboratory, California Institute of Technology; 2000

[11] Downs DS. Interplanetary Navigation using Pulsating Radio Sources. In: NASA Technical Reports TR N74-34150; October 1974. pp. 1-12

[12] Sheikh SI, Pines DJ, Ray PS, Wood KS, Lovellette MN, Wolff MT. Spacecraft navigation using X-ray pulsars. Journal of Guidance, Control, and Dynamics. 2006;**29**(1):49-63

[13] Sheikh SI, Golshan AR, PinesDJ. Absolute and relative position determination using variable celestial X-ray sources. In: 30th Annual AAS Guidance and Control Conference; 2007. pp. 855-874

[14] Ashby N, Howe DA. Relativity and Timing in X-ray Pulsar Navigation. IEEE; Contribution of the U.S. Government; 1-4244-0074-0/06; 2006

[15] Ma C et al. The international celestial reference frame as realized by very long baseline interferometry. The Astronomical Journal. 1998;**116**(1): 516-546

[16] Fey AL, Ma C, Arias EF, Charlot P, Feissel-Vernier M, Gontier AM, et al. The second extension of the international celestial reference frame: ICRF-EXT.1. Astronomical Journal. 2004;**127**:3587-3608

[17] Ma C, Arias EF, Bianco G, Boboltz DA, Bolotin SL, Charlot P et al. The second realization of the international celestial reference frame by very long baseline interferometry. IERS Technical Note No. 35; International Earth Rotation and Reference System Service (Frankfurt am Main: Verlag des Bundsesamts fur Kartographie und Geodasie) 2009

[18] Semyonov OG. Radiation hazard of relativistic interstellar flight. Acta Astronautica. 2009;**64**:644-653

[19] Hewitt A, Burbidge G. A revised and updated catalogue of quasi stellar objects. The Astrophysical Journal Supplement Series. 1993;**87**:451-947

[20] Green RM. Spherical Astronomy. New York, USA: Cambridge University Press; 1985

[21] Calabrò E. Interacting winds in eclipsing symbiotic systems. The case study of EG Andromedae. Journal of Astrophysics and Astronomy (JOAA). 2014;**35**(1):69-85

[22] Harwit M. Astrophysical Concepts. New York, USA: Springer-Verlag; 1988

Chapter 6

Analysis and Control of Nonlinear Attitude Motion of Gravity-Gradient Stabilized Spacecraft via Lyapunov-Floquet Transformation and Normal Forms

Peter M.B. Waswa and Sangram Redkar

Abstract

This chapter demonstrates analysis and control of the attitude motion of a gravity-gradient stabilized spacecraft in eccentric orbit. The attitude motion is modeled by nonlinear planar pitch dynamics with periodic coefficients and additionally subjected to external periodic excitation. Consequently, using system state augmentation, Lyapunov-Floquet (L-F) transformation, and normal form simplification, we convert the unwieldy attitude dynamics into relatively more amenable schemes for motion analysis and control law development. We analyze the dynamical system's periodicity, stability, resonance, and chaos via numerous nonlinear dynamic theory techniques facilitated by intuitive system state augmentation and Lyapunov-Floquet transformation. Versal deformation of the normal forms is constructed to investigate the bifurcation behavior of the dynamical system. Outcome from the analysis indicates that the motion is quasi-periodic, chaotic, librational, and undergoing a Hopf bifurcation in the small neighborhood of the critical point-engendering locally stable limit cycles. Consequently, we demonstrate the implementation of linear and nonlinear control laws (i.e., bifurcation and sliding mode control laws) on the relatively acquiescent transformed attitude dynamics. By employing a two-pronged approach, the quasiperiodic planar motion is independently shown to be stabilizable via the nonlinear control approaches.

Keywords: gravity gradient, nonlinear attitude control, sliding mode, Lyapunov-Floquet transformation, normal forms

1. Introduction

Ever since the launch of *Sputnik*—the first artificial satellite put into earth's orbit in 1957—mankind has increasingly become dependent on space-based technology in many areas of our daily lives. For instance, space-based technology performs an indispensable role in telecommunications, navigation, personal entertainment, weather forecasting, farming, security, defense, scientific exploration, research, innovation, etc. Undoubtedly, the prominence of space technology in shaping humanity's future is unequivocal. The success or failure of a given space mission is largely contingent upon the complex system analysis and design methodologies exerted in converting the initial idea into an elaborate functioning enterprise [1]. It is for this reason that reliable and efficacious methodologies and tools are consistently utilized in space mission formulation and implementation. Thus, there is a need to continuously examine the effectiveness of prevailing space mission analysis and design methodologies. This is in order to improve prevailing tools and approaches that shall expedite relatively simpler, more reliable, and accurate mission modeling and analysis.

Space systems are required to function nominally in their designated orbital locations, maintain appropriate orientation, and conform to planned trajectories despite the ambient perturbing space environment. Strict mission pointing requirements normally constrain spacecraft in orbit around a large body to maintain a fixed stable orbital position and orientation during operation. However, perturbing space-environment torques act to dislodge positioned spacecraft and disorient stabilized ones [2, 3].

Modeling, analyzing, and controlling dynamics of space systems are therefore a crucial component of space mission design. The quest for relatively simpler, more accurate, and more reliable analytical methodologies and tools to represent, scrutinize, and manipulate the dynamics of space systems is therefore a worthwhile undertaking.

Inopportunely, dynamics of space systems tend to be commonly represented by coupled analytical models that possess complex structures encompassing nonlinearity, parameter-variant coefficients, and periodic external excitation terms [4–9]. The requisite analysis essential to fathom such motion is not a trivial undertaking—except for few special cases, the general solution for such dynamical systems cannot be found. The complex structures of the motion's analytical models characteristically point to nondeterministic and potentially chaotic systems over a range of initial conditions and system parameters. Therefore, to analyze dynamical space systems, we often have to be content with nonautonomous, nonlinear, and periodic differential equations [10, 11]. This presents an immense analysis challenge. For instance, time-varying eigenvalues of the periodic linear system matrix cannot determine the system stability. Consequently, methods such as linearization [12], averaging [13], and perturbation techniques [14, 15] have been consistently used to analyze such complex nonlinear, periodic motion. However, the two latter approaches tend to be limited to minimally excited systems (parameter multiplying the periodic terms is small), while linearization is restricted to small domains about the operating point. Further, such methods are inclined to be relatively cumbersome and normally augmented with numerical approaches to analyze dynamical systems and accomplish real-life applications [16–18].

The presence of perturbing torques in the ambient space environment tends to disorient an already stabilized spacecraft and further alters the orbital motion [2, 3]. This is contrary to the prevailing strict pointing requirements that constrain the satellite in orbit to maintain a stable orbital motion and fixed orientation during mission operation. A number of strategies are employed to stabilize spacecraft attitude motion and maintain a desired orientation despite the presence of perturbing torques in the space environment. The most common attitude control and stabilization approaches are three-axis stabilization, spin stabilization, and gravity-gradient stabilization. To provide the determined control input required to offset undesired attitude deviations, these methods employ either active control systems (e.g., thrusters, magnetic torquers, reaction wheels) or passive control systems (e.g., booms). Unlike passive controllers, active controllers utilize an external source of energy to drive the attitude control actuators [17, 19].

Among the stated attitude stabilization methods, gravity-gradient stabilization of spacecraft attitude is attractive due to its relatively intrinsic simplicity, reliability, and low cost [20]. However, it is mostly feasible in low earth orbit due to its principle of operation as discussed in Section 2.

The motion about COM of a rigid gravity-gradient stabilized spacecraft is libratory about the pitch axis. This axis is normal to the orbital plane in an inverse-square gravity field. The satellite will oscillate about a position of stable relative equilibrium if the work done by external perturbing forces is greater than the rotational kinetic energy. The sufficient conditions for stability of relative equilibrium are explained in Section 2. The complete formulation of COM motion for a gravity-gradient stabilized satellite in eccentric orbit consists of six coupled, nonlinear second-order differential equations. This system of equations is considered analytically unsolvable in closed form [5, 7, 21–23].

This chapter aims to first investigate the periodicity, quasi-periodicity, and chaotic behavior of the gravity-gradient stabilized attitude motion. Moreover, the motion stability, resonances and bifurcation behavior will also be examined. Subsequently in Section 4, we synthesize suitable controllers to adequately offset the attitude perturbations experienced by the gravity-gradient stabilized spacecraft in an eccentric orbit. Requisite assumptions made to facilitate the attitude motion analyses will be explicitly stated and qualified.

To model, analyze, and control the nonlinear motion with parameter-variant coefficients and periodic forcing terms, we intend to use approaches based on:

- System state augmentation
- Lyapunov-Floquet (L-F) transformations
- Normal form (NF) theory

The fitting use of the aforementioned transformations and techniques enables dynamical system analysis and control law development in transformed, parameter-invariant, and more tractable coordinates that preserve the original system Lyapunov stability properties [24, 25]. Consequently, we intend to exploit this propitious attribute in our investigation. Applications of L-F transformations in spacecraft dynamics have been previously investigated by authors such as [8, 26]. The former demonstrates how L-F theory enhances the representation of relative spacecraft dynamics in elliptical orbits, while the latter further proposes an orbit control law based on L-F theory. On the other hand, this chapter focuses on the dynamics of a rigid body about its COM while in elliptical orbit around a central large mass.

2. Gravity-gradient attitude stabilization in eccentric orbit

A gravity-gradient stabilized spacecraft attains a state of stable relative equilibrium when its I_z points along the radius vector, I_y points along the normal to the orbit plane, and I_x is along the tangent to the orbit in the LVLH frame (Hill frame) as shown in **Figure 1a**. In addition, the condition $I_y > I_x > I_z$ must similarly be satisfied.

If the work done by external perturbing torques is greater than the rotational kinetic energy of the spacecraft about its COM, motion of the spacecraft in an elliptical orbit will be libratory as illustrated in **Figure 1b**. Equations representing the

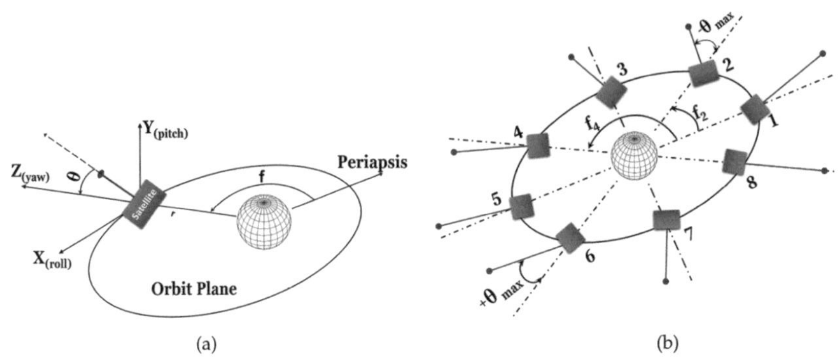

Figure 1.
Gravity-gradient stabilization. (a) Geometry of orbit and attiutde parameters, and (b) Pitch angle librations.

spacecraft orbital motion are identical with those of a point mass in an inverse-square law force field. To analyze the attitude dynamics, the spacecraft orbital motion (motion of COM) can be reasonably assumed to be independent of the spacecraft attitude motion (motion about COM). This assumption is justifiable because the satellite is small compared to the dimensions of the orbit. Under this assumption, the spacecraft's orbital motion can hence transfer energy to the attitude motion, but the converse is assumed not to be possible. Thus, orbital parameters as determined functions of time are considered in analyzing attitude motion [7, 27].

When the spacecraft is considered as a rigid body in an inverse-square force field along an elliptical orbit, a complete formulation of equations of motion can be derived [5, 7, 21]. The resulting six second-order differential equations of motion are nonlinear and coupled. These equations of motion cannot be solved analytically in this exact form.

Ignoring other torques such as aerodynamic, magnetic, thermal bending, and solar radiation pressure, we can derive the equations of spacecraft attitude motion under the influence of inverse-square force field in an elliptical orbit. Additional assumptions are an ideal, perfect sphere earth without oblateness; largest spacecraft dimensions are extremely small compared to the orbit radius, and the spacecraft mass is negligible compared to the mass of the central body [27].

We further assume that the exact equations of motion can be linearized in small-angle motion characterization. Subsequently, the attitude dynamic models may be considered to consist of two equations with coupled roll-yaw angles and a third uncoupled equation describing the pitch angle dynamics. The pitch motion equation is hence independent of roll-yaw motion. The coupled roll-yaw (Ψ-Ω) equations are homogeneous and can be solved for $\Psi = \dot{\Psi} = \Omega = \dot{\Omega} = 0$ [5, 7, 21, 27].

Consequently, the exact problem is reduced to the equation of pitch motion with orbital parameters as functions of time and spacecraft mass parameters shown in Eq. (1):

$$\ddot{\Theta} + 3\frac{\mu}{r^3}\sigma \sin\Theta \cos\Theta = -\dot{\omega} \tag{1}$$

where $0 \le \sigma \le 1$ is a dimensionless ratio of the spacecraft's principal moments of inertia given by

$$\sigma = \frac{I_x - I_z}{I_y} = \frac{I_{roll} - I_{yaw}}{I_{pitch}} \tag{2}$$

To analyze attitude motion in eccentric orbit, we substitute time with true anomaly, where f is the independent variable. Moreover, the COM will obey the following Keplerian orbit relations:

$$r = \frac{P}{1 + e\cos f} \tag{3}$$

$$\omega = \frac{df}{dt} = \frac{\sqrt{\mu P}}{r^2} = \frac{\sqrt{\mu P}}{P^2}(1 + e\cos f)^2 \tag{4}$$

Thus

$$\dot{\omega} = -2\frac{\mu}{r^3}e\sin f \tag{5}$$

$$\ddot{\Theta} = (1 + e\cos f)\frac{\mu}{r^3}\left(\frac{d^2\Theta}{df^2}\right) - 2\frac{\mu}{r^3}e\sin f\left(\frac{d\Theta}{df}\right) \tag{6}$$

Substituting Eqs. (5) and (6) into Eq. (1) yields

$$(1 + e\cos f)\Theta'' - 2e\sin f\Theta' + 3\sigma\sin\Theta\cos\Theta = 2e\sin f \tag{7}$$

This is the well-known equation of plane pitch angle libratory motion in elliptical orbit [5, 7, 21, 27]. The primes indicate differentiation with respect to f. The planar pitch attitude motion equation is hence nonlinear with periodic coefficients in f. Analysis of this motion and subsequent synthesis of a fitting controller is not a trivial task. We hence intend to analyze this motion and synthesize suitable controllers to stabilize the system.

3. Attitude motion analysis

In general, L-F transformation techniques facilitate obtaining solutions of dynamical systems with periodic coefficients, evaluate periodically forced responses, and design feedback control laws. We shall augment these capabilities with normal form techniques (simplifies nonlinearity) and state augmentation (converts nonautonomous to autonomous system). Subsequently, the emanating synergies serve to accomplish the objectives of this chapter.

Floquet theory enables stability and response analysis of linear systems with periodic coefficicnts, i.e., $\dot{\boldsymbol{x}}(t) = \boldsymbol{A}(t)\boldsymbol{x}(t),\ s.t\ \boldsymbol{A}(t) = \boldsymbol{A}(t+T)$. Floquet theory utilizes knowledge of characteristic exponents of the state transition matrix (STM) to infer that if the solution of a system is obtained over a full principle period, then the solution is known for all time [28, 29].

Further, the the Lyapunov-Floquet (L-F) transformation $\boldsymbol{x}(t) = \boldsymbol{Q}(t)\boldsymbol{z}(t)$ reduces the original nonautonomous linear differential system to an autonomous one with the form $\dot{\boldsymbol{z}}(t) = \boldsymbol{R}\boldsymbol{z}(t)$ where $\boldsymbol{R}$ is an $n \times n$ constant matrix. In the LFT matrix, $\boldsymbol{Q}(t)$ can be approximated via the methodology described by [25, 30] using shifted Chebyshev polynomials of the first kind. Chebyshev polynomials are chosen because they produce better approximation and convergence than other special orthogonal polynomials [31]. The approximation of LFT matrix via Chebyshev polynomials contains elements $Q_{ij}(t)$ with truncated Fourier representations as shown in Eq. (8). $Q_{ij}^{-1}(t)$ has similar Fourier representation:

$$Q_{ij}(t) \approx \frac{a_0}{2} + \sum_{n=1}^{q} a_n\cos\left(\frac{\pi nt}{T}\right) + \sum_{n=1}^{q} b_n\sin\left(\frac{\pi nt}{T}\right) \tag{8}$$

Time-independent normal form (TINF) simplification facilitates construction of relatively lesser complex but qualitatively equivalent models of the original nonlinear dynamical systems. On the other hand, time-dependent normal form (TDNF) simplification considers original nonlinear dynamical systems with periodic coefficients by utilizing Lyapunov-Floquet transformation. This simplification is generally implemented on equations arising from Taylor series expansion via successive application of nonlinear near-identity transformations. Such a transformation entails preservation of the original system's stability and bifurcation characteristics by the transformed models. The fundamental concept behind normal forms methodology is to simplify the system by eliminating as many nonlinear terms as possible. This is accomplished via application of successive series of near-identity transformations on the original system. The near-identity coordinate transformations are nonlinear and local. The reader is directed to the well-documented literature on normal form theory found in works by authors such as [32–35].

To normalize nonlinear systems subjected to external periodic excitation, several authors such as [34–37] either utilize approaches that introduce equation variables and/or detuning parameters or incorporate a bookkeeping parameter in their methodology. However, the augmenting parameters involved seemingly lack a uniform explicit connection to the terms in the dynamic equations under consideration. Consequently, we shall utilize a relatively more straightforward and intuitive approach that involves augmenting the system states by converting the periodic external excitation into a system state. The state augmentation approach here is intuitive because the augmented states directly emanate from the periodic forcing term(s); hence, they are neither ad hoc nor arbitrary. Moreover, neither a detuning parameter nor a bookkeeping parameter is required.

Strictly speaking the behavior of the attitude motion as shown in Eq. (7) is characterized in terms variation in true anomaly, f. However, the true anomaly similarly varies with time; hence, the reference as implicit time history is preferred. To demonstrate implicit time history behavior, we initially select $e = 0.2$ and $\sigma = 0.3$. Later on in Section 3.3, we shall analyze the impact of different e-σ pair values on the motion.

Both the motion in original coordinates, Eq. (7) and the corresponding state augmented system, will be scrutinized. **Figure 2** shows the implicit time history behavior of the motion in original coordinates.

Similarly, we shall scrutinize the history behavior of the system in Eq. (7) after augmenting the system states. In accordance with binomial expansion theorem [38], since $|e\cos f| < 1$ and $|(-1)e\cos f| \ll 1$, then the magnitude of the terms in the binomial series progressively become smaller. Therefore, the binomial expansion of the term $(1 + e\cos f)^{-1}$ can be approximated as $(1 - e\cos f)$. Eq. (7) becomes

$$\Theta'' = (1 - e\cos f)(2e\sin f\Theta' - 3\sigma\sin\Theta\cos\Theta + 2e\sin f) \tag{9}$$

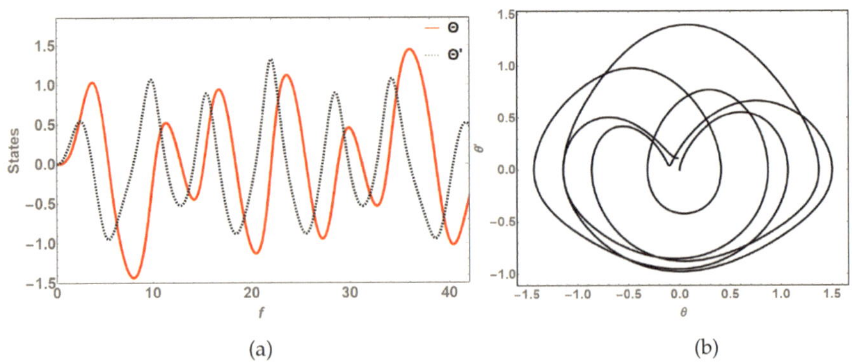

Figure 2.
Motion history behavior in original coordinates for seven complete orbits. (a) Θ and Θ′, (b) Phase portrait.

Let

$$\left.\begin{aligned} p &= \cos f \\ p' &= -sinf = -q \\ q' &= \cos f = p \end{aligned}\right\} \tag{10}$$

After further substituting the trigonometric product term with its series approximation given in Eq. (13) to the 7th order, the motion in Eq. (9) can be expressed as

$$\Theta'' = (1 - ep)\left[2eq\Theta' - 3\sigma\left(\Theta - \frac{2}{3}\Theta^3 + \frac{2}{15}\Theta^5 - \frac{4}{315}\Theta^7\right) + 2eq\right] \tag{11}$$

Therefore, the augmented system state-space representation with $(\Theta_1{}' = \Theta_2, \Theta_2{}' = \Theta_1{}'')$ is shown in Eq. (12):

$$\begin{bmatrix} \Theta_1' \\ \Theta_2' \\ q' \\ p' \end{bmatrix} = \begin{bmatrix} 0 & 1 & 0 & 0 \\ -3\sigma & 0 & 2e & 0 \\ 0 & 0 & 0 & 1 \\ 0 & 0 & -1 & 0 \end{bmatrix} \begin{bmatrix} \Theta_1 \\ \Theta_2 \\ q \\ p \end{bmatrix} + \begin{bmatrix} 0 \\ 2eq[ep(1-\Theta_2)+\Theta_2] - 3\sigma ep\Theta_1 - 3\sigma(1+ep)\left(-\frac{2}{3}\Theta_1^3 + \frac{2}{15}\Theta_1^5 - \frac{4}{315}\Theta_1^7\right) \\ 0 \\ 0 \end{bmatrix} \tag{12}$$

Figure 3 shows the augmented state system history behavior. Allowing for the expected minor discrepancies due to series and binomial expansion approximations, the system state response is comparable to that of the system in original coordinates shown in **Figure 2**.

From **Figures 2** and **3a and b**, the attitude motion is quasiperiodic as characterized by the absence of closed trajectory attractors in the phase space. The pitch angle librates roughly between $-1.5 < \Theta < +1.5$ *radians*, while the pitch angle rate of change varies between $-1.0 < \Theta' < +1.5$. The orbits generally appear to follow a "heart-shaped" path starting at the origin with two non-closing lobes on either side. Conversely, the augmented states are periodic as characterized by the closed circular limit cycle attractor centered at the origin in **Figure 3d**. The motion behavior discussed here indicates that Θ and Θ' are susceptible to instability, unpredictability, and chaos. Consequently, these aspects of the motion flow will be investigated next. Neither the eigenvalues of the linear periodic term of Eq. (7) nor of the state augmented system Eq. (12) can be used to determine stability. Consequently, we'll have to construct the Floquet transformation matrix (FTM) to analyze the dynamical system's stability.

3.1 Stability and chaos

Stability analysis is preceded by computation of the dynamical system's state transition matrix, $\mathbf{\Phi}(t)$. We first prepare the motion in Eq. (7) for expansion via shifted Chebyshev polynomials of the first kind by normalizing the principal

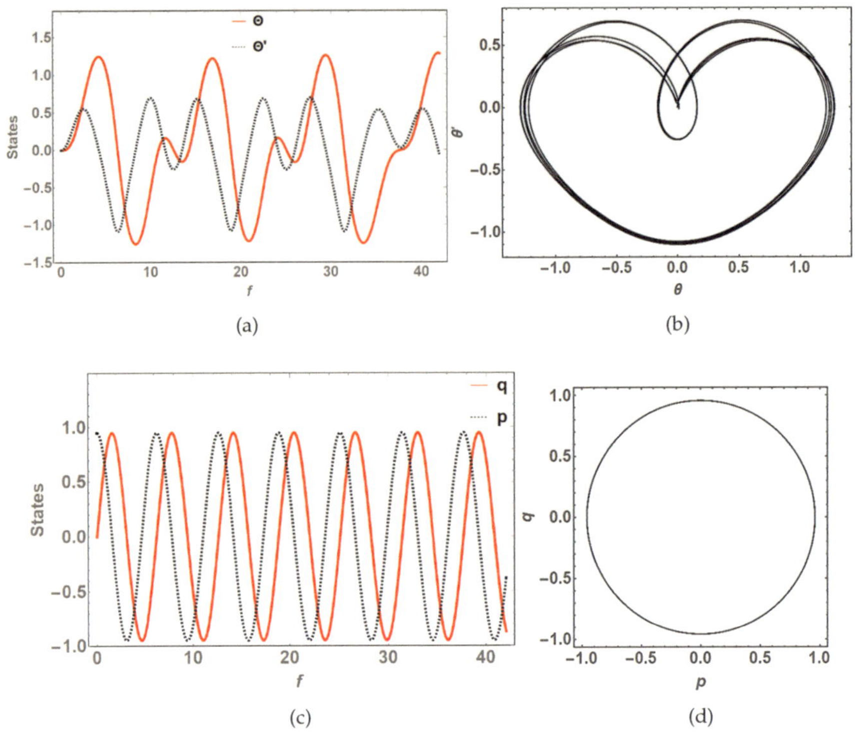

Figure 3.
Motion history behavior of the augmented state system for seven complete orbits. (a) Θ and Θ′ response, (b) Θ and Θ′ phase portrait, (c) q(f) and p(f) response.

period. This is because shifted Chebyshev polynomials are only valid for the period interval $[0, 1]$.

The pitch angle trigonometric product term in Eq. (7) can be represented as a product of the respective Taylor series expansion of sine and cosine as shown in Eq. (13):

$$\sin\Theta\cos\Theta = \Theta - \frac{2}{3}\Theta^3 + \frac{2}{15}\Theta^5 - \frac{4}{315}\Theta^7 + \ldots + a_k\Theta^{2k-1} \tag{13}$$

where $k = 1, 2, 3, 4\ldots$, also series coefficient $a_k \to 0$, as $k \to \infty$.

We can ignore the terms of order greater than 7 in Eq. (13) without significant loss of accuracy because the follow-on terms have relatively small successive coefficients that rapidly approach zero. For instance, the 9th order term has the coefficient $a_5 = 6.7791 \times 10^{-4}$, while the 11th order term's coefficient is $a_6 = 1.98412 \times 10^{-5}$. Substituting the expanded trigonometric product in Eq. (7), we obtain

$$\Theta'' = \frac{1}{(1 + e\cos f)}\left(2e\sin f\Theta' - 3\sigma\left(\Theta - \frac{2}{3}\Theta^3 + \frac{2}{15}\Theta^5 - \frac{4}{315}\Theta^7\right) + 2e\sin f\right) \tag{14}$$

To normalize the principal period, let $f = 2\pi\zeta$. Equivalently, $\zeta \in [0, 1]$ represents duration within the principal period. Let Z represent the principal period, and then this implies that for a periodic term, $\boldsymbol{A}(\zeta) = \boldsymbol{A}(\zeta + Z)$. It follows

$$\Theta' = \frac{d\Theta}{d\zeta}\cdot\frac{1}{2\pi}, \qquad \Theta'' = \frac{1}{4\pi^2}\frac{d^2\Theta}{d\zeta^2} \tag{15}$$

After substituting Θ' and Θ'' from Eq. (15) into Eq. (14), the obtained state-space representation of the normalized attitude motion is given in Eq. (16), where $d\Theta/d\zeta = \mathring{x}_1$ and $d^2\Theta/d\zeta^2 = \mathring{x}_2$. Further, $\left\{\Theta(0) = 0,\ \mathring{\Theta}(0) = 0\right\}$ constitutes the initial conditions of the represented motion which correspond to pitch librations at position 1 of **Figure 1b**:

$$\begin{bmatrix} \mathring{x}_1 \\ \mathring{x}_2 \end{bmatrix} = \begin{bmatrix} 0 & 1 \\ \dfrac{-12\pi^2\sigma}{(1+e\cos 2\pi\zeta)} & \dfrac{4\pi e\sin 2\pi\zeta}{(1+e\cos 2\pi\zeta)} \end{bmatrix} \begin{bmatrix} x_1 \\ x_2 \end{bmatrix} + \frac{12\pi^2\sigma}{(1+e\cos 2\pi\zeta)} \begin{bmatrix} 0 \\ \dfrac{2}{3}x_1^3 - \dfrac{2}{15}x_1^5 + \dfrac{4}{315}x_1^7 \end{bmatrix} + \begin{bmatrix} 0 \\ \dfrac{8\pi^2 e\sin 2\pi\zeta}{(1+e\cos 2\pi\zeta)} \end{bmatrix} \tag{16}$$

It is clear that Eq. (16) is of the form

$$\mathring{\boldsymbol{x}}(\zeta) = \boldsymbol{A}(\zeta)\boldsymbol{x}(\zeta) + \boldsymbol{f}(\zeta,\boldsymbol{x}) + \boldsymbol{F}(\zeta) \tag{17}$$

3.1.1 Floquet multipliers and exponents

To facilitate computation of STM, FTM, and L-F transformation matrices using Chebyshev polynomials, we utilized the Chebfun software package on MATLAB™ [39]. Summarily, Chebfun applies piecewise Chebyshev polynomial interpolation to construct smooth functions over the interval $[-1, +1]$. Recall that $\zeta = f/2\pi$ and Z are the normalized principal period; hence, the computed $FTM = \boldsymbol{\Phi}(Z)$ is

$$\boldsymbol{\Phi}(Z) = \begin{bmatrix} 0.9462 & -0.0529 \\ 1.9796 & 0.9462 \end{bmatrix} \tag{18}$$

The computed Floquet multipliers are critical since they lie on the unit circle with values of $(0.9462 \pm 0.3236i)$ as shown in **Figure 4**. Consequently, this reveals a marginally stable system for the chosen e-σ pair. It follows that the corresponding

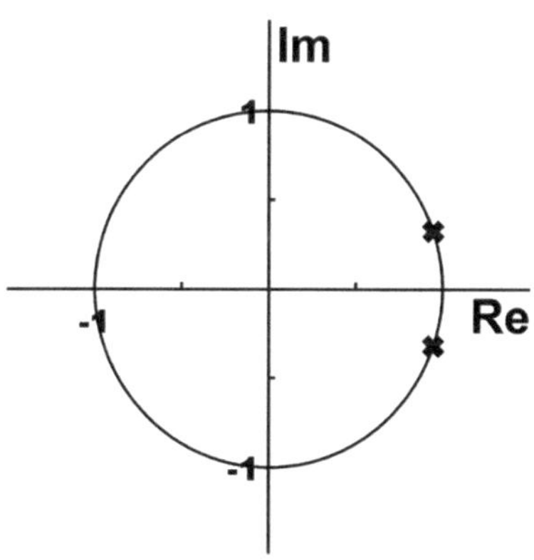

Figure 4.
Floquet multiplier location.

Floquet exponents ($0 \pm 0.3295i$) are purely imaginary. This is consistent with the quasiperiodic system phase portraits that illustrate a librational motion.

The motion is hence stable in the sense of Lyapunov, but the inherent oscillations are disruptively significant to jeopardize nominal execution of the spacecraft mission.

3.1.2 Poincaré map

Figure 5 shows the constructed Poincaré section of the flow. There is a discernible main cluster of points in close proximity to the origin but restricted to the positive side of Θ'. Relatively scanty, isolated discrete points occupy the lower bottom half of the plot bound by $-1 < \Theta < 2$.

The Poincaré section composition suggests two possible flow behaviors. The groupings suggest a quasiperiodic trajectory. On the other hand, the scanty random points devoid of clustering could be due to transient behavior or chaos. A chaotic motion can briefly dwell on a near periodic trajectory before changing to a disparate trajectory with a period that is a multiple of the preceding motion. Consequently, it is needed to further investigate the presence of chaos in the attitude motion dynamics.

3.1.3 Chaos

We define chaos as a bounded aperiodic steady-state motion behavior that is not in equilibrium and is sensitive to initial conditions. A minuscule divergence in the input rapidly grows to spawn an overwhelming difference in the system response. We begin investigating chaos in the attitude motion by plotting the system implicit time history with a minute divergence to the initial condition of the state, Θ. We set the first initial condition to zero; the divergent second initial condition is obtained

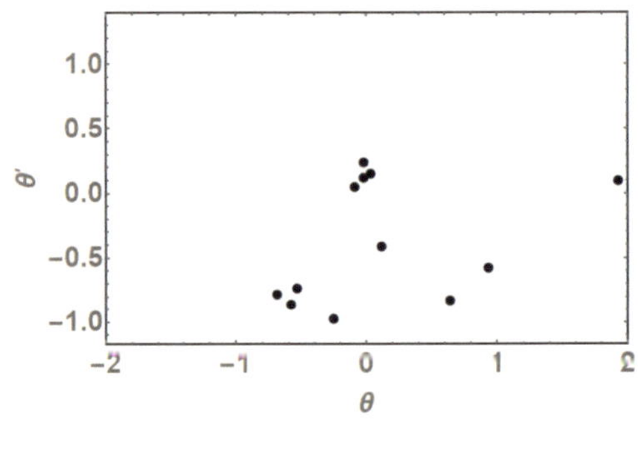

Figure 5.
Poincaré section of the attitude motion.

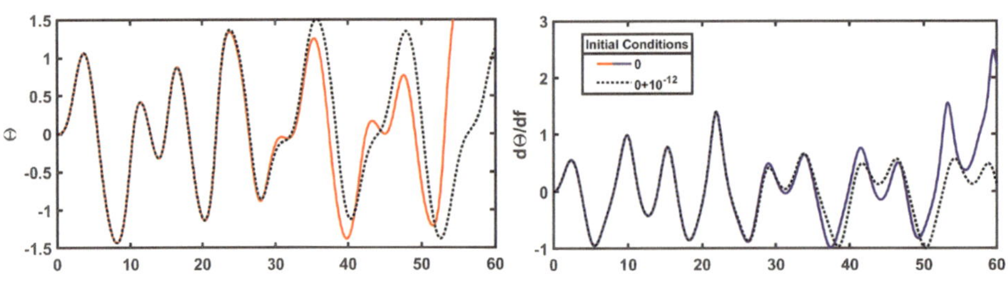

Figure 6.
Chaos: attitude motion sensitivity to initial conditions.

by adding $\epsilon = 10^{-12}$ to the zero initial condition. The obtained implicit time history of the two curves reveal the onset and progression of an overwhelming difference in the system response as illustrated in **Figure 6**. The slightly divergent initial condition results in an overwhelming difference in response that begins in the second half of the 4th orbit and then rapidly grows in the subsequent orbits. The attitude motion is hence chaotic.

3.1.4 Lyapunov exponents

To determine the average rate of divergence between the initially neighboring trajectories defined locally in the state space, we shall scrutinize the dynamic behavior of the motion's Lyapunov exponents. Lyapunov exponent stability analysis affords a means of quantifiable expression for initial conditions sensitivity and dependence (i.e., chaos), by describing the exponential rate of growth or decay of a perturbation to a trajectory as time progresses [40]. Lyapunov exponent λ is numerically expressed as

$$\lambda = \lim_{t \to \infty} \frac{1}{t} \ln \left(\left\| \frac{\boldsymbol{\delta y}(t)}{\boldsymbol{\delta y}(0)} \right\| \right) \tag{19}$$

where $\boldsymbol{\delta y}(t)$ is the tiny separating perturbation vector between the trajectories. The value of Lyapunov exponent will distinguish the nature of the trajectory according to the following criteria: (i) $\lambda < 0$: Trajectory is stable and the motion is asymptotically stable. (ii) $\lambda = 0$: Trajectory is neutral and the motion is characterized by some sort of steady-state. (iii) $\lambda > 0$: Trajectory is unstable and chaotic. The Lyapunov exponent behavior for the motion given Eq. (16) is illustrated in **Figure 7**.

The computed Lyapunov exponents are equal in magnitude but opposite in sign with increasing periods because the flow in Eq. (16) is nonautonomous Hamiltonian. Hamiltonian systems are conservative. Therefore, the magnitude of λ_1 which measures expansion in one direction is equal to the magnitude of λ_2 which measures contraction in another direction. Since $\lambda_1 > 0$ always, then as prescribed by the above distinction for λ, the attitude motion is chaotic. This outcome is consistent with the preceding chaos analyses that scrutinized motion sensitivity to initial conditions.

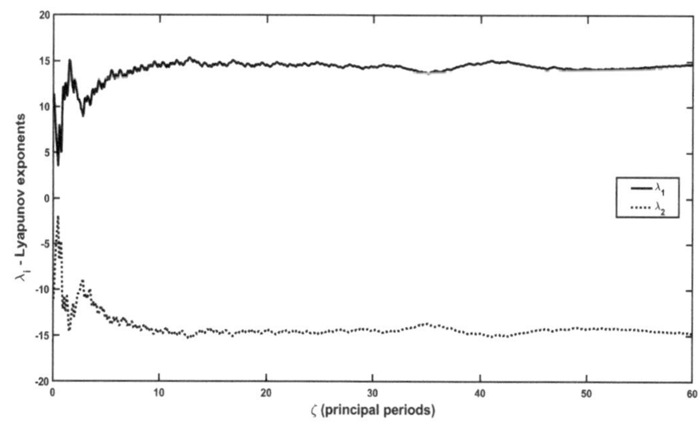

Figure 7.
Dynamics of Lyapunov exponents for the attitude motion.

3.1.5 Stability charts

The orbit eccentricity and spacecraft's ratios of principal moments of inertia are, respectively, defined as $e \in \{0, 1\}$ and $\sigma \in \{0, 1\}$. On the other hand, our stability analysis so far has been illustrated in the courtesy of arbitrarily set values of $(e = 0.2\ , \sigma = 0.3)$. Consequently is it essential to holistically scrutinize the motion behavior for all possible values of e and σ. Constructing stability charts which partition the e-σ plane into stable and unstable regions enables scrutiny of motion stability as e and σ vary simultaneously. Transition curves in stability charts constitute frontiers that separate stable regions from unstable regions. We can derive transition curves in closed form via the FTM. Floquet theory requires a stable system to have a Floquet multiplier of magnitude, $|\rho_k| \ \leq 1$. It can hence be proven that the transition from stability to instability occurs when both Floquet multipliers are equal to 1 or both are equal to -1 (see **Figure 4**). Therefore, the transition curves in the e-σ plane where the solution to the linear periodic term of Eq. (17) changes from stable to unstable (or vice versa) are determined by the conditions $Trace[FTM] = \pm 2$ or $|\rho_k| \ = 1$.

Even though we can only construct transition curves associated with the linear term of the attitude motion equation, the outcome provides insightful perusal into the behavior of the whole equilibrium solution. The complete solution behavior can be arrived at by augmenting the evaluated linear periodic stability behavior with the combined effect of the nonlinear term $\boldsymbol{f}(x, \zeta)$ and forcing term $\boldsymbol{F}(\zeta)$ in Eq. (17). For instance, if a given e-σ pair is initially located in the unstable region of the FTM-dependent stability chart, then the nonlinear and forcing terms will tend to exacerbate this instability, rendering the complete solution unstable for that particular $e - \sigma$ combination. A similar argument can be made for an e-σ pair located in a stable region. The constructed stability chart of the attitude motion is shown in **Figure 8**.

The darker regions constitute stable points while the lighter regions are unstable. The stability chart has a slightly larger stable region than the unstable region.

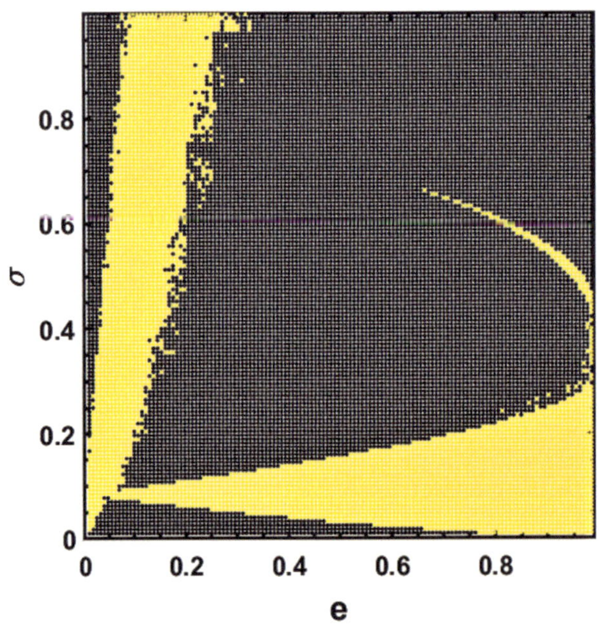

Figure 8.
Stability chart of the attitude motion.

Unstable solutions appear to be dominated by two regions approximately defined by (i) $\sigma \in \{0, 0.2\}$ and increasing values of e plus (ii) $e \in \{0, 0.3\}$ and increasing values of σ. Essentially, a spacecraft with mass distribution such that $\sigma > 0.2$ is amenable to a wider range of eccentricity values above 0.2 to achieve an intended stable pitch angle motion.

For instance, we have previously considered the pair $\{e = 0.2, \sigma = 0.3\}$; this pair is located in the stable region corresponding to critical Floquet multipliers confirmed in **Figure 4**. The motion at this location is stable in the sense of Lyapunov. The stability chart further accords a means of scrutinizing generalized stability behavior trends or commonalities between disparate e-σ pairs. By picking representative e-σ pairs from different regions, we tabulate the resultant illustrative Floquet multipliers as shown in **Table 1**.

From **Table 1**, we note that both selected marginal and stable regions are associated with critical Floquet multipliers. This implies that the pertinent e-σ pair characterizes a motion stable in the sense of Lyapunov; i.e., the pitch angle librations are bound by $\Theta \in \{-\pi, +\pi\}$. This is consistent with the dynamics presented in **Figures 1** and 3. However, in the unstable regions, Floquet multipliers have magnitudes $|\rho_k| > 1$, implying that the pitch angle wanders beyond $\pm\pi$.

3.2 Resonance

The attitude dynamics are dominated by the linear and forcing terms delineated in Eq. (17). This is because if we consider the motion composed of only the linear and forcing terms, i.e., $\mathring{\boldsymbol{x}}(\zeta) = \boldsymbol{A}(\zeta)\boldsymbol{x}(\zeta) + \boldsymbol{F}(\zeta)$, the numerical solution is unbounded as shown in **Figure 9**. This is not the case if we consider motion composed of any of the following term combinations: $\mathring{\boldsymbol{x}}(\zeta) = \boldsymbol{A}(\zeta)\boldsymbol{x}(\zeta)$, $\mathring{\boldsymbol{x}}(\zeta) = \boldsymbol{A}(\zeta)\boldsymbol{x}(\zeta) + \boldsymbol{f}(x, \zeta)$, or $\mathring{\boldsymbol{x}}(\zeta) = \boldsymbol{f}(x, \zeta) + \boldsymbol{F}(\zeta)$.

Moreover, if we consider the L-F transformed motion, Floquet exponents (eigenvalues of $\boldsymbol{R}$) represent frequencies associated with the linear term. Periodic elements of the nonlinear matrix are a product of the truncated Fourier series matrices $\boldsymbol{Q}(\zeta)$ and $\boldsymbol{Q}^{-1}(\zeta)$. On the other hand, periodic elements of the forcing matrix are likewise multiplied by $\boldsymbol{Q}^{-1}(\zeta)$. Subsequently, resonance between the Floquet exponents and any of the periodic terms in the forcing matrix elements will trigger instability in the motion as well.

Bifurcation triggers the system's equilibrium solutions to transition between the disparate regions of the stability chart. The orbit eccentricity, e, is the bifurcation parameter for the attitude motion (see Eq. (16)). This is because the general on-orbit spacecraft mass properties represented by σ typically tend to be constant for a

Stability region	e	σ	Floquet multipliers (ρ_1, ρ_2)	$\lvert\rho_1\rvert$	$\lvert\rho_2\rvert$
	0.82	0.6	$-1.06522, \ -0.93877$	1.06	0.94
Unstable	0.6	0.1	$-0.20126, \ -4.96854$	0.20	4.96
	0.8	0.14	$-0.10212, \ -9.79244$	0.10	9.79
Marginal	0.2	0.5	$0.13463 \pm 0.99099i$	1	1
	0.4	0.4	$0.79713 \pm 0.60382i$	1	1
Stable	0.6	0.5	$-0.1117 \pm 0.99373i$	1	1
	0.8	0.9	$0.71565 \pm 0.69844i$	1	1

Table 1.
Stability of representative e-σ pairs.

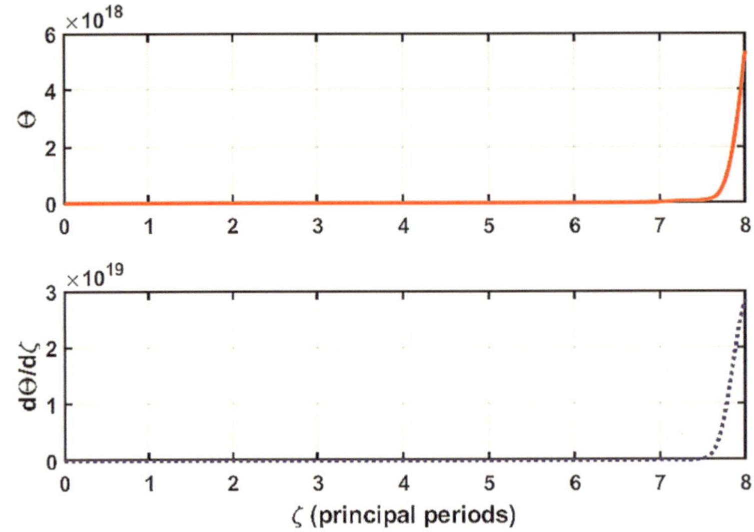

Figure 9.
Resonance in linear and forcing terms.

gravity-gradient stabilized spacecraft. Therefore, it is essential to analyze the equilibrium solution dynamics as small increments are applied on the bifurcation parameter. We develop the normal form of our dynamics in the next section to facilitate bifurcation behavior analysis. Normal forms are not unique, consequently near identity transformation for the state-augmented system, and the L-F transformed system will be undertaken separately.

3.3 Versal deformation of the normal form and bifurcation analysis

Versal deformation refers to embedding the system in a parameterized family of systems containing all possible dynamics that can occur near the bifurcation point. Moreover, the family of systems should be transverse to the bifurcation surface with the number of parameters equal to the codimension of the bifurcation [41]. The attitude motion undergoes a codimension-one bifurcation because only one parameter e is responsible for the loss of stability (for gravity-gradient stabilization to be maintained, σ has to remain fixed). Because our critical Floquet multipliers are complex and lie on the unit circle, this system will experience a Hopf bifurcation. Further, it is "well-known" that Hopf bifurcation is a codimension-one bifurcation. Firstly, we shall formulate the normal forms of nonlinearities up to the cubic order in Eqs. (16) and (12) to demonstrate the intended approach. Normalization of dynamics with higher order nonlinearities can be accomplished via the same techniques.

3.3.1 State-augmented system

To normalize the state-augmented system, we first apply the modal transformation $\boldsymbol{\Theta} = \boldsymbol{M}\boldsymbol{y}$ to Eq. (12) and obtain Eq. (20), where $\boldsymbol{\Theta} = [\Theta_1 \quad \Theta_2 \quad q \quad p]^T$:

$$y' = \boldsymbol{J}y + \boldsymbol{M}^{-1}\begin{bmatrix} 0 \\ 2eq[ep(1-\Theta_2)+\Theta_2] - 3\sigma ep\Theta_1 - 3\sigma(1+ep)\left(-\frac{2}{3}\Theta_1^3\right) \\ 0 \\ 0 \end{bmatrix} \tag{20}$$

This system now possesses 4th order nonlinearity. $\boldsymbol{J}$ is in the Jordan canonical form. The normal form is evaluated by successive application of near identity transformation of the form

$$\boldsymbol{y} = \boldsymbol{v} + \boldsymbol{h}_4(\boldsymbol{v}, f) \tag{21}$$

where $\boldsymbol{h}_r(\boldsymbol{v})$ is an $n \times 1$ homogeneous vector of monomials in $\boldsymbol{v}$ of degree r. The state-augmented system is independent of periodic coefficients; hence, we solely obtain the TINF given in Eq. (22):

$$\begin{bmatrix} v_1' \\ v_2' \\ v_3' \\ v_4' \end{bmatrix} = \begin{bmatrix} -iv_1 \\ iv_2 \\ -i0.948683v_3 + i30.4002v_1v_2v_3 + i1.05409v_3^2v_4 \\ i0.948683v_4 - i30.4002v_1v_2v_4 - i1.05409v_3v_4^2 \end{bmatrix} \tag{22}$$

When the external forcing term is augmented as a system state, the magnitude of the external forcing frequency appears as solitary, linear, imaginary conjugate coefficients in the normal form, i.e., 1 (see Eq. (9)). Eigenvalues of the linear matrix in Eq. (12) constitute the conjugate coefficients in the linear terms of the reduced nonlinearity normal forms (i.e., $\pm i0.948683$).

Moreover, after obtaining the straightforward solutions of $v_1(f)$ and $v_2(f)$ and then substituting these values in the equations for $v_3'(f)$ and $v_4'(f)$, the last two equations of the normal form can be expressed as

$$v_3' = i(-0.948683 + 30.4002C_1C_2)v_3 + i1.05409v_3^2v_4$$
$$v_4' = i(0.948683 - 30.4002C_1C_2)v_4 - i1.05409v_3v_4^2$$

where C_1 and C_2 are the integration constants originating from the analytical solutions of $v_1(f)$ and $v_2(f)$, respectively.

We shall investigate the bifurcation of Eq. (12) via its normal form given above. Because the periodic system maintains the same general structure, we may treat the respective limit cycles as equilibria and study their bifurcations. We utilize the versal deformation of the normal form to investigate the change in the stability structure of the dynamics in the neighborhood of the critical point of the bifurcation parameter. Essentially, construction of versal deformation of the normal form facilitates characterization of system dynamics at the critical point and its small neighborhood. Therefore, we handily gain complete understanding of the qualitative phase space dynamics of the dynamical system in the neighborhood of the critical point.

We define the normal form versal deformation parameter as μ_1. The parameter μ_1 represents a small change in the eigenvalues of the normal form corresponding to a small change in the bifurcation parameter in the original system coordinates. It is a prerequisite condition to obtain a relationship between the versal deformation parameter μ_1 and the original system bifurcation parameter e. Incorporating μ_1 in Eq. (22), we obtain Eq. (23):

$$\begin{bmatrix} v_1' \\ v_2' \\ v_3' \\ v_4' \end{bmatrix} = \begin{bmatrix} \mu_1 - i & 0 & 0 & 0 \\ 0 & \mu_1 + i & 0 & 0 \\ 0 & 0 & \mu_1 - \varpi & 0 \\ 0 & 0 & 0 & \mu_1 + \varpi \end{bmatrix} \begin{bmatrix} v_1 \\ v_2 \\ v_3 \\ v_4 \end{bmatrix} + \begin{bmatrix} 0 \\ 0 \\ i1.05409v_3^2v_4 \\ -i1.05409v_3v_4^2 \end{bmatrix} \tag{23}$$

where $\varpi = i(0.948683 - 30.4002C_1C_2)$.

By defining small increments on the bifurcation parameter as η, we can write $e_k = e_c + \eta_k$ to represent the k disparate sets of bifurcation parameter in the neighborhood of the critical parameter $e_c = 0.2$. We employ the least squares, curve fitting technique proposed by [42] to obtain the relationship between μ_1 and η, as $\mu_1 = (1.47476 + i0.301628)\eta - (1.82052 + i0.414608)\eta^2$. NB values of C_1 and C_2 in Eq. (23) were evaluated by forward action transformations of initial conditions in the original coordinates, i.e., $\Theta_1 = \Theta_2 = 0, p = 1, q = 0$.

The closed-form analytical solutions for $v_1(f)$ and $v_2(f)$ in the versal deformation in Eq. (23) are straightforward. To obtain $v_3(f)$ and $v_4(f)$, we introduce the complex changes of variable $v_3(f) = u_1 - iu_2$ and $v_4(f) = u_1 + iu_2$ followed by the polar coordinates $u_1 = R\cos\theta$ and $u_2 = Rsin(\theta)$. The last two equations in Eq. (23) become

$$\left.\begin{aligned} R' &= \mathrm{Re}(\mu_1)R \\ \theta' &= 0.948683 - 30.4002C_1C_2 - 1.05409R^2 \end{aligned}\right\} \tag{24}$$

After solving Eq. (24), we utilize the results to complete the closed-form analytical solution of Eq. (23) as given in Eq. (24):

$$\begin{aligned} v_1(f) &= e^{\mu_1 f} C_1 \exp(-if) \\ v_2(f) &= e^{\mu_1 f} C_2 \exp(if) \\ v_3(f) &= e^{\mu_1 f} C_3 \exp - \left((0.948683 - 30.4C_1C_2)f - \frac{0.52705e^{2\mu_1 t} C_3^2}{\mu_1} + C_4 \right) i \\ v_4(f) &= e^{\mu_1 f} C_3 \exp\left((0.948683 - 30.4C_1C_2)f - \frac{0.52705e^{2\mu_1 f} C_3^2}{\mu_1} + C_4 \right) i \end{aligned} \tag{25}$$

Similarly, C_3 and C_4 are the integration constants originating from the analytical solutions of $v_3(f)$ and $v_4(f)$, respectively. The values of these integration constants are evaluated from the initial conditions specified in the original coordinates. After back transformation of the normal form closed-form analytical solutions above, we obtain the motion in the original coordinates.

The back transformed $v1(f)$ and $v_2(f)$ constitute the augmented states given in **Figure 3c**. Moreover, $\mu_1 \neq 0$ but is generally small in the order of magnitude 10^{-4}. The integration constants on evaluation are imaginary whose magnitudes are close to identity. Consequently, from Eq. (25), back transformation of the sinusoidal $v1(f)$ and $v_2(f)$ will result in the trigonometric augmented states whose amplitude is determined by the magnitude of the integration constants. From Eq. (24), we can express the transient solution of R as $R = e^{i(\mp i\mu_1 \pm |C|)}$. Since $\mu_1 \neq 0$, the motion is characterized by a locally stable limit cycle in the neighborhood of the bifurcation point. The limit cycle is stable in the sense of Lyapunov but not asymptotically stable. Post-bifurcation attractors that transform into quasiperiodic attractors portraying a limit cycle in the original coordinates are obtained via back transformation and subsequently shown in **Figure 10** ($\eta = 0.00001$).

3.3.2 L-F transformed system

Here, we also demonstrate analysis of bifurcation behavior subject to different values of e-σ pair. Consequently, by utilizing the developed stability chart (**Figure 8**), we select $\{e = 0.1, \sigma = 0.2\}$. This e-σ pair lies on a transition curve; hence, the system is guaranteed to be bifurcating. Again, by considering up to the cubic nonlinearity, the history behavior from Eq. (16) is shown in **Figure 11**.

The system likewise possesses critical Floquet multipliers that lie on the unit circle of values $(0.1435 \pm 0.9896i)$ and purely imaginary Floquet exponents, $\pm 1.4268i$. The computed FTM and $\boldsymbol{R}$ matrices are given in Eq. (26):

$$\boldsymbol{\Phi}(Z) = \begin{bmatrix} 0.1435 & -0.1914 \\ 5.1161 & 0.1435 \end{bmatrix}, \qquad \mathbf{R} = \begin{bmatrix} 0 & -0.276 \\ 7.376 & 0 \end{bmatrix} \tag{26}$$

The computed periodic L-F transformation $\mathbf{Q}(\zeta)$ matrix and $\mathbf{Q}^{-1}(\zeta)$ are plotted in **Figure 12**.

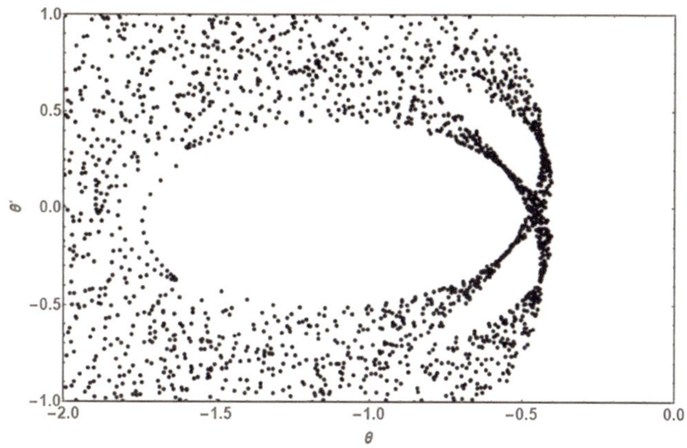

Figure 10.
Poincaré map of state augmented motion post-bifurcation behavior.

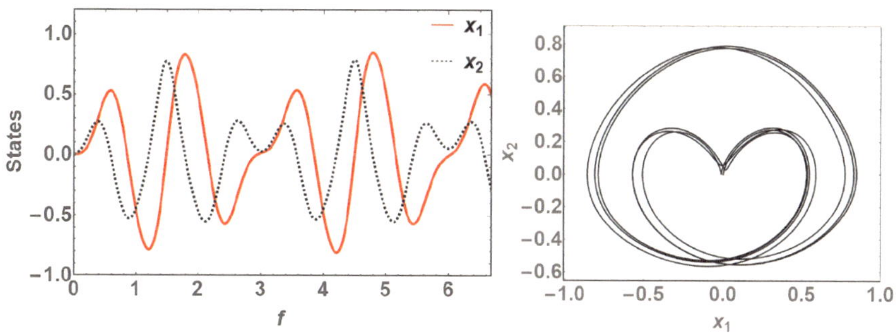

(a) (b)

Figure 11.
L-F transformed system implicit time history response for seven complete orbits. (a) x_1 and x_2, and (b) Phase portrait

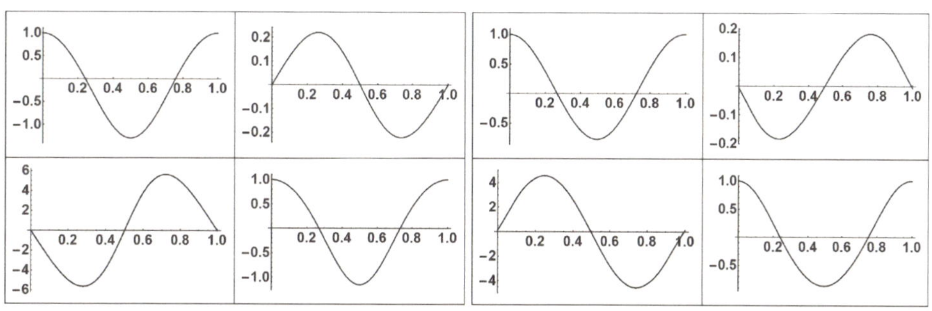

(a) (b)

Figure 12.
Plot of elements of the LFT matrix and its inverse. (a) Q_{ij} (ζ), (b) Q_{ij}^{-1} (ζ).

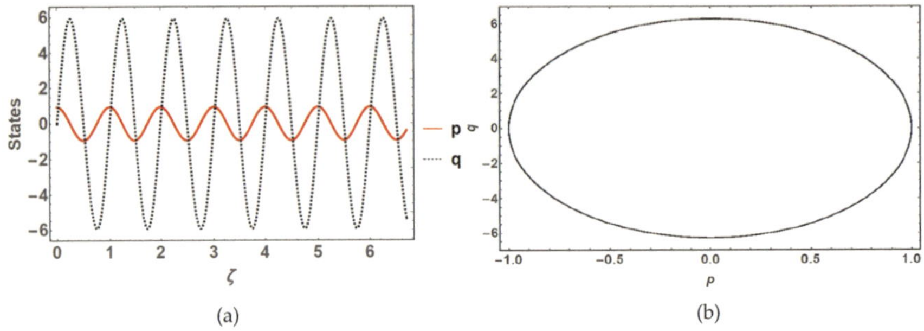

Figure 13.
Augmented system states behavior. (a) $p(\zeta)$ and $q(\zeta)$, (b) Phase portrait.

We consider the L-F transformed dynamics in Eq. (27) up to the cubic nonlinearity. After applying the L-F transformation given in Eq. (19) on the attitude motion in Eq. (16), we obtain the system in Eq. (27):

$$\mathring{z} = Rz + Q^{-1}\begin{bmatrix} 0 \\ \dfrac{12\pi^2\sigma}{(1+e\cos 2\pi\zeta)}\left\{\dfrac{2}{3}(Q_{11}z_1 + Q_{12}z_2)^3\right\} \end{bmatrix} + Q^{-1}\begin{bmatrix} 0 \\ \dfrac{8\pi^2 e \sin 2\pi\zeta}{(1+e\cos 2\pi\zeta)} \end{bmatrix} \tag{27}$$

To normalize this externally excited motion, the system states are augmented to convert the system from nonautonomous to autonomous. We define additional states shown in Eq. (28) and plotted in **Figure 13**:

$$\left.\begin{aligned} p &= \cos(2\pi\zeta) \\ \mathring{p} &= -2\pi sin(2\pi\zeta) = -q \\ \mathring{q} &= 4\pi^2\cos(2\pi\zeta) = 4\pi^2 p \end{aligned}\right\} \tag{28}$$

After substituting the above augmented states into Eq. (27), we obtain the system shown in Eq. (29)—whose order of nonlinearity increases to four. The transformed denominator term $(1+e\cos(2\pi\zeta))^{-1}$ has been approximated by the binomial expansion equivalent, i.e., $(1+ep)^{-1} \approx (1-ep)$.

Apart from raising the order of nonlinearity, state augmentation further introduces periodic linear terms, $4\pi e Q_{12}^{-1} q$ and $4\pi e Q_{22}^{-1} q$ in Eq. (29). Consequently, the augmented dynamics with linear parameter-variant coefficients necessitate a second L-F transformation to convert the linear term to parameter invariant:

$$\begin{bmatrix} \mathring{z}_1 \\ \mathring{z}_2 \\ p^\circ \\ q^\circ \end{bmatrix} = \begin{bmatrix} 0 & -0.276 & 0 & 0 \\ 7.376 & 0 & 0 & 0 \\ 0 & 0 & 0 & -1 \\ 0 & 0 & 4\pi^2 & 0 \end{bmatrix}\begin{bmatrix} z_1 \\ z_2 \\ p \\ q \end{bmatrix} + 8\pi^2\sigma\begin{bmatrix} (1-ep)(Q_{11}z_1 + Q_{12}z_2)^3 Q_{12}^{-1} \\ (1-ep)(Q_{11}z_1 + Q_{12}z_2)^3 Q_{22}^{-1} \\ 0 \\ 0 \end{bmatrix} + 4\pi e\begin{bmatrix} q(1-ep)Q_{12}^{-1} \\ q(1-ep)Q_{22}^{-1} \\ 0 \\ 0 \end{bmatrix} \tag{29}$$

The computed parameters corresponding to the second L-F transformation are as follows: critical Floquet multipliers, $0.1435 \pm 0.9897i$, and purely imaginary

Floquet exponents, $0 \pm 1.4268i$. The computed second FTM and constant $\boldsymbol{R}$ matrices are given in Eqs. (30) and (31), respectively:

$$\boldsymbol{\Phi}^*(Z) = \begin{bmatrix} 0.1435 & -0.1914 & -0.4343 & -0.054 \\ 5.1161 & 0.1435 & -2.0844 & 0.307 \\ 0 & 0 & 1 & 1 \\ 0 & 0 & 0 & 1 \end{bmatrix} \tag{30}$$

$$\boldsymbol{R}^* = \begin{bmatrix} 0 & -0.276 & -0.6456 & -0.0022 \\ 7.376 & 0 & -0.1165 & 0.4524 \\ 0 & 0 & 0 & 0 \\ 0 & 0 & 0 & 0 \end{bmatrix} \tag{31}$$

The computed second periodic L-F transformation matrix, $\boldsymbol{Q}^*(\zeta)$, and its inverse, $\boldsymbol{Q}^{*-1}(\zeta)$, are similarly presented as plots in **Figure 14**.

We designate the second L-F transformation as $\boldsymbol{z} = \boldsymbol{Q}^*\boldsymbol{w}$. Here, $\boldsymbol{z} = [z_1\ z_2\ p\ q]^T$. After applying the second L-F transformation to the state augmented periodic system, we obtain

$$\mathring{\boldsymbol{w}} = \boldsymbol{R}^*\boldsymbol{w} + \boldsymbol{Q}^{*-1}(8\pi^2\sigma)\begin{bmatrix} (1-ep)(Q_{11}z_1 + Q_{12}z_2)^3 Q_{12}^{-1} \\ (1-ep)(Q_{11}z_1 + Q_{12}z_2)^3 Q_{22}^{-1} \\ 0 \\ 0 \end{bmatrix} - \boldsymbol{Q}^{*-1}(4\pi e^2)\begin{bmatrix} qpQ_{12}^{-1} \\ qpQ_{22}^{-1} \\ 0 \\ 0 \end{bmatrix} \tag{32}$$

Applying the modal transformation $\boldsymbol{w} = \boldsymbol{M}^*\boldsymbol{y}$ to Eq. (32) transmutes this system into Eq. (33):

$$\mathring{\boldsymbol{y}} = \boldsymbol{J}\boldsymbol{y} + \boldsymbol{M}^{*-1}\boldsymbol{Q}^{*-1}(8\pi^2\sigma)\begin{bmatrix} (1-ep)(Q_{11}z_1 + Q_{12}z_2)^3 Q_{12}^{-1} \\ (1-ep)(Q_{11}z_1 + Q_{12}z_2)^3 Q_{22}^{-1} \\ 0 \\ 0 \end{bmatrix} - \boldsymbol{M}^{*-1}\boldsymbol{Q}^{*-1}(4\pi e^2)\begin{bmatrix} qpQ_{12}^{-1} \\ qpQ_{22}^{-1} \\ 0 \\ 0 \end{bmatrix} \tag{33}$$

where

$$z_1 = \sum_{i=1}^{4} Q_{1i}^* w_i, \quad z_2 = \sum_{i=1}^{4} Q_{2i}^* w_i, \quad p = \sum_{i=1}^{4} Q_{3i}^* w_i, \quad q = \sum_{i=1}^{4} Q_{4i}^* w_i$$

and

$$w_1 = \sum_{i=1}^{4} M_{1i}^* y_i, \quad w_2 = \sum_{i=1}^{4} M_{2i}^* y_i, \quad w_3 = \sum_{i=1}^{4} M_{3i}^* y_i, \quad w_4 = \sum_{i=1}^{4} M_{4i}^* y_i$$

$\boldsymbol{J}, \boldsymbol{y}$, and $\mathring{\boldsymbol{y}}$ take the form previously described in Eqs. (20) and (21).

We first evaluate the TDNF and then average out the periodic terms to obtain the simplified TINF. The closed-form analytical solutions for $v_1(\zeta)$ and $v_2(\zeta)$ are constants. Variables $v_1(\zeta)$ and $v_2(\zeta)$ in the $\mathring{v}_3$ and $\mathring{v}_4$ differential equations are substituted by their respective computed constants. This computation is carried out

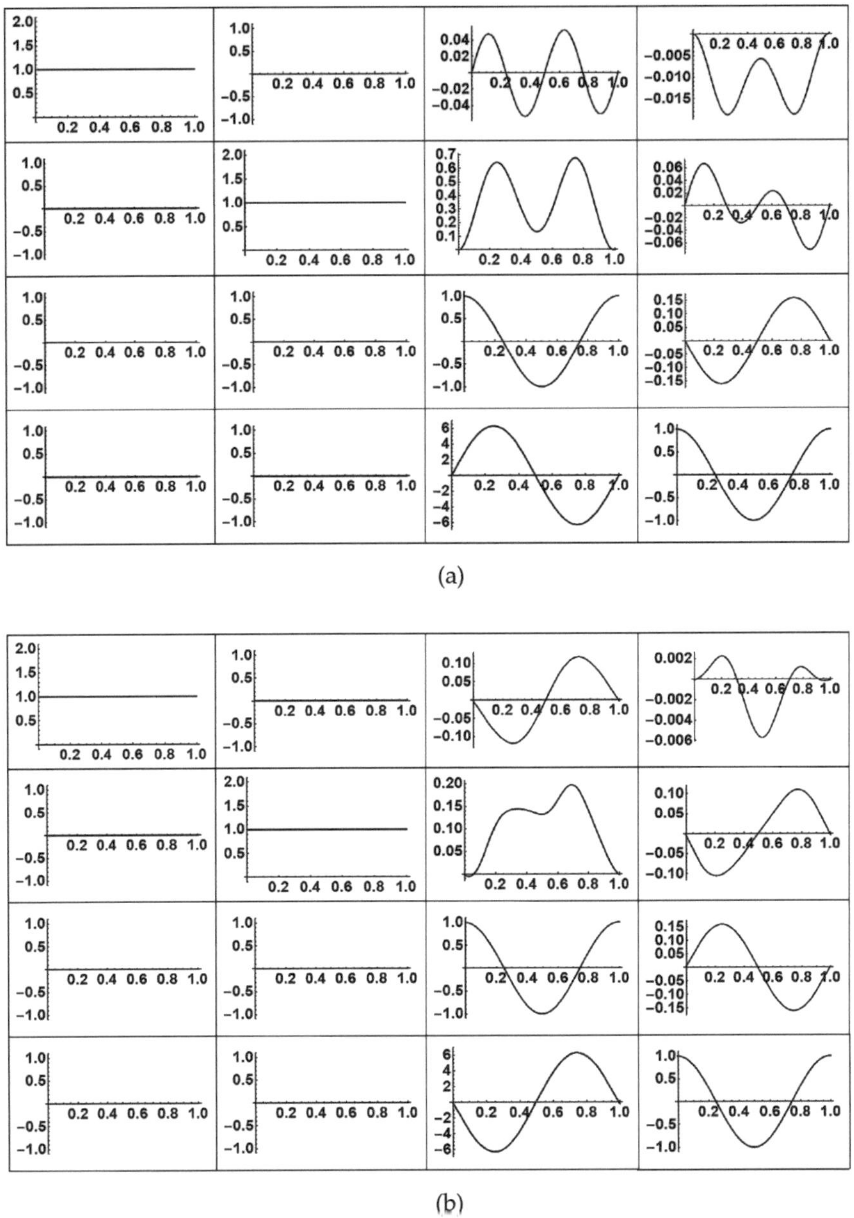

Figure 14.
Plot of elements of the second LFT matrix and its inverse. (a) $Q^{*}_{ij}(\zeta)$, (b) $Q^{*-1}_{ij}(\zeta)$.

through the forward action transform of the L-F, modal, and near-identity transformations of the initial conditions declared in the original coordinates:

$$\begin{bmatrix} \mathring{v}_1 \\ \mathring{v}_2 \\ \mathring{v}_3 \\ \mathring{v}_4 \end{bmatrix} = \begin{bmatrix} 0 \\ 0 \\ -(0.0106945 - i2.12186)v_3 + i0.0005599v_3^2 v_4 \\ -(0.0106945 + i2.12186)v_4 - i0.0005599v_3 v_4^2 \end{bmatrix} \tag{34}$$

The Floquet exponents are conjugate coefficients in the linear terms of the normal forms before being multiplied by the substituted constant values equal to v_1 and v_2.

Similarly, we define the normal form versal deformation parameter as μ_2 and incorporate it into Eq. (34):

$$\begin{bmatrix} \mathring{v}_1 \\ \mathring{v}_2 \\ \mathring{v}_3 \\ \mathring{v}_4 \end{bmatrix} = \begin{bmatrix} 0 & 0 & 0 & 0 \\ 0 & 0 & 0 & 0 \\ 0 & 0 & \mu_2 - \lambda_3 & 0 \\ 0 & 0 & 0 & \mu_2 - \lambda_4 \end{bmatrix} \begin{bmatrix} v_1 \\ v_2 \\ v_3 \\ v_4 \end{bmatrix} + \begin{bmatrix} 0 \\ 0 \\ -i0.0005599 v_3^2 v_4 \\ i0.0005599 v_3 v_4^2 \end{bmatrix} \tag{35}$$

where $\lambda_3 = (0.0106945 + i2.12186)$ and $\lambda_4 = (0.0106945 - i2.12186)$.

After defining small increments on the bifurcation parameter again as η, we express the k disparate sets of bifurcation parameter in the neighborhood of the critical parameter $e_c = 0.1$ as, $e_k = e_c + \eta$. The relationship between μ_2 and η is evaluated via the procedure previously stated in Section 3.3.1 to yield $\mu_2 = (0.132784 - i0.842528)\eta - (3.04196 - i7.13545)\eta^2$.

The closed-form analytical solutions for $v_1(\zeta)$ and $v_2(\zeta)$ of the versal deformation normal form are straightforward. To obtain $v_3(\zeta)$ and $v_4(\zeta)$, we introduce the complex changes of variable, $v_3(\zeta) = u_1 - iu_2$ and $v_4(\zeta) = u_1 + iu_2$ followed by the polar coordinates $u_1 = R\cos\theta$ and $u_2 = R sin(\theta)$. The last two equations in (35) become

$$\left.\begin{aligned} \mathring{R} &= [\mathrm{Re}(\mu_2) - 0.0106945]R \\ \mathring{\theta} &= 2.12187 + 0.0005599R^2 \end{aligned}\right\} \tag{36}$$

We solve Eq. (36) and use the results to obtain the remaining analytical solutions of Eq. (35) as shown in Eq. (37):

$$\begin{aligned} v_1(\zeta) &= \mu_2\zeta + C_1 \\ v_2(\zeta) &= \mu_2\zeta + C_2 \\ v_3(\zeta) &= e^{(-0.01069+\mu_2)\zeta} C_3 e^{-\Gamma i} \\ v_4(\zeta) &= e^{(-0.01069+\mu_2)\zeta} C_3 e^{\Gamma i} \end{aligned} \tag{37}$$

where $\Gamma = \left(2.1219\zeta + \frac{0.00028e^{(-0.02138+2\mu_2)\zeta}\mathrm{Re}\left(C_3^2\right)}{\mu_2 - 0.01069} + \mathrm{Re}(C_4)\right)$.

C_i $(i = 1, 2, 3, 4)$ are the respective constants of integration whose value is evaluated from the initial conditions specified in the original coordinates. C_1 and C_2 are real, whereas C_3 and C_4 are complex. μ_2 is similarly small in the order of magnitude

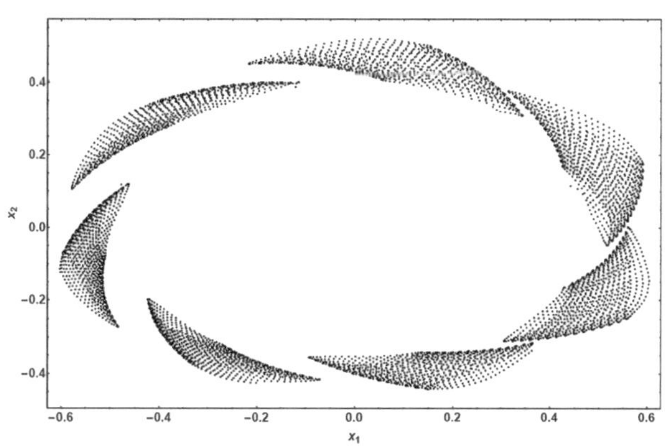

Figure 15.
Poincaré map of L-F transformed motion post-bifurcation behavior.

10^{-4}. $v_1(\zeta)$ and $v_2(\zeta)$ back transformation via inverse near-identity, modal, and single L-F transformations forms the augmented states given in Eq. (28) and is plotted in **Figure 13**. Eq. (36) yields a steady-state solution of the limit cycle amplitude as $R = \frac{\mathrm{Re}(\mu_2)}{0.0106945}$. When $\mu_2 \neq 0$, the solution of $v_3(\zeta)$ and $v_4(\zeta)$ results in locally stable limit cycle with amplitude corresponding to $R = \frac{\mathrm{Re}(\mu_2)}{0.0106945}$. Consequently, the quasiperiodic attractors in the original coordinates delineating a limit cycle are obtained after back transformation as shown in **Figure 15** ($\eta = 0.0001$).

Solutions of the versal deformation equations enable investigation of the post-bifurcation steady-state behavior in the small neighborhood of the bifurcation point. However, as observed by [42], this method is only useful for local analysis. This is because minor errors introduced by back transformation close to the bifurcation points significantly grow as you move further away.

4. Attitude motion feedback control

After setting $e = 0.2$ and $\sigma = 0.3$, the motion in Eqs. (9) and (16) is then numerically integrated to obtain the uncontrolled responses shown in **Figure 16a** and **b**, respectively (NB **Figure 16a** is the same as **Figure 2a**). The slight difference in the long-term motion behavior in **Figure 16b** may be attributed to the approximation of the trigonometric product term by a truncated series in Eq. (13) and possibly fidelity of the numerical integrator used in the computation. As established in Section 3.1, the attitude motion for the considered (e-σ) pair is quasiperiodic, marginally stable, and chaotic. Despite the system being stable in the sense of Lyapunov, the inherent oscillations are disruptively significant and require stabilization if the spacecraft is expected to successfully conduct its mission.

Motion controller design shall be undertaken on augmented state system, L-F transformed, and near-identity transformed coordinates. We shall hence first transform the system dynamics into these more amenable but topologically equivalent dynamical structures that retain the Lyapunov stability and bifurcation properties of the original system. Augmentation of the attitude dynamic states has been conducted in Section 3. Control law development will be first considered in L-F transformed coordinates and then followed by near-identity transformed coordinates of Eq. (16).

4.1 Lyapunov-Floquet transformation

Prior to computing the L-F transformation matrix $\boldsymbol{Q}(\zeta)$ and its inverse, $\boldsymbol{Q}^{-1}(\zeta)$ matrices (for the $e = 0.2$, $\sigma = 0.3$ case), we computed the FTM and $\boldsymbol{R}$ matrices over the interval $\zeta \in [0, 1]$ via shifted Chebyshev polynomials of the first kind for the

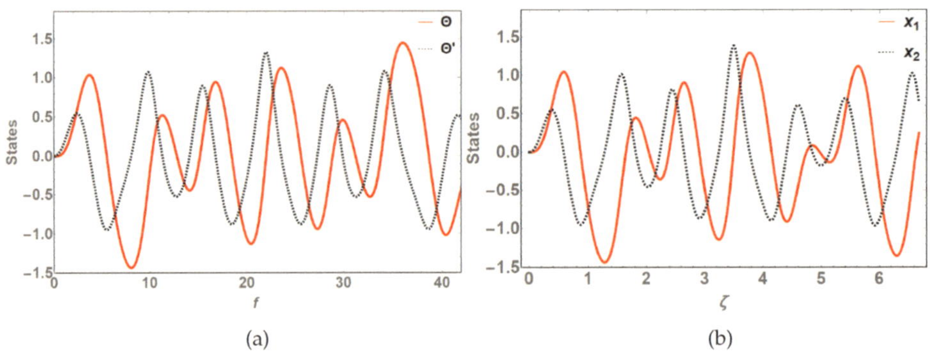

Figure 16.
Uncontrolled attitude motion.

system in Eq. (16). The evaluated aforementioned matrices are shown below. Here we alternatively present elements of $\boldsymbol{Q}(\zeta)$ and $\boldsymbol{Q}^{-1}(\zeta)$ as truncated Fourier series described in Eq. (8). Previously in **Figures 12** and **14**, we have presented the periodic plots of these series. Further, recall that $\zeta = \frac{f}{2\pi}$ and Z are the normalized principal period; hence, $FTM = \boldsymbol{\Phi}(Z)$:

$$\boldsymbol{\Phi}(Z) = \begin{bmatrix} 0.9462 & -0.0529 \\ 1.9796 & 0.9462 \end{bmatrix}, \qquad \boldsymbol{R} = \begin{bmatrix} 0 & -0.0539 \\ 2.0159 & 0 \end{bmatrix} \tag{38}$$

$$\boldsymbol{Q}(\zeta) = \begin{bmatrix} Q_{11} & Q_{12} \\ Q_{21} & Q_{22} \end{bmatrix}, \qquad \boldsymbol{Q}^{-1}(\zeta) = \begin{bmatrix} Q_{11}^{-1} & Q_{12}^{-1} \\ Q_{21}^{-1} & Q_{22}^{-1} \end{bmatrix} \tag{39}$$

where

$$\begin{aligned} Q_{11} &= -0.138896 + 1.24968\cos(2\pi\zeta) - 0.121824\cos(4\pi\zeta) + 0.0121464\cos(4\pi\zeta) \\ &\quad - 0.00121854\cos(8\pi\zeta) + 0.000122567\cos(10\pi\zeta) \\ Q_{12} &= 0.201902\sin(2\pi\zeta) - 0.0196812\sin(4\pi\zeta) + 0.00196227\sin(6\pi t) \\ &\quad - 0.000196857\sin(8\pi\zeta) \\ Q_{21} &= -7.44496\sin(2\pi\zeta) + 1.49121\sin(4\pi\zeta) - 0.224998\sin(6\pi\zeta) \\ &\quad + 0.0302284\sin(8\pi\zeta) \\ Q_{22} &= 0.0074811 + 1.20128\cos(2\pi\zeta) - 0.24076\cos(4\pi\zeta) + 0.0363337\cos(6\pi\zeta) \\ &\quad - 0.00488192\cos(8\pi\zeta) + 0.000615458\cos(10\pi\zeta) \end{aligned}$$

$$\begin{aligned} Q_{11}^{-1} &= 0.178684 + 0.831759\cos(2\pi\zeta) - 0.00416167\cos(4\pi\zeta) \\ &\quad - 0.00201274\cos(6\pi\zeta) - 0.001088\cos(8\pi\zeta) - 0.000706808\cos(10\pi\zeta) \\ &\quad - 0.000523275\cos(12\pi\zeta) - 0.000466034\cos(14\pi\zeta) \\ &\quad - 0.0258801\sin(2\pi\zeta) + 0.000259231\sin(4\pi\zeta) + 0.000188365\sin(6\pi\zeta) \\ &\quad + 0.000136072\sin(8\pi\zeta) + 0.000110821\sin(10\pi\zeta) + 0.000103106\sin(14\pi\zeta) \end{aligned}$$

$$\begin{aligned} Q_{12}^{-1} &= -0.0043031\cos(2\pi\zeta) - 0.000772303\cos(4\pi\zeta) + 0.00012908\cos(6\pi\zeta) \\ &\quad + 0.000115479\cos(8\pi\zeta) + 0.000107006\cos(10\pi\zeta) + 0.000110097\cos(12\pi\zeta) \\ &\quad + 0.000112704\cos(14\pi\zeta) - 0.138297\sin(2\pi\zeta) - 0.0123985\sin(4\pi\zeta) \\ &\quad + 0.00137926\sin(6\pi\zeta) + 0.000923347\sin(8\pi\zeta) + 0.00068248\sin(10\pi\zeta) \\ &\quad + 0.000583058\sin(12\pi\zeta) + 0.000509416\sin(14\pi\zeta) \end{aligned}$$

$$\begin{aligned} Q_{21}^{-1} &= 0.155206\cos(2\pi\zeta) - 0.00365144\cos(4\pi\zeta) - 0.00363534\cos(6\pi\zeta) \\ &\quad - 0.00338305\cos(8\pi\zeta) - 0.00343186\cos(10\pi\zeta) - 0.00327608\cos(12\pi\zeta) \\ &\quad - 0.00335147\cos(14\pi\zeta) + 4.98815\sin(2\pi\zeta) - 0.0586199\sin(4\pi\zeta) \\ &\quad - 0.0388447\sin(6\pi\zeta) - 0.0270502\sin(8\pi\zeta) - 0.0218882\sin(10\pi\zeta) \\ &\quad - 0.0173497\sin(12\pi\zeta) - 0.0151485\sin(14\pi\zeta) \end{aligned}$$

$$\begin{aligned} Q_{22}^{-1} &= 0.0835005 + 0.844365\cos(2\pi\zeta) + 0.0815938\cos(4\pi\zeta) - 0.00321618\cos(6\pi\zeta) \\ &\quad - 0.00173261\cos(8\pi\zeta) - 0.0010434\cos(10\pi\zeta) - 0.000742632\cos(12\pi\zeta) \\ &\quad - 0.00062635\cos(14\pi\zeta) - 0.0262723\sin(2\pi\zeta) - 0.00508249\sin(4\pi\zeta) \\ &\quad + 0.000300991\sin(6\pi\zeta) + 0.00021669\sin(8\pi\zeta) + 0.000163595\sin(10\pi\zeta) \\ &\quad + 0.000140229\sin(12\pi\zeta) + 0.000138575\sin(14\pi\zeta) \end{aligned}$$

After applying the L-F transformation, $\boldsymbol{x}(\zeta) = \boldsymbol{Q}(\zeta)\boldsymbol{z}(\zeta)$, to the attitude motion in Eq. (16), it becomes

$$z^{\circ} = Rz + Q^{-1}\begin{bmatrix} 0 \\ \dfrac{12\pi^2\sigma}{(1+e\cos 2\pi\zeta)}\left\{\dfrac{2}{3}k^3 - \dfrac{2}{15}k^5 + \dfrac{4}{315}k^7\right\} \end{bmatrix} + Q^{-1}\begin{bmatrix} 0 \\ \dfrac{8\pi^2 e\sin 2\pi\zeta}{(1+e\cos 2\pi\zeta)} \end{bmatrix} \tag{40}$$

where $k = (Q_{11}z_1 + Q_{12}z_2)$.

The Lyapunov stability properties are preserved in the new coordinates after the system is transformed by the L-F transformation matrix. The L-F transformation theory guarantees that a suitable controller realized in the L-F transformed coordinates will be correspondingly efficacious after back transformation into the original system coordinates. Consequently, we shall endeavor to systematically synthesize suitable controllers to stabilize the motion in the transformed coordinates. Our control synthesis strategy will first consider linear control laws before exploring nonlinear control strategies.

In order to formulate appropriate control laws that would stabilize the quasiperiodic motion analyzed in Section 3, we introduce a control input $u(t)$ in Eq. (1) as shown below:

$$\ddot{\Theta} = -3\frac{\mu}{r^3}\sigma\sin\Theta\cos\Theta - \dot{\omega} + u(t) \tag{41}$$

Using Eqs. (5) and (6), we perform a change of independent variable from time (t) to true anomaly (f). The closed-loop attitude dynamics hence will be

$$\Theta'' = \frac{1}{(1+e\cos f)}(2e\sin f\Theta' - 3\sigma\sin\Theta\cos\Theta + 2e\sin f) + u(f) \tag{42}$$

The control action $u(f)$ will generally represent torque per unit moment of inertia as a function of true anomaly. Eq. (42) is first used to synthesize linear control laws followed by nonlinear control law development.

4.2 Linear control

Though linear control law principles are conventionally intended for controlling linear parameter invariant systems [43], we initially consider them to control our nonlinear dynamics as an initial analysis step. Since most linear control methods tend to be relatively simpler to analyze and implement compared to nonlinear control methods, it is prudent to ascertain the suitability of linear control prior to embarking on relatively more complicated techniques. To implement linear control, we shall consider pole-placement approach to determine the negative feedback gain required to stabilize the system.

4.2.1 State-augmented system

The autonomous state-augmented system in Eq. (12) can be represented in abbreviated form as shown in Eq. (43):

$$\boldsymbol{\Theta}' = \boldsymbol{A\Theta}(f) + \boldsymbol{f}(\boldsymbol{\Theta}, f) \tag{43}$$

where $\boldsymbol{f}(\boldsymbol{\Theta}, f)$ constitutes the nonlinear terms, $\boldsymbol{A}$ is the linear matrix, and $\boldsymbol{\Theta} = [\Theta_1 \ \Theta_2 \ q \ p]^T$ is the state vector. To synthesize the parameter-invariant linear state feedback controller, Eq. (42) becomes

$$\boldsymbol{\Theta}' = \boldsymbol{A}\boldsymbol{\Theta}(f) + \boldsymbol{f}(\boldsymbol{\Theta},f) + \boldsymbol{G}_1 u(f) \quad (44)$$

The linear state feedback controller is of the form $u = -\boldsymbol{K}\boldsymbol{\Theta}(f)$, and the control input scaling vector is $\boldsymbol{G}_1 = [0\ \ 1\ \ 0\ \ 0]^T$. Though $\boldsymbol{A}$ is full rank, the linear pair $\{\boldsymbol{A}\ ,\boldsymbol{G}_1\}$ is not controllable. This is because the system controllability matrix, $\boldsymbol{CM}$, shown in Eq. (45) has a rank of 2, instead of rank 4:

$$\boldsymbol{CM} = \begin{bmatrix} 0 & 1 & 0 & -0.9 \\ 1 & 0 & -0.9 & 0 \\ 0 & 0 & 0 & 0 \\ 0 & 0 & 0 & 0 \end{bmatrix} \quad (45)$$

Consequently, a linear state feedback controller cannot stabilize the system dynamics associated with the e-σ pair considered. It is to be noted that the two states, (p, q) in Eq. (44), are virtual states serving to simplify the system but are not accessible in the actual system dynamics. This is illustrated by the fact that controllability matrix does not have full rank.

4.2.2 L-F transformed system

In this case, the parameter-invariant linear state feedback controller is similarly of the form $u(\zeta) = -\boldsymbol{K}\boldsymbol{z}(\zeta)$. The control input is scaled by the matrix $\boldsymbol{G}_2 = [1\ \ 1]^T$ in L-F transformed coordinates. Back transformation of the $\boldsymbol{G}_2 u(\zeta)$ product via inverse L-F transformation will guarantee a single control input in the system original coordinates as will be demonstrated in Eqs. (54) and (55). $\boldsymbol{R}$ is full rank and the linear pair $\{\boldsymbol{R}, \boldsymbol{G}_2\}$ is controllable. The L-F transformed Eq. (42) will be shown in Eq. (46):

$$\mathring{\boldsymbol{z}} = \boldsymbol{R}\boldsymbol{z}(\zeta) + \boldsymbol{Q}^{-1}\boldsymbol{f}(\boldsymbol{z},\zeta) + \boldsymbol{Q}^{-1}\boldsymbol{F}(\zeta) + \boldsymbol{G}_2 u(\zeta) \quad (46)$$

Therefore, the system closed-loop dynamics subjected to a linear control law will be of the form

$$\mathring{\boldsymbol{z}} = [\boldsymbol{R} - \boldsymbol{G}_2\boldsymbol{K}]\boldsymbol{z}(\zeta) + \boldsymbol{Q}^{-1}\boldsymbol{f}(\boldsymbol{z},\zeta) + \boldsymbol{Q}^{-1}\boldsymbol{F}(\zeta) \quad (47)$$

We initially place poles at $(-1\ ,0)$. Then we evaluate the corresponding matrix $\boldsymbol{K} = [K_1\ \ K_2]$ to realize this pole placement. Several stable double pole locations with a decreasing factor of 10 (i.e., -1, -10, -100, -1000...) were considered. None of these pole-placement locations demonstrated notable success in stabilizing the system. For instance, poles placed at $\{p_1 = -0.1\ ,\ p_2 = -0.2\}$ produce a response for a duration of slightly beyond 1.5 cycles before the states abruptly become indeterminate at about $\zeta \approx 1.74$ as shown in **Figure 17**.

In this analysis, the system response in the original coordinates is realized via the back transformation $\boldsymbol{z}(\zeta) = \boldsymbol{Q}^{-1}(\zeta)\boldsymbol{x}(\zeta)$. Therefore, similar to the state augmentation case, the L-F transformed nonlinear system demonstrates inability to be stabilized by a linear control law. The presence of periodic coefficients (elements of $\boldsymbol{Q}^{-1}(\zeta)$) associated with the nonlinear and forcing terms renders the system untenable to be controlled via LTI system control approaches. In general, the "region of application" of linear control for nonlinear systems is dependent on magnitude of nonlinearity and initial conditions. Many times, linear control may stabilize nonlinear systems locally, but this is not guaranteed.

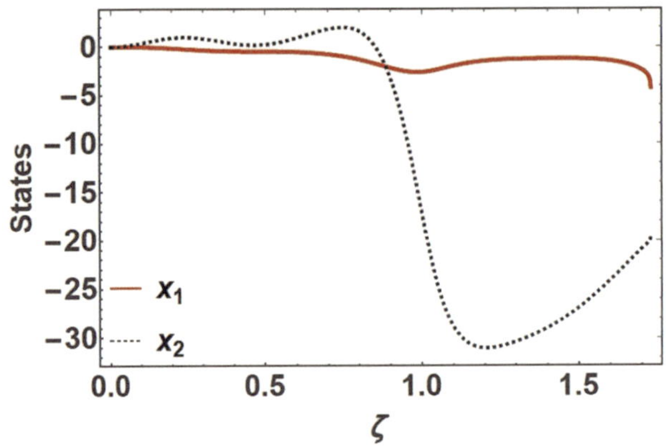

Figure 17.
Linear control of L-F transformed state.

4.3 Nonlinear control

Nonlinear control appears more suitable than linear control to stabilize the attitude motion. However, conventional nonlinear techniques are not readily amenable to dynamics with periodic coefficients and periodic external excitation. Hence, requisite system state augmentation, L-F, or near-identity transformations will be undertaken prior to controller design. We shall first consider sliding mode control (SMC), and then bifurcation control will be implemented on the marginally stable system to stabilize post-bifurcation response.

4.3.1 Sliding mode control

Sliding mode control is a robust nonlinear feedback control methodology that is suitable for achieving accurate tracking for a class of nonlinear systems. SMC methodology is based on variable structure control law that results in the state trajectory "sliding" along a discontinuity surface in the state space [44, 45]. Though SMC is deterministic, nonlinear, and robust, its implementation is prone to undesirable "chattering" along the sliding surface [46]. Design of SMC involves (i) selection of the switching function (stable hyperplane in the state space on which the dynamics will be restricted) and (ii) control law synthesis.

4.3.1.1 State-augmented system

Here, we implement a SMC that tracks a desired null pitch angle via a negative rate of growth. Dynamics in the original coordinates possess periodic coefficients rendering the dynamics unwieldy and unfavorable to synthesize a SMC. Therefore, we develop the SMC law based on the augmented state dynamics—which are liberated from periodic coefficients. To design a sliding mode controller for the state augmented systems, we designate the switching function as given in Eq. (48). The switching function represents the actual system state (i.e., attitude pitch angle) reference error (difference between desired and actual pitch angle) that the controller desires to maintain at zero. Therefore, when $s_1 = 0$, $\Theta_1 \to 0$ as $\Theta_2 \to 0$:

$$s_1 = (\beta\Theta_1 + \Theta_2)^2 \tag{48}$$

Subsequently, the closed-loop dynamics of the controlled system are similar to Eq. (42) as shown in Eq. (49):

$$\left.\begin{aligned}
&\Theta_1' = \Theta_2\\
&\Theta_2' = 2eq + 2e^2pq(1-\Theta_2) + 2eq\Theta_2 - 3\Theta_1\sigma - 3ep\Theta_1\sigma\\
&\qquad - 3(1+ep)\left\{-\frac{2}{3}\Theta_1^3 + \frac{2}{15}\Theta_1^5 - \frac{4}{315}\Theta_1^7\right\}\sigma + u(f)\\
&q' = p\\
&p' = -q
\end{aligned}\right\} \tag{49}$$

where $u(f)$ represents the control input. Derivative of the switching function with Θ_1' and Θ_2' substituted from Eq. (49) is

$$\begin{aligned} s_1' = &\frac{2}{105}(\beta\Theta_1 + \Theta_2)\left((1+ep)\left(210eq + \sigma\Theta_1\left[-315 + 210\Theta_1^2 - 42\Theta_1^4 + 4\Theta_1^6\right]\right)\right.\\ &\left.+ 105(\beta - 2e(-1+ep)q)\Theta_2\right) + u(f) \end{aligned} \tag{50}$$

Setting $\beta = 1$, we derive the following control input:

$$\begin{aligned} u(f) = &-\frac{2}{105}(\Theta_1 + \Theta_2)\left((1+ep)\left(210eq + \sigma\Theta_1\left[-315 + 210\Theta_1^2 - 42\Theta_1^4 + 4\Theta_1^6\right]\right)\right)\\ &- 105(1 - 2e(-1+ep)q)\Theta_2) - \rho\ \mathrm{sgn}\,(s_1) \end{aligned} \tag{51}$$

where

$$\begin{aligned} \rho(\boldsymbol{\Theta}) > \bigg| &-\frac{2}{105}(\beta\Theta_1 + \Theta_2)\left((1+ep)\left(210eq + \sigma\Theta_1\left[-315 + 210\Theta_1^2 - 42\Theta_1^4 + 4\Theta_1^6\right]\right)\right)\\ &-105(\beta - 2e(-1+ep)q)\Theta_2)\bigg| \end{aligned} \tag{52}$$

A sigmoid function, $\frac{s_1}{|s_1|+\varepsilon}$ is preferred instead of the signum function to reduce chattering around the sliding surface typical of sliding mode controllers. ε is generally small. Employing a direct Lyapunov approach, stability of the sliding mode controller applied here is ascertained by setting $V = \frac{1}{2}s_1^2$ to be the Lyapunov function. Hence, $V' = s_1 s_1'$. The guaranteed negative definiteness of the Lyapunov function derivative demonstrated in Eq. (53) points to a stable controller. This equation is obtained after subsequent substitution for s_1' and $u(f)$ in the equation of V':

$$V' = -2s_1^{\frac{3}{2}}\left[\varrho\frac{s_1}{|s_1| + \epsilon}\right] < 0 \quad \forall s_1 \neq 0 \tag{53}$$

Figure 18 shows the sliding mode controlled system response.

It is observable that the sliding mode controller in the state-augmented system achieves stabilization of the motion throughout any number of orbits. Both Θ_1 and Θ_2 are adequately confined to zero. The augmented states p and q remain unaffected.

4.3.1.2 L-F transformed system

In this case, we similarly assume a control input $u(\zeta)$ first applied to Eq. (16), prior to L-F transformation as shown in Eq. (54). $\boldsymbol{G}_3 = [0 \quad 1]^T$ is the control input scaling vector:

$$\begin{bmatrix} \mathring{x}_1 \\ \mathring{x}_2 \end{bmatrix} = \begin{bmatrix} 0 & 1 \\ \dfrac{-12\pi^2\sigma}{(1+e\cos 2\pi\zeta)} & \dfrac{4\pi e \sin 2\pi\zeta}{(1+e\cos 2\pi\zeta)} \end{bmatrix} \begin{bmatrix} x_1 \\ x_2 \end{bmatrix} + \frac{12\pi^2\sigma}{(1+e\cos 2\pi\zeta)} \begin{bmatrix} 0 \\ \frac{2}{3}x_1^3 - \frac{2}{15}x_1^5 + \frac{4}{315}x_1^7 \end{bmatrix} + \begin{bmatrix} 0 \\ \dfrac{8\pi^2 e\sin 2\pi\zeta}{(1+e\cos 2\pi\zeta)} \end{bmatrix} + \mathbf{G}_3 u(\zeta) \tag{54}$$

It then follows from Eq. (40) that the controlled L-F transformed system is as shown in Eq. (55):

$$\left.\begin{aligned} \mathring{z} &= R_{11}z_1 + R_{12}z_2 + Q_{12}^{-1}\Bigg(\frac{12\pi^2\sigma}{(1+e\cos 2\pi\zeta)}\left[\frac{2}{3}k^3 - \frac{2}{15}k^5 + \frac{4}{315}k^7\right] \\ &\quad + \left[\frac{8\pi^2 e\sin 2\pi\zeta}{(1+e\cos 2\pi\zeta)}\right] + u(\zeta)\Bigg) \\ \mathring{z} &= R_{21}z_1 + R_{22}z_2 + Q_{22}^{-1}\Bigg(\frac{12\pi^2\sigma}{(1+e\cos 2\pi\zeta)}\left[\frac{2}{3}k^3 - \frac{2}{15}k^5 + \frac{4}{315}k^7\right] \\ &\quad + \left[\frac{8\pi^2 e\sin 2\pi\zeta}{(1+e\cos 2\pi\zeta)}\right] + u(\zeta)\Bigg) \end{aligned}\right\} \tag{55}$$

where $k = Q_{11}z_1 + Q_{12}z_2$. We define the sliding function according to Eq. (56) to ensure that when $s_2 = 0, z_1 \to 0$ as $z_2 \to 0$. The sliding surface represents the reference pitch angle error. The controller attempts to maintain a zero error throughout, i.e., $s_2 = 0, \ \forall f > 0$:

$$s_2 = \alpha z_1 + z_2 \tag{56}$$

After obtaining the derivative of the sliding function, we substitute for $\mathring{z}_1$ and $\mathring{z}_2$ from Eq. (55) to obtain Eq. (57). Moreover, from Eq. (38), $R_{11} = R_{22} = 0$:

$$\begin{aligned} \mathring{s}_2 &= R_{21}z_1 + \alpha R_{12}z_2 + \left(\alpha Q_{12}^{-1} + Q_{22}^{-1}\right)\Bigg(\frac{12\pi^2\sigma}{(1+e\cos 2\pi\zeta)}\left[\frac{2}{3}k^3 - \frac{2}{15}k^5 + \frac{4}{315}k^7\right] \\ &\quad + \left[\frac{8\pi^2 e\sin 2\pi\zeta}{(1+e\cos 2\pi\zeta)}\right] + u(\zeta)\Bigg) \end{aligned} \tag{57}$$

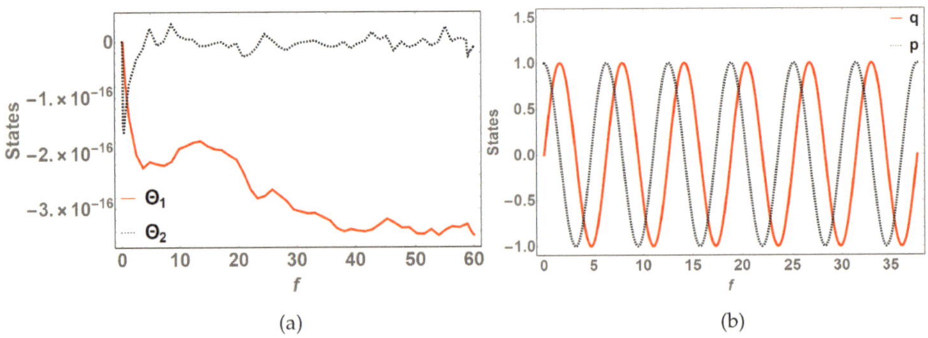

Figure 18.
Sliding mode controlled actual and augmented states. (a) Θ_1 and Θ_2, (b) q and p.

From Eq. (57), we set $\alpha = 1$ and derive the following sliding mode control law:

$$u(\zeta) = \left(-R_{21}z_1 - R_{12}z_2 - \left(Q_{12}^{-1} + Q_{22}^{-1}\right)\left\{\frac{12\pi^2\sigma}{(1+e\cos 2\pi\zeta)}\left[\frac{2}{3}k^3 - \frac{2}{15}k^5 + \frac{4}{315}k^7\right] - \left[\frac{8\pi^2 e \sin 2\pi\zeta}{(1+e\cos 2\pi\zeta)}\right]\right\}\right)\frac{1}{\left(Q_{12}^{-1} + Q_{22}^{-1}\right)} - \rho\,\mathrm{sgn}\,(s_2) \tag{58}$$

where

$$\rho(z) > \left|\left(-R_{21}z_1 - R_{12}z_2 - \left(Q_{12}^{-1} + Q_{22}^{-1}\right)\left\{\frac{12\pi^2\sigma}{(1+e\cos 2\pi\zeta)}\left[\frac{2}{3}k^3 - \frac{2}{15}k^5 + \frac{4}{315}k^7\right] - \left[\frac{8\pi^2 e \sin 2\pi\zeta}{(1+e\cos 2\pi\zeta)}\right]\right\}\right)\frac{1}{\left(Q_{12}^{-1} + Q_{22}^{-1}\right)} - \rho\,\mathrm{sgn}\,(s_2)\right| \tag{59}$$

To reduce chattering around the sliding surface typical of sliding mode controllers due to fast switching of the signum function, a sigmoid function is similarly preferred. We again apply the direct Lyapunov approach to analyze the sliding mode controller stability by selecting $V = \frac{1}{2}s_2^2$ as the Lyapunov function. Asymptotic stability will be guaranteed if the sliding function derivative is negative definite. Hence, the switching function derivative is $\mathring{V} = s_2\mathring{s}_2$. Substituting for $\mathring{s}_2$ with the control input likewise substituted, we obtain the stability-criteria satisfying relationship below:

$$\mathring{V} = -s_2\left[\rho\frac{s_2}{|s_2| + \epsilon}\right] < 0, \quad \forall s_2 \neq 0 \tag{60}$$

Figure 19 shows the sliding mode controlled system in L-F transformed coordinates. The response in **Figure 19** is back transformed via the inverse L-F transformation resulting in controlled states in the original coordinates shown in **Figure 20**.

We observe that, similar to the state augmented case, the sliding mode controller stabilizes the L-F transformed motion as well by invariably confining the states to zero as desired. Though specific values for the *e*-σ were used to demonstrate this technique, stabilization of the planar pitch motion by SMC approach is independent of the assigned *e*-σ values. However, the possibility of a synthesized sliding mode

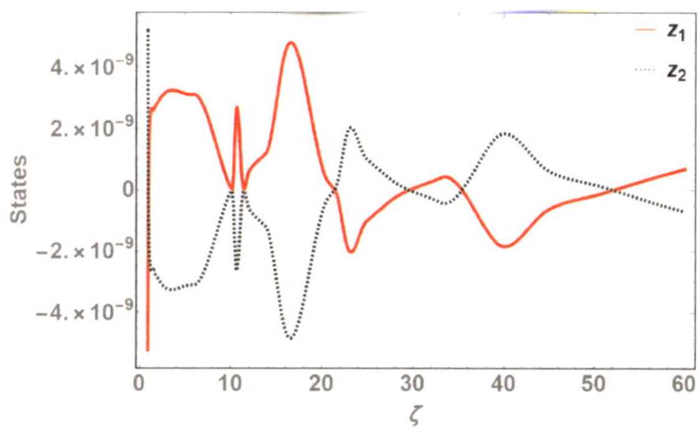

Figure 19.
Sliding mode controlled states in L-F transformed coordinates.

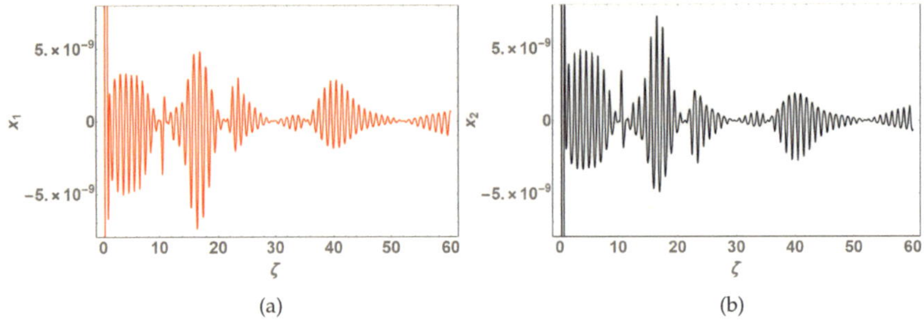

Figure 20.
Sliding mode controlled states in original coordinates. (a) x_1 state, (b) x_2 state.

controller being impractical to implement exists if the required control effort is colossally prohibitive.

4.3.2 Bifurcation control

The critical Floquet multipliers corresponding to purely imaginary Floquet exponents indicate that the system is in the stability boundary. Consequently, it is essential to stabilize the system post bifurcation apart from modifying other motion characteristics such as rate of growth. To achieve these objectives, we engage nonlinear bifurcation control with full state feedback. Synthesis of such a controller is facilitated by the normalized dynamics which are relatively more tractable compared to the dynamics as represented in the original coordinates. Periodic coefficients and complexity in structure of the dynamic equations in the original coordinates drastically convolute synthesis of bifurcation control law. Dynamics of the states in the original coordinates will eventually be obtained via back transformation of the normal form, modal, and L-F transformations. Location of the complex Floquet multipliers on the unit circle (**Figure 4**) indicates that the pitch attitude motion is undergoing a Hopf bifurcation with a limit cycle attractor of controllable radius. Therefore, the structure of the normal form will also verify a Hopf bifurcation occurring in the neighborhood of the critical point of the bifurcation parameter (i.e., orbit eccentricity).

To illustrate the intended approach, we shall formulate the normal forms of nonlinearities up to the cubic order in Eqs. (12) and (16). Normalization of dynamics with higher order nonlinearities can be accomplished through the same techniques. Similar to the preceding cases, we'll consider the augmented states and L-F transformed systems separately.

4.3.2.1 State-augmented system

In Section 3.1.1, we demonstrated how to obtain the TINF of the state-augmented system—shown in Eq. (23). Obtaining the closed-form analytical solutions for $v_1(f)$ and $v_2(f)$ in Eq. (23) is straightforward. On the other hand to evaluate $v_3(f)$ and $v_4(f)$, we introduce the complex changes of variable, $v_3 = u_1 - iu_2$ and $v_4 = u_1 + iu_2$ followed by the polar coordinates $u_1 = Rcos\theta$ and $u_2 = Rsin(\theta)$. The last two equations in Eq. (23) become

$$\left.\begin{aligned} R' &= 0 \\ \theta' &= 0.948683 - 30.4002C_1C_2 - 1.05409R^2 \end{aligned}\right\} \quad (61)$$

where C_1 and C_2 are the integration constants obtained when solving for $v_1(f)$ and $v_2(f)$, respectively.

We solve Eq. (61) and then utilize the results to complete the closed-form analytical solution of Eq. (23). The closed-form solutions of $v_1(f)$, $v_2(f)$, $v_3(f)$, and $v_4(f)$ are then back transformed to the original coordinates producing the uncontrolled motion behavior shown in **Figure 20**. The system response in **Figure 21** is a cognate approximation of the originally obtained numerical solution in **Figures 3a** and **16**. Again, a quasiperiodic motion is characterized by non-closed curves and is observed in the corresponding phase portrait. Moreover, a codimension-one Hopf Bifurcation is verified by the normal form structure.

To synthesize a bifurcation control law of the normal form, we first add a control input in Eq. (62):

$$\begin{bmatrix} v_1' \\ v_2' \\ v_3' \\ v_4' \end{bmatrix} = \begin{bmatrix} -i & 0 & 0 & 0 \\ 0 & i & 0 & 0 \\ 0 & 0 & i0.948683 & 0 \\ 0 & 0 & 0 & -i0.948683 \end{bmatrix} \begin{bmatrix} v_1 \\ v_2 \\ v_3 \\ v_4 \end{bmatrix} + \begin{bmatrix} 0 \\ 0 \\ i30.4002v_1v_2v_3 + i1.05409v_3^2v_4 \\ -i30.4002v_1v_2v_4 - i1.05409v_3v_4^2 \end{bmatrix} + \boldsymbol{G_4u} \tag{62}$$

Let the scaling matrix and control input, respectively, be of the form

$$\boldsymbol{G_4} = \begin{bmatrix} 0 & 0 & 0 & 0 \\ 0 & 0 & 0 & 0 \\ 0 & 0 & 1 & 0 \\ 0 & 0 & 0 & 1 \end{bmatrix}, \quad \boldsymbol{u} = \gamma_1 \begin{bmatrix} 0 \\ 0 \\ K_1v_1v_2v_3 + K_2v_3^2v_4 \\ K_1v_1v_2v_3 + K_2v_3v_2^4 \end{bmatrix} \tag{63}$$

Back transformation of the $\boldsymbol{G_4u}$ product via inverse normal form and modal transformations will guarantee a single control input in the system original coordinates as demonstrated in Eqs. (54) and (55). The proportional gains are custom tuned to $K_1 = -5$ and $K_2 = -10$. $\gamma_1 = 1$ is a scalable parameter meant to suppress

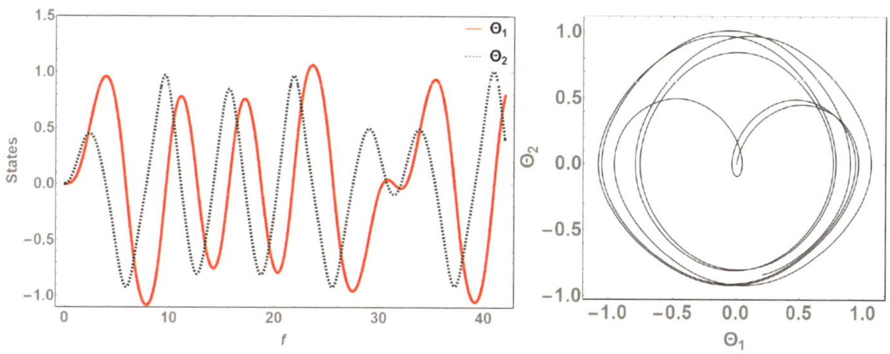

(a) (b)

Figure 21.
Uncontrolled dynamics of the normalized state augmented system for seven complete orbits. (a) Θ and Θ′ response, (b) Phase portrait.

strange trajectory behavior according to Poincaré-Bendixson theorem in the system phase space. The resulting response of the bifurcation-controlled augmented state system is shown in **Figure 22**. The augmented states remain unaffected as previously shown in **Figure 18b**.

The libratory amplitude of the quasiperiodic pitch angle motion is tremendously stabilized and confined to a significantly diminished limit cycle attractor as illustrated in **Figure 21**.

4.3.2.2 L-F transformed system

As already indicated in Section 3.3.2, in addition to synthesizing bifurcation control law via L-F transformed dynamics, we shall also demonstrate analysis of the spacecraft attitude dynamics due to different values of e and σ. Therefore, $e = 0.1$ and $\sigma = 0.2$ is once again considered in this section. L-F transformation analysis of the attitude dynamics associated with these values of e and σ has been comprehensively conducted in Section 3.3.2. Subsequently, the normalized TINF system was obtained in Eq. (34).

In Eq. (34), the closed-form analytical solutions for $v_1(\zeta)$ and $v_2(\zeta)$ are constants. Variables v_1 and v_2 in the $\mathring{v}_3$ and $\mathring{v}_4$ differential equations are substituted by their respective computed constants. This computation is carried out through the forward action transform of the L-F, modal, and near-identity transformations of the initial conditions declared in the original coordinates.

The Floquet exponents are conjugate coefficients in the linear terms of the normal forms before being multiplied by the substituted constant values equal to v_1 and v_2.

To obtain $v_3(\zeta)$ and $v_4(\zeta)$, we introduce the complex changes of variable $v_3 = u_1 - iu_2$ and $v_4 = u_1 + iu_2$ followed by the polar coordinates $u_1 = R\cos\theta$ and $u_2 = Rsin(\theta)$. The last two equations in Eq. (34) become

$$\left.\begin{aligned}\mathring{R} &= -0.0106945R\\ \mathring{\theta} &= 2.12186 + 0.0005599R^2\end{aligned}\right\} \tag{64}$$

Results from Eq. (64) which is easier to solve are then used to obtain the closed-form analytical solutions to Eq. (34). Then, $[v_1(\zeta) \quad v_2(\zeta) \quad v_3(\zeta) \quad v_4(\zeta)]^T$ are then back transformed to the original coordinates, producing the uncontrolled motion shown in **Figure 23**. The system response in **Figure 23** (with nonzero initial

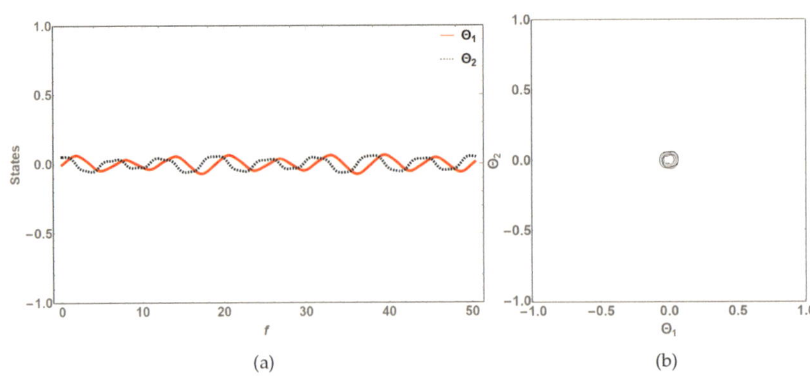

Figure 22.
Dynamics of the bifurcation-controlled state-augmented system. (a) Θ and Θ′ response, (b) Phase portrait.

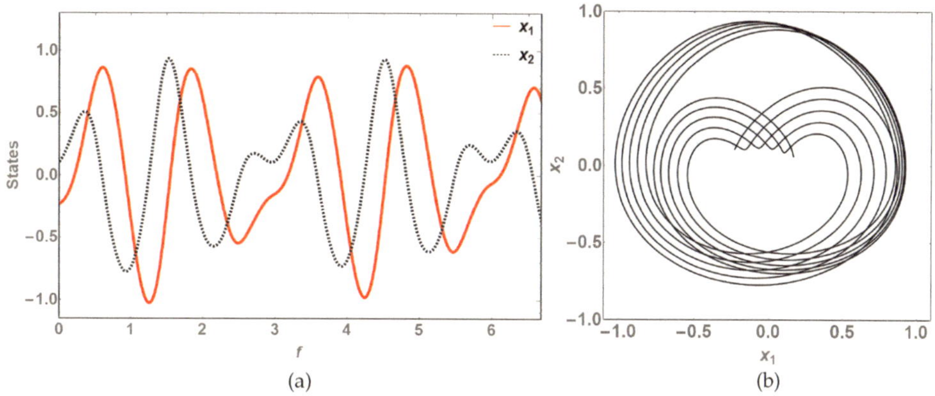

Figure 23.
Behavior of the normalized L-F transformed system states for seven complete orbits. (a) Original back transformed normalized states, (b) Phase portrait.

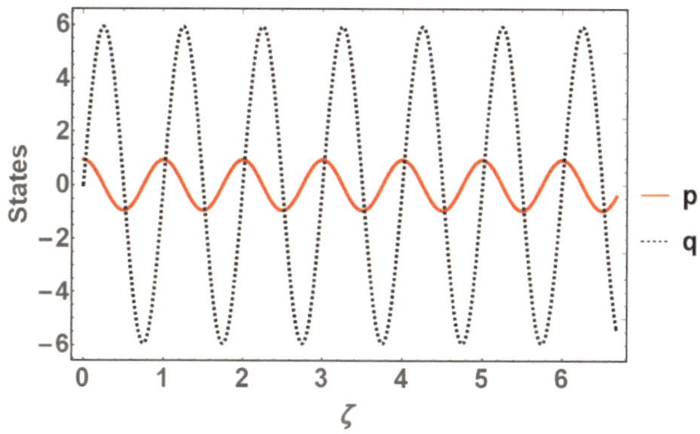

Figure 24.
Behavior of back-transformed normalized augmented states for seven complete orbits.

conditions) is a cognate approximation of the originally obtained numerical integration solution in **Figure 11**. The back transformed augmented states are similarly shown in **Figure 24** corresponding to Eq. (28) where the amplitude of $q(\zeta)$ is 2π times that of p.

The normal form in Eq. (34) verifies that this is a system undergoing a codimension-one Hopf bifurcation. To synthesize a bifurcation control law, a control input is added to Eq. (34) as shown below:

$$\begin{bmatrix} \overset{\circ}{v}_1 \\ \overset{\circ}{v}_2 \\ \overset{\circ}{v}_3 \\ \overset{\circ}{v}_4 \end{bmatrix} = \begin{bmatrix} 0 & 0 & 0 & 0 \\ 0 & 0 & 0 & 0 \\ 0 & 0 & -0.0106945 - i2.12186 & 0 \\ 0 & 0 & 0 & -0.0106945 + i2.12186 \end{bmatrix} \begin{bmatrix} v_1 \\ v_2 \\ v_3 \\ v_4 \end{bmatrix} + \begin{bmatrix} 0 \\ 0 \\ -i0.0005599 v_3^2 v_4 \\ i0.0005599 v_3 v_4^2 \end{bmatrix} + \mathbf{G}_5 \boldsymbol{u} \tag{65}$$

Let the scaling matrix and control input be of the form

$$G_5 = \begin{bmatrix} 0 & 0 & 0 & 0 \\ 0 & 0 & 0 & 0 \\ 0 & 0 & 1 & 0 \\ 0 & 0 & 0 & 1 \end{bmatrix}, \quad u = \gamma_2 \begin{bmatrix} 0 \\ 0 \\ K_1 v_3^2 v_4 \\ K_2 v_3 v_4^2 \end{bmatrix} \tag{66}$$

The proportional gains are custom-tuned to $K_1 = K_2 = -2$ and $\gamma_2 = 1$. **Figure 25** shows dynamic behavior of the implemented bifurcation control in original coordinates with nonzero initial conditions.

The oscillating amplitude of the quasiperiodic pitch angle motion is tremendously stabilized relative to the initial behavior illustrated in **Figure 23**. This hence demonstrates successful control of the post-bifurcation attitude dynamics about the spacecraft center of mass.

Bifurcation control is a nonlinear control technique that affects the behavior of the closed-loop system by modifying nonlinearity and post-bifurcation behavior. Therefore, the location of Floquet multipliers (exponents) is generally preserved post-bifurcation control. **Figure 26** shows the preserved locations of the Floquet multipliers before and after bifurcation control ($e = 0.1$ and $\sigma = 0.2$). This location of Floquet multipliers is consistent with the limit cycle shown in **Figure 25b** corresponding to a simply stable system with relatively subdued librations.

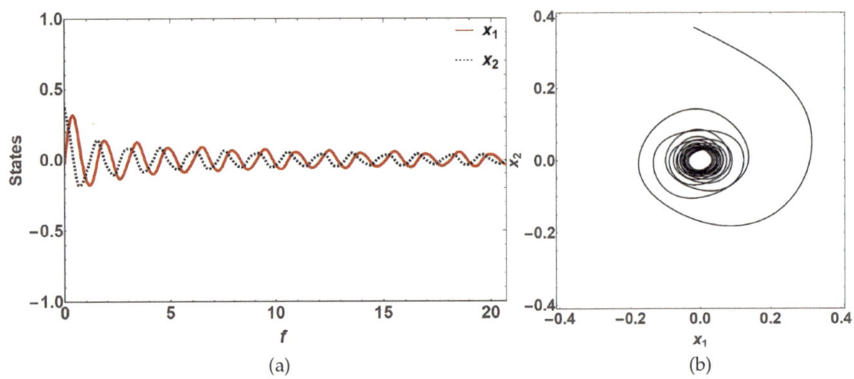

Figure 25.
State response of the bifurcation-controlled L-F transformed system. (a) x_1 and x_2 response, (b) Phase portrait.

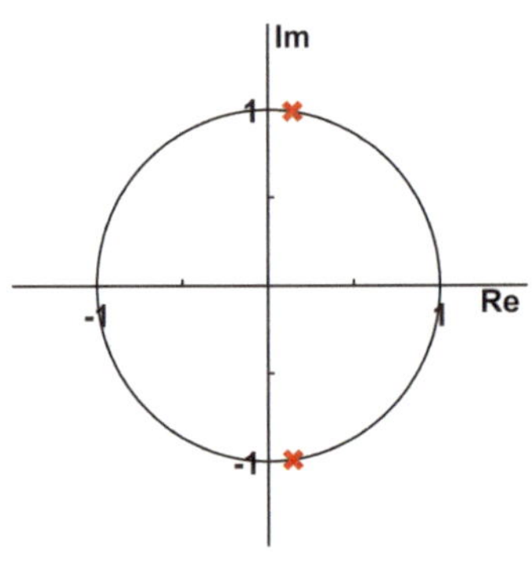

Figure 26.
Preserved locations of Floquet multipliers after bifurcation control.

5. Conclusions

In this chapter, we illustrated techniques for analyzing and stabilizing the attitude motion of a gravity-gradient stabilized spacecraft. The motion dynamics are shown to be nonlinear with periodic coefficients and subjected to external periodic excitation. Methodologies employed here utilize state augmentation, Lyapunov-Floquet transformation theory, and normal forms to realize relatively more tractable dynamical systems that are amenable to conventional controller synthesis techniques. Floquet theory was used to investigate system stability. State augmentation facilitated analysis via normal forms by transforming the dynamical system from nonautonomous to an autonomous one.

Outcome from the analysis showed that the attitude motion is quasiperiodic, chaotic, and stable in the sense of Lyapunov for the particular e-σ pairs considered. Subsequently, the motion stability chart that was constructed facilitated prediction of e-σ combination leading to stable or unstable dynamics. The stable regions of the stability curves were found to predict marginal and not asymptotically stable dynamics. However, the emanating librations need to be stabilized for nominal mission operations to be realized. Conversely, the e-σ combinations located in the unstable regions resulted in aperiodic unstable dynamics. The computed Lyapunov exponents indicate that the chaotic dynamics also depend on initial values of $\{\Theta, \ \Theta'\}$ pair, not just on the magnitudes of e-σ pairs.

Both outcomes of the twofold versal deformation analyses (disparate values of e-σ pairs considered) indicate establishment of locally stable limit cycles by the quasiperiodic flow post bifurcation. Since the eccentricity varies as $0 < e < 1$, relatively small deviations from the critical point e_c of the order $10^{-4} < \eta < 10^{-3}$ trigger a significant topological change in the structure of the motion flow.

The quasiperiodic, nonlinear, and periodically forced pitch attitude motion is challenging to control. The synthesized linear controller served as starting point for developing more adept control laws. Not surprisingly, the linear controller failed to stabilize the complexly structured nonlinear dynamical system. As stated, in general the "region of application" of linear control for nonlinear systems is dependent on magnitude of nonlinearity and initial conditions. Many times, linear control may stabilize nonlinear systems locally, but this is not guaranteed.

On-orbit perturbations cause disturbing torques that bifurcate the attitude motion; it is hence imperative to stabilize the system attitude dynamics in the small neighborhood of the bifurcation parameter's critical point. Local nonlinear bifurcation control law implemented on the attitude motion undergoing a Hopf bifurcation was shown to stabilize the attitude motion. The bifurcation controller which modifies the nonlinearity and post-bifurcation behavior further prevents the attitude motion from becoming chaotic because bifurcation is the path to chaos. Implemented in the TINF, the bifurcation control law would subsequently stabilize the secular and periodic attitude perturbations experienced by a spacecraft in elliptical orbit about its nominal operating point.

Sliding mode control law was based on driving both system states to zero on the sliding surface when the sliding surface reference error is equal to zero. The SMC law was similarly shown to be successful by invariably restricting the pitch angle to zero.

Future work will consider torques generated by sources such as magnetism and oblateness of the earth, atmospheric drag, solar radiation pressure, thermal bending, etc. Further, nonlinearities beyond the cubic term in the L-F transformation and TDNF case of near identity transformation would also be analyzed. In addition, physical implementation of the controllers and derivation of TDNF-based control laws requires future scrutiny. As demonstrated, all the control effort inputs are single torques per unit moment of inertia which for instance can be implemented via thrusters. Consequently, sizing and implementation of the control effort are an essential task.

Acknowledgements

Financial support for this research was partially provided by the Interplanetary Initiative of Arizona State University.

Conflict of interest

The authors declare no conflict of interest.

Nomenclature

e	eccentricity
f	true anomaly, *radians*
r	orbit radius, m
COM	center of mass
$\boldsymbol{I}$	identity matrix
I_x, I_y, I_z	principal moment of inertia about the roll, pitch, and yaw axes, respectively, kgm^2
LVLH	local-vertical/local-horizontal
P	semilatus rectum
$\mathbf{Q}(t)$	Lyapunov-Floquet transformation matrix (LFT)
TDNF	time-dependent normal forms
TINF	time-independent normal forms
μ	earth gravitational parameter, $3.986004418 \times 10^{14}$ $m^3 s^{-2}$
σ	$\frac{(I_x - I_z)}{I_y}$
ω	spacecraft angular velocity, rad/s
Θ	spacecraft pitch angle, *radians*
Ψ	spacecraft roll angle, *radians*
Ω	spacecraft yaw angle, *radians*
$\boldsymbol{\Phi}(t)$	state transition matrix (STM)
$\boldsymbol{\Phi}(T)$	Floquet transition matrix (FTM)

Author details

Peter M.B. Waswa*† and Sangram Redkar†
Arizona State University, Mesa, Arizona, USA

*Address all correspondence to: pwaswa@asu.edu

†These authors contributed equally.

IntechOpen

© 2020 The Author(s). Licensee IntechOpen. Distributed under the terms of the Creative Commons Attribution - NonCommercial 4.0 License (https://creativecommons.org/licenses/by-nc/4.0/), which permits use, distribution and reproduction for non-commercial purposes, provided the original is properly cited. (cc) BY-NC

References

[1] Waswa PMB, Redkar S. A survey of space mission architecture and system actualisation methodologies. International Journal of Space Science and Engineering. 2017;**4**(3):234-252

[2] Larson WJ, Wertz JR, editors. Space Missions Analysis and Design. El Segundo, CA: Microcosm Press; 2006

[3] Wie B. Space Vehicle Dynamics and Control. Reston, VA: American Institute of Aeronautics and Astronautics; 2008

[4] Suleman A. Multibody dynamics and nonlinear control of flexible space structures. Modal Analysis. 2004;10(11):1639-1661. Available from: https: //doi.org/10.1177/10775463040 42049

[5] Beletskii VV. Librations on an eccentric orbit: NASA Technical Translation F-8504. Washington, DC, USA: National Aeronautics and Space Administration; 1963

[6] Fujii HA, Ichiki W. Nonlinear dynamics of the tethered subsatellite system in the station keeping phase. Journal of Guidance, Control, and Dynamics. 2011;**20**(2):403-406

[7] Beletskii VV. Motion of an artificial satellite about its center of mass: NASA Technical Translation F-429. U.S. Department Commerce: Springfield, VA, USA; 1963

[8] Sherrill RE, Sinclair AJ, Sinha SC, Lovell TA. Lyapunov-Floquet control of satellite relative motion in elliptic orbits. IEEE Transactions on Aerospace and Electronic Systems. 2015;**51**(4): 2800-2810

[9] Mingori DL. Effects of energy dissipation on the attitude stability of dual-spinsatellites. AIAA Journal. 1969; 7(1):20-27

[10] Kojima H, Iwasaki M, Fujii HA, Blanksby C, Trivailo P. Nonlinear control of librational motion of tethered satellites in elliptic orbits. Journal of Guidance, Control, and Dynamics. 2004;**27**(2):229-239. American Institute of Aeronautics and Astronautics

[11] Sinha S, Wu D, Juneja V, Joseph P. Analysis of dynamic systems with periodically varying parameters via Chebyshev polynomials. Journal of Vibration and Acoustics. 1993;**115**(1): 96-102

[12] Vincent TL, Grantham WJ. Nonlinear and Optimal Control Systems. New York: John Wiley & Sons; 1997

[13] Yakubovich VA, Starzhinskii VM. Linear Differential Equations with Periodic Coefficients. Vol. I and II. New York: John Wiley & Sons; 1975

[14] Johnson W. Perturbation solutions for the influence of forward flight on helicopter rotor flapping stability. NASA TM X-62; 1974. p. 165

[15] Jordan DW, Smith P. Nonlinear Ordinary Differential Equations: An Introduction for Scientists and Engineers. 4th ed. Oxford: Oxford University Press; 2007

[16] Lovera M, Marchi ED, Bittanti S. Periodic attitude control techniques for small satellites with magnetic actuators. IEEE Transactions on Control Systems Technology. 2002;**10**(1):90-95

[17] Rossa FD, Dercole F, Lovera M. Attitude stability analysis for an earth pointing magnetically controlled spacecraft. IFAC Proceedings Volumes. 2013;**46**(19):518-523. 19th IFAC Symposium on Automatic Control in Aerospace

[18] Sheng J, Elbeyli O, Sun JQ. Stability and optimal feedback controls for time-

delayed linear periodic systems. American Institute of Aeronautics and Astronautics Journal. 2004;**42**(5): 908-911

[19] Wisniewski R, Blanke M. Fully magnetic attitude control for spacecraft subject to gravity gradient. Automatica. 1999;**35**(7):1201-1214

[20] Arduini C, Baiocco P. Active magnetic damping attitude control for gravity gradient stabilized spacecraft. Journal of Guidance, Control, and Dynamics. 1997;**20**(1):117-122. American Institute of Aeronautics and Astronautics

[21] Chernous'ko FL. On the motion of a satellite about its center of mass under the action of gravitational moments. Journal of Applied Mathematics and Mechanics. 1963; **27**(3):708-722

[22] Kane T. Attitude stability of earth-pointing satellites. AIAA Journal. 1965 Apr;**3**:726-731

[23] Breakwell J, Pringle Jr R. Nonlinear resonance affecting gravity-gradient stability. In: 16th International Astronautical Congress; Astrodynamics. Athens, Greece; 1965. pp. 305-325

[24] D'Angelo H. Linear Time-Varying Systems: Analysis and Synthesis. Boston: Allyn and Bacon; 1970

[25] Sinha S, Joseph P. Control of general dynamic systems with periodically varying parameters via Lyapunov-Floquet transformation. Transanctions-American Society of Mechanical Engineers Journal of Dynamic Systems Measurement and Control. 1994;**116**: 650-650

[26] Sherrill RE, Sinclair AJ, Sinha SC, Lovell TA. Time-varying transformations for Hill-Clohessy-Wiltshire solutions in elliptic orbits. Celestial Mechanics and Dynamical Astronomy. 2014;**119**(1):55-73. DOI: 10.1007/s10569-014-9543-x

[27] Fleig A Jr. On the libration of a gravity gradient stabilized spacecraft in an eccentric orbit. NASA TECHNICAL REPORT R-350; Washington, DC, USA: National Aeronautics and Space Administration; 1970

[28] Coddington A, Levinson N. Theory of Ordinary Differential Equations. International Series in Pure and Applied Mathematics. New York, NY: McGraw-Hill; 1955

[29] Floquet G. Sur les équations différentielles linéaires á coefficients périodiques. Annales scientifiques de l'École Normale Supérieure. 1883;**12**: 47-88. Available from: http://eudml.org/doc/80895

[30] Sinha S. On the analysis of time-periodic nonlinear dynamical systems. Sadhana. 1997;**22**:411-434

[31] Mason JC, Handscomb DC. Chebyshev Polynomials. Boca Raton, Florida: CRC Press; 2003

[32] Poincaré H. Les méthodes Nouvelles de la mécanique céleste. Vol. 1. Paris: Gauthier-Villars; 1892

[33] Arnold VI, Vogtmann K, Weinstein A. Mathematical Methods of Classical Mechanics. Graduate Texts in Mathematics. New York: Springer; 2013. Available from: https://books.google.com/books?id=UOQlBQAAQBAJ

[34] Nayfeh AH. Method of Normal Forms. Wiley Series in Nonlinear Science. New York: Wiley; 1993

[35] Jezequel L, Lamarque CH. Analysis of non-linear dynamical systems by the normal form theory. Journal of Sound and Vibration. 1991;**149**(3):429-459

[36] Gabale AP, Sinha SC. A direct analysis of nonlinear systems with

external periodic excitations via normal forms. Nonlinear Dynamics. 2009;**55** (1-2):79-93

[37] Neild SA, Champneys AR, Wagg DJ, Hill TL, Cammarano A. The use of normal forms for analysing nonlinear mechanical vibrations. Philosophical Transactions of the Royal Society of London A: Mathematical, Physical and Engineering Sciences. 2015;**373**(2051):1-22

[38] Herman EA. Convergence, Approximation, and Differential Equations. Hoboken, NJ: Wiley; 1986

[39] Driscoll T A NH, Trefethen LN, editors. Chebfun Guide: For Chebfun version 5. www.chebfun.org; 2014

[40] Nayfeh AH, Balachandran B. Applied Nonlinear Dynamics: Analytical, Computational and Experimental Methods. Wiley Series in Nonlinear Science. Wiley & Sons Inc: New York, NY; 1995

[41] Wiggins S. Introduction to Applied Nonlinear Dynamical Systems and Chaos. Texts in Applied Mathematics. New York: Springer; 2003. Available from: https://books.google.com/books?id=RSI4RGdwnU4C

[42] Dávid A, Sinha SC. Versal deformation and local bifurcation analysis of time-periodic nonlinear systems. Nonlinear Dynamics. 2000;**21** (4):317-336. Available from: https: //doi. org/10.1023/A:1008330023291.

[43] Franklin GF, Powell J, Emami-Naeini A. Feedback Control of Dynamic Systems. 3rd ed. Reading, MA: Addison-Wesley; 1994

[44] Utkin VI. Sliding Modes and their Application in Variable Structure Systems. Rev. from the 1974 ed. Moscow: Mir Publishers; 1978

[45] Slotine JJ, Sastry SS. Tracking control of non-linear systems using sliding surfaces with application to robot manipulators. In: 1983 American Control Conference; 1983. pp. 132-135

[46] Kachroo P, Tomizuka M. Chattering reduction and error convergence in the sliding-mode control of a class of nonlinear systems. IEEE Transactions on Automatic Control. 1996;**41**(7): 1063-1068

Chapter 7

Optimal Trajectory Synthesis and Tracking Control for Spacecraft Large Attitude Manoeuvers

Ranjan Vepa

Abstract

The classical approach to the problem of synthesizing an optimal attitude manoeuver trajectory, involves the use of the calculus of variations and the use Lagrange multipliers or co-states. The nonlinear large attitude manoeuver trajectory is controlled by a set of nonlinear evolving co-states. In this paper, following a review of the methodologies available for trajectory synthesis followed by tracking control, the optimal trajectory for a typical optimal attitude manoeuver is synthesized by solving for the states and co-states defined by a two point boundary value problem. Gravity gradient torques are included as a matter of course. Following the synthesis of the optimal attitude-rate trajectory, tracking control laws are synthesized by re-formulating the optimal control as a feedback law. The approximate linear tracking feedback controls are evaluated by relating the optimal state and co-state vector by a linear relation. The control laws are synthesized numerically. The problem of optimal attitude orientation trajectory synthesis is also addressed. The methodologies are applied to typical sample problems and results are presented. Quantitative comparisons of the results of the methods are made to the results obtained by the application of other linear and nonlinear methods, to illustrate the key features of the methodologies.

Keywords: attitude manoeuvers, optimal manoeuver trajectory, trajectory optimization, trajectory tracking, feedback control laws

1. Introduction

The need for designing fast attitude and angular rate acquisition manoeuvers for a spacecraft with restricted or low actuation torqueing capacity arises in many space recent applications. Spacecraft are usually equipped with an attitude control system (ACS), which operates in one of two modes; the first mode involves maneuvering for attitude or angular rate acquisition while the second is to ensure stability. In the first mode, the ACS is responsible for acquiring and tracking an attitude or an angular rate state trajectory which could include a steady state. The requirements for this mode are set by the need to remotely capture an orbiting body, de-tumble a spacecraft, synchronize with another orbiting body or point at a specific direction in space. Although a large number of publications have appeared before the end of the last millennium on the subject of attitude stabilization and feedback control, a few recent papers have focused on the construction of optimal maneuvering trajectories synthesis for attitude or angular rate acquisition. There have been some publications related to the synthesis of optimal maneuvering trajectories for attitude or angular rate acquisition during the last two decades of the preceding millennium. Yet some significant advances have been made in the early part of this century. This includes papers by Lee et al. [1], Yoshida et al. [2], Aghili [3, 4], Yang and Wu [5], Liu et al. [6], and Zhang et al. [7], who have considered the maneuvering for attitude or angular rate acquisition problems using classical methodologies. Sharma and Tewari [8] have addressed the issue of nonlinear tracking of spacecraft attitude manoeuvers while Hegrenas et al. [9] have considered the maneuvering for attitude or angular rate acquisition problem by means of explicit model predictive control, via a nonlinear programming approach.

In dealing with the optimal attitude trajectory synthesis, several real world effects such as gravity gradient torques are generally neglected. Neglecting the gravity gradient torques can have a serious effect on the trajectory synthesis problem as (i) the gravity gradient torques can influence the stability of the spacecraft and (ii) they tend to couple the attitude rate dynamics with the quaternion kinematics. For this reason, it is not always advisable to ignore these torques on grounds of "smallness" as even the smallest of these perturbations can not only trigger instability but also induce the bifurcation of the orbit. It is the gravity that is primarily responsible for the orbital motion and the attitude stability of a spacecraft. The importance of gravity gradient torques has also been underscored by Lobo et al. [10].

The classical methodologies for trajectory synthesis compare well with other nonlinear and deterministic artificial intelligence approaches such as those developed by Sands et al. [11], Nakatani and Sands [12] and Baker et al. [13].

In this paper, the optimal trajectory for a typical attitude rate manoeuver is synthesized by solving for the states and co-states defined by a two point boundary value problem. Gravity gradient torques are included as a matter of course. Following the synthesis of the optimal attitude-rate trajectory, tracking control laws are synthesized by re-formulating the optimal control as a feedback law. The approximate linear tracking feedback controller gains are evaluated by relating the optimal state and co-state vector by a linear relation. The feedback control laws are synthesized numerically. The problem of optimal attitude orientation trajectory synthesis is also addressed. The optimization methodologies are applied to typical sample problems and results are presented. Quantitative comparisons of the results of the optimization method are made to the results obtained by the application of other linear and nonlinear control methods, to illustrate the key features of the methodologies.

2. Spacecraft attitude dynamics and quaternion kinematics

In matrix form, when the inertia matrix is not diagonal the equations of attitude motion of chaser spacecraft are

$$\mathbf{I}\dot{\boldsymbol{\omega}} + \boldsymbol{\Omega}\mathbf{I}\boldsymbol{\omega} = \mathbf{M} + \mathbf{M}_{gg} + \mathbf{M}_{d}, \tag{1}$$

where $\mathbf{I}$ is the moment of inertia matrix which is assumed to be

$$\mathbf{I} = \begin{bmatrix} I_{11} & I_{12} & I_{13} \\ I_{12} & I_{22} & I_{23} \\ I_{13} & I_{23} & I_{33} \end{bmatrix}, \boldsymbol{\omega} \equiv \begin{bmatrix} \omega_1 \\ \omega_2 \\ \omega_3 \end{bmatrix}, \boldsymbol{\Omega} = \begin{bmatrix} 0 & -\omega_3 & \omega_2 \\ \omega_3 & 0 & -\omega_1 \\ -\omega_2 & \omega_1 & 0 \end{bmatrix} \tag{2}$$

$\mathbf{M}_d$ are the disturbance torques and $\mathbf{M}_{gg}$ are the gravity gradient torques.

It is important to emphasize that the targets dynamics are irrelevant to us as there is little or no chance of acquiring the target's inertia properties. However the target's angular velocity vector is assumed to be given by $\boldsymbol{\omega}_d$, its attitude quaternion relative to the chaser's body frame is assumed to be $\mathbf{q}_d$ or its relative attitude quaternion $\Delta\mathbf{q}$, relative to the chaser's body frame, can in principle be measured from within the chaser spacecraft.

Expressions for the gravity gradient moment are obtained assuming that z axis of the spacecraft body is nominally pointing to the Earth. The direction vector the center of gravity of the spacecraft pointing to the Earth is given by the last column of $\mathbf{T}_{BR}$, the transformation from the Earth orbiting frame to the body fixed frame of the spacecraft as

$$\mathbf{c} = \begin{bmatrix} c_1 & c_2 & c_3 \end{bmatrix}^T. \tag{3}$$

The corresponding cross product operator $\mathbf{c}_\times$ is defined as

$$\mathbf{c}_\times = \begin{bmatrix} 0 & -c_3 & c_2 \\ c_3 & 0 & -c_1 \\ -c_2 & c_1 & 0 \end{bmatrix}. \tag{4}$$

Hence the gravity gradient moments acting on the spacecraft and manipulator body are:

$$\mathbf{M}_{gg} = 3n^2\mathbf{c}_\times\mathbf{Ic} \equiv 3n^2\mathbf{c}_\times\mathbf{Ic} = \begin{bmatrix} L_{gg} & M_{gg} & N_{gg} \end{bmatrix}^T. \tag{5}$$

Thus,

$$\mathbf{M}_{gg} = \begin{bmatrix} L_{gg} \\ M_{gg} \\ N_{gg} \end{bmatrix} = 3n^2 \begin{bmatrix} c_2c_1I_{31} + c_2^2I_{32} + c_2c_3(I_{33} - I_{22}) - c_3c_1I_{21} - c_3^2I_{23} \\ c_3c_1(I_{11} - I_{33}) + c_2c_3I_{12} + c_3^2I_{13} - c_1^2I_{31} - c_1c_2I_{32} \\ c_1^2I_{21} + c_1c_2(I_{22} - I_{11}) + c_1c_3I_{23} - c_2^2I_{12} - c_2c_3I_{13} \end{bmatrix}. \tag{6}$$

If we express the transformation from the orbiting frame to the body coordinates in terms of an attitude quaternion of the chaser spacecraft with components ε_1, ε_2, ε_3 and η as

$$\mathbf{T}_{BR}(\mathbf{q}) = \begin{bmatrix} \eta^2 + \varepsilon_1^2 - \varepsilon_2^2 - \varepsilon_3^2 & 2(\varepsilon_1\varepsilon_2 + \varepsilon_3\eta) & 2(\varepsilon_1\varepsilon_3 - \varepsilon_2\eta) \\ 2(\varepsilon_1\varepsilon_2 - \varepsilon_3\eta) & \eta^2 - \varepsilon_1^2 + \varepsilon_2^2 - \varepsilon_3^2 & 2(\varepsilon_2\varepsilon_3 + \varepsilon_1\eta) \\ 2(\varepsilon_1\varepsilon_3 + \varepsilon_2\eta) & 2(\varepsilon_2\varepsilon_3 - \varepsilon_1\eta) & \eta^2 - \varepsilon_1^2 - \varepsilon_2^2 + \varepsilon_3^2 \end{bmatrix}, \tag{7}$$

then from the last column, the Earth pointing direction vector is:

$$\mathbf{c} = \begin{bmatrix} c_1 \\ c_2 \\ c_3 \end{bmatrix} = \begin{bmatrix} 2(\varepsilon_1\varepsilon_3 - \varepsilon_2\eta) \\ 2(\varepsilon_2\varepsilon_3 + \varepsilon_1\eta) \\ \eta^2 - \varepsilon_1^2 - \varepsilon_2^2 + \varepsilon_3^2 \end{bmatrix}. \tag{8}$$

The quaternion kinematics satisfies

$$\frac{d\mathbf{q}}{dt} = \frac{1}{2}\mathbf{A}_\omega(\boldsymbol{\omega})\mathbf{q}, \ \mathbf{q}^T \cdot \mathbf{q} = \mathbf{1} \tag{9}$$

where the quaternion $\mathbf{q} = \begin{bmatrix} \varepsilon_1 & \varepsilon_2 & \varepsilon_3 & \eta \end{bmatrix}^T$, consists of a vector part, $\boldsymbol{\varepsilon} = \begin{bmatrix} \varepsilon_1 & \varepsilon_2 & \varepsilon_3 \end{bmatrix}^T$ and the scalar η so,

$$\mathbf{q}=\begin{bmatrix}\boldsymbol{\varepsilon}\\ \eta\end{bmatrix}\text{ and }\mathbf{A}_{\omega}=\begin{bmatrix}-\boldsymbol{\Omega}(\boldsymbol{\omega}) & \boldsymbol{\omega}\\ -\boldsymbol{\omega}^{T} & 0\end{bmatrix},\ \boldsymbol{\Omega}(\boldsymbol{\omega})=\begin{bmatrix}0 & -\omega_3 & \omega_2\\ \omega_3 & 0 & -\omega_1\\ -\omega_2 & \omega_1 & 0\end{bmatrix}. \tag{10}$$

The quaternion kinematics may also be compactly expressed as

$$\frac{d\mathbf{q}}{dt}=\frac{d}{dt}\begin{bmatrix}\boldsymbol{\varepsilon}\\ \eta\end{bmatrix}=\frac{1}{2}\boldsymbol{\Gamma}(\mathbf{q})\boldsymbol{\omega}, \tag{11}$$

$$\boldsymbol{\Gamma}(\mathbf{q})=\begin{bmatrix}\eta\mathbf{I}_{3\times3}+\mathbf{S}(\boldsymbol{\varepsilon})\\ -\boldsymbol{\varepsilon}^{T}\end{bmatrix},\ \mathbf{S}(\boldsymbol{\varepsilon})=\begin{bmatrix}0 & -\varepsilon_3 & \varepsilon_2\\ \varepsilon_3 & 0 & -\varepsilon_1\\ -\varepsilon_2 & \varepsilon_1 & 0\end{bmatrix}, \tag{12}$$

where $\mathbf{I}_{3\times3}$ is the 3×3 unit matrix. These relations may be inverted as

$$\boldsymbol{\omega}=2\begin{bmatrix}\eta\mathbf{I}_{3\times3}+\mathbf{S}^{T}(\boldsymbol{\varepsilon}) & -\boldsymbol{\varepsilon}\end{bmatrix}\begin{bmatrix}\dot{\boldsymbol{\varepsilon}} & \dot{\eta}\end{bmatrix}^{T}=2\boldsymbol{\Gamma}^{-1}(\mathbf{q})\begin{bmatrix}\dot{\boldsymbol{\varepsilon}} & \dot{\eta}\end{bmatrix}^{T}. \tag{13}$$

The desired attitude quaternion relative to the chaser's body frame which is assumed to be $\mathbf{q}_d$ and the relative attitude quaternion $\Delta\mathbf{q}$, relative to the chaser's body frame are related to the chasers attitude by

$$\mathbf{q}_d=\Delta\mathbf{q}\otimes\mathbf{q}. \tag{14}$$

Given two quaternions, $\Delta\mathbf{q}=\begin{bmatrix}\Delta q_1 & \Delta q_2 & \Delta q_3 & \Delta q_4\end{bmatrix}^{T}$, $\mathbf{q}=\begin{bmatrix}\varepsilon_1 & \varepsilon_2 & \varepsilon_3 & \eta\end{bmatrix}^{T}$, the quaternion product $\mathbf{q}_d=\Delta\mathbf{q}\otimes\mathbf{q}$ is defined as

$$\mathbf{q}_d=\begin{bmatrix}q_{1d}\\ q_{2d}\\ q_{3d}\\ q_{4d}\end{bmatrix}=\begin{bmatrix}\eta & \varepsilon_3 & -\varepsilon_2 & \varepsilon_1\\ -\varepsilon_3 & \eta & \varepsilon_1 & \varepsilon_2\\ \varepsilon_2 & -\varepsilon_1 & \eta & \varepsilon_3\\ -\varepsilon_1 & -\varepsilon_2 & -\varepsilon_3 & \eta\end{bmatrix}\begin{bmatrix}\Delta q_1\\ \Delta q_2\\ \Delta q_3\\ \Delta q_4\end{bmatrix}=\begin{bmatrix}\Delta q_4 & -\Delta q_3 & \Delta q_2 & \Delta q_1\\ \Delta q_3 & \Delta q_4 & -\Delta q_1 & \Delta q_2\\ -\Delta q_2 & \Delta q_1 & \Delta q_4 & \Delta q_3\\ -\Delta q_1 & -\Delta q_2 & -\Delta q_3 & \Delta q_4\end{bmatrix}\begin{bmatrix}\varepsilon_1\\ \varepsilon_2\\ \varepsilon_3\\ \eta\end{bmatrix}. \tag{15}$$

Hence it is expressed in matrix form as

$$\mathbf{q}_d=\mathbf{C}_0\Delta\mathbf{q},\ \mathbf{C}_0=\begin{bmatrix}\eta & \varepsilon_3 & -\varepsilon_2 & \varepsilon_1\\ -\varepsilon_3 & \eta & \varepsilon_1 & \varepsilon_2\\ \varepsilon_2 & \varepsilon_1 & \eta & \varepsilon_3\\ -\varepsilon_1 & -\varepsilon_2 & -\varepsilon_3 & \eta\end{bmatrix},\ \mathbf{C}_0=\begin{bmatrix}\eta\mathbf{I}_{3\times3}+\mathbf{S}^{T}(\boldsymbol{\varepsilon}) & \mathbf{q}_{1:3}\\ -\mathbf{q}_{1:3} & \eta\end{bmatrix}. \tag{16}$$

Similarly,

$$\mathbf{q}_d=\begin{bmatrix}q_{1d}\\ q_{2d}\\ q_{3d}\\ q_{4d}\end{bmatrix}=\begin{bmatrix}\Delta q_4 & -\Delta q_3 & \Delta q_2 & \Delta q_1\\ \Delta q_3 & \Delta q_4 & -\Delta q_1 & \Delta q_2\\ -\Delta q_2 & \Delta q_1 & \Delta q_4 & \Delta q_3\\ -\Delta q_1 & -\Delta q_2 & -\Delta q_3 & \Delta q_4\end{bmatrix}\begin{bmatrix}\varepsilon_1\\ \varepsilon_2\\ \varepsilon_3\\ \eta\end{bmatrix}=\begin{bmatrix}\Delta q_4\mathbf{I}_{3\times3}+\mathbf{S}(\Delta\mathbf{q}_{1:3}) & \Delta\mathbf{q}_{1:3}\\ -\Delta\mathbf{q}_{1:3} & \Delta q_4\end{bmatrix}\mathbf{q}. \tag{17}$$

The traditional method of defining the attitude of a spacecraft is by the use of Euler angle sequences. The conversion of Euler angles defined as Euler angle sequences, may be converted to an equivalent quaternion set, using well-known conversion formulae, such as, those given by Smeresky et al. [14].

3. Formulation of the optimal angular rate trajectory synthesis problem

The first task is to formulate the optimal control problem, so it can be solved numerically. This is briefly reviewed. The attitude equations of the spacecraft may be expressed in state space form as

$$\frac{d\mathbf{q}}{dt} = \frac{1}{2}\mathbf{A}_{\omega}(\boldsymbol{\omega})\mathbf{q}, \mathbf{q}^T \cdot \mathbf{q} = \mathbf{1}, \tag{18}$$

$$\mathbf{I}\dot{\boldsymbol{\omega}} + \boldsymbol{\Omega}\mathbf{I}\boldsymbol{\omega} = \mathbf{M}_c + \mathbf{M}_{gg} + \mathbf{M}_d, \tag{19}$$

where $\mathbf{M}_c$ is the control torque vector acting on the spacecraft. For our purposes it will be assumed that it can be expressed as, $\mathbf{M}_c = \mathbf{I}T\mathbf{W}\boldsymbol{u}$, where T is the scalar magnitude of the specific torque or torque per unit inertia, $\mathbf{W}$ is a symmetric, positive-definite, torque transformation weighting matrix and $\boldsymbol{u}$ defines the direction of the torque vector. It is the attitude steering control input to the spacecraft. Thus Eq. (19) is

$$\dot{\boldsymbol{\omega}} = \mathbf{I}^{-1}\mathbf{M}_{gg}(\mathbf{q}) - \mathbf{I}^{-1}\boldsymbol{\Omega}\mathbf{I}\boldsymbol{\omega} + T\mathbf{W}\boldsymbol{u}, \tag{20}$$

with

$$\boldsymbol{u} = \begin{bmatrix} \sin\alpha\cos\beta & \cos\alpha\cos\beta & \sin\beta \end{bmatrix}^T. \tag{21}$$

When one is interested in the problem of finding the steering control

$$\boldsymbol{u} = \boldsymbol{u}(t), t_0 \le t \le t_f, \tag{22}$$

the torque direction time history is sought, such that it minimizes the cost functional:

$$J = 0.5(\boldsymbol{\omega}(t) - \boldsymbol{\omega}_d)^T \mathbf{Q}_f(\boldsymbol{\omega}(t) - \boldsymbol{\omega}_d)\Big|_{t=t_f} = \Phi\{\boldsymbol{\omega}(t)\}\Big|_{t=t_f}, \tag{23}$$

subject to, Eqs. (18), (20) and (21).

Introducing the single state vector, $\mathbf{x} = \begin{bmatrix} \mathbf{q}^T & \boldsymbol{\omega}^T \end{bmatrix}^T$, so the Eqs. (18), (20) and (22) are expressed as

$$\frac{d}{dt}\mathbf{x}^T = \frac{d}{dt}\begin{bmatrix} \mathbf{q}^T & \boldsymbol{\omega}^T \end{bmatrix} = \boldsymbol{f}^T, \mathbf{q}^T \cdot \mathbf{q} - \mathbf{1} \equiv g = 0 \tag{24}$$

To solve the optimization problem, seven Lagrangian multipliers or co-states are introduced given by the two vectors $\boldsymbol{\lambda}_{\mathbf{q}}(t)$ and $\boldsymbol{\lambda}_{\omega}(t)$, and a scalar λ_c denoted by a single column vector, $\boldsymbol{\lambda}(t)$. Following Bryson and Ho [15], a Hamiltonian function is defined as

$$H = \boldsymbol{\lambda}^T(t)\begin{bmatrix} \boldsymbol{f}^T(t) & g \end{bmatrix}^T. \tag{25}$$

Hence,

$$H = \boldsymbol{\lambda}_{\mathbf{q}}^T\left(\frac{1}{2}\mathbf{A}_{\omega}(\boldsymbol{\omega})\mathbf{q}\right) + \boldsymbol{\lambda}_{\boldsymbol{\omega}}^T\left(\mathbf{I}^{-1}\mathbf{M}_{gg}(\mathbf{q}) - \mathbf{I}^{-1}\boldsymbol{\Omega}\mathbf{I}\boldsymbol{\omega}\right) + \boldsymbol{\lambda}_{\boldsymbol{\omega}}^T T\mathbf{W}\boldsymbol{u} + \lambda_c g. \tag{26}$$

The necessary conditions (Bryson and Ho [15], Conway [16]) for the first variation of J to be zero include the co-state differential equations

$$\frac{d}{dt}\boldsymbol{\lambda}^T(t) = -\frac{\partial H}{\partial \mathbf{x}}. \tag{27}$$

Explicitly the co-state equations are

$$\dot{\lambda}_{q_i}(t) = -\frac{\partial H}{\partial q_i} = -\left\{\boldsymbol{\lambda}_{\mathbf{q}}^T\left(\frac{1}{2}\mathbf{A}_\omega(\boldsymbol{\omega})\frac{\partial}{\partial q_i}\mathbf{q}\right) + \boldsymbol{\lambda}_{\boldsymbol{\omega}}^T\left(\mathbf{I}^{-1}\frac{\partial}{\partial q_i}\mathbf{M}_{gg}(\mathbf{q})\right)\right\}, \tag{28}$$

$$\dot{\lambda}_{\omega_i}(t) = -\frac{\partial H}{\partial \omega_i} = -\left\{\boldsymbol{\lambda}_{\mathbf{q}}^T\left(\frac{1}{2}\frac{\partial}{\partial \omega_i}\mathbf{A}_\omega(\boldsymbol{\omega})\mathbf{q}\right) - \boldsymbol{\lambda}_{\boldsymbol{\omega}}^T\mathbf{I}^{-1}\frac{\partial}{\partial \omega_i}(\boldsymbol{\Omega}\mathbf{I}\boldsymbol{\omega})\right\}, \tag{29}$$

$$\dot{\lambda}_c = 0. \tag{30}$$

The optimality conditions are

$$\frac{\partial H}{\partial \alpha} = \boldsymbol{\lambda}_{\boldsymbol{\omega}}^T \cdot T\mathbf{W}[\cos\alpha\cos\beta \quad -\sin\alpha\cos\beta \quad 0]^T = 0, \tag{31}$$

and

$$\frac{\partial H}{\partial \beta} = \boldsymbol{\lambda}_{\boldsymbol{\omega}}^T \cdot T\mathbf{W}[-\sin\alpha\sin\beta \quad -\cos\alpha\sin\beta \quad \cos\beta]^T = 0. \tag{32}$$

Hence,

$$\mathbf{W}\boldsymbol{\lambda}_{\boldsymbol{\omega}} = |\mathbf{W}\boldsymbol{\lambda}_{\boldsymbol{\omega}}|[\sin\alpha\cos\beta \quad \cos\alpha\cos\beta \quad \sin\beta]^T = -|\mathbf{W}\boldsymbol{\lambda}_{\boldsymbol{\omega}}|\boldsymbol{u}. \tag{33}$$

Thus the two-parameter control vector $\boldsymbol{u}$, can be expressed as

$$\boldsymbol{u} = -\mathbf{W}\boldsymbol{\lambda}_{\boldsymbol{\omega}}/|\mathbf{W}\boldsymbol{\lambda}_{\boldsymbol{\omega}}|. \tag{34}$$

The choice of the sign in Eq. (34) will depend on the direction of the desired torque, forward or reverse torque. Thus the closed-loop equations of motion are

$$\frac{d}{dt}\mathbf{x} = \boldsymbol{f}, \mathbf{q}^T \cdot \mathbf{q} - \mathbf{1} \equiv g = 0, \boldsymbol{u} = -\mathbf{W}\boldsymbol{\lambda}_{\boldsymbol{\omega}}/|\mathbf{W}\boldsymbol{\lambda}_{\boldsymbol{\omega}}|. \tag{35}$$

To complete the definition of the optimal solution, the boundary conditions at $t = t_f$ for the co-state system are found by applying the transversality conditions. The transversality conditions ensure that the initial and final states are selected optimally within the feasible regions of the states. For the transversality conditions, one may write

$$\boldsymbol{\lambda}_{\mathbf{q}}(t_f) = \left.\frac{\partial \Phi\{\boldsymbol{\omega}(t)\}}{\partial \mathbf{q}}\right|_{t=t_f} = 0, \tag{36}$$

$$\boldsymbol{\lambda}_{\boldsymbol{\omega}}(t_f) = \left.\frac{\partial \Phi\{\boldsymbol{\omega}(t)\}}{\partial \boldsymbol{\omega}}\right|_{t=t_f} = \mathbf{Q}_f(\boldsymbol{\omega}(t_f) - \boldsymbol{\omega}_d). \tag{37}$$

$$\lambda_c(t_f) = 0. \tag{38}$$

Thus, $\lambda_c(t) = 0$ is a feasible solution. If it can be ensured that the constraint on the quaternion, $\mathbf{q}^T \cdot \mathbf{q} - \mathbf{1} \equiv g = 0$ is satisfied at each and every integration time step, $\lambda_c(t)$ could be set to zero for all time.

The solution to the state and co-state equations, subject to the optimal control defined by Eq. (34) and the boundary conditions defined by Eqs. (36) and (37), may be found by solving a two point boundary value problem (TPBVP). This can be done using MATLAB's function, *bvp4c.m*.

Re-considering Eq. (34), the control input vector may be expressed as

$$\boldsymbol{u} = -\mathbf{W}\boldsymbol{\lambda}_{\boldsymbol{\omega}}/|\mathbf{W}\boldsymbol{\lambda}_{\boldsymbol{\omega}}| = -r^{-1}\mathbf{B}^T\boldsymbol{\lambda}, \tag{39}$$

where $\mathbf{B}^T$ is a projection matrix relating $\boldsymbol{\lambda}$ to $\mathbf{W}\boldsymbol{\lambda}_{\boldsymbol{\omega}}$ and $r = |\mathbf{W}\boldsymbol{\lambda}_{\boldsymbol{\omega}}|$. It is important to recognize that the weighting matrix $\mathbf{W}$ also needs to be chosen. On one hand it provides a set of free parameters so one can construct an optimal solution, while on the other it makes the solution much harder to obtain. Its choice is discussed further in Section 8.

4. Feedback implementation of optimal co-states

Of interest at this stage is to be able to implement the controller, obtained in the last section, as a feedback control law. Thus, inspired by linear optimal control, it can be assumed that, locally, the co-state vector $\boldsymbol{\lambda}$ is proportional to the state vector $\mathbf{x}$. Hence one may express

$$\boldsymbol{\lambda} = \mathbf{P}\mathbf{x}. \tag{40}$$

Furthermore it is assumed that the matrix P is slowly varying and hence does not change appreciably as the time t changes from the current time t to $t \pm \Delta t$ and to $t \pm n\Delta t$, $n = 2, 3, 4$. Thus the matrix P may be obtained by evaluating $\boldsymbol{\lambda}$ and $\mathbf{x}$ in Eq. (40) at the times $t \pm n\Delta t$, $n = 0, 1, \ldots, 4$, where Δt is a reasonably small time step, over which both $\boldsymbol{\lambda}$ and $\mathbf{x}$ change appreciably. The solution for the matrix $\mathbf{P}$ is obtained by differencing the data and solving an over determined system of linear equations by a least squares approach over a moving time window. The matrix $\mathbf{P}$ is also constrained to be a symmetric non-negative definite matrix. Moreover $\mathbf{x}$ is expressed as

$$\mathbf{x} = \left(\mathbf{x} - \mathbf{x}_{ref}\right) + \mathbf{x}_{ref}, \tag{41}$$

where $\mathbf{x}_{ref}$ is the optimal transfer trajectory for $\mathbf{x}$ obtained by solving the system of equations for the states, parameters and co-states defined in Section 3, which together constitute a TPBVP. By solving the system of equations for the states, parameters and co-states defined in Section 3, one also seeks the trajectory coordinates of the reference trajectory and the total transfer time. It is important that the spacecraft's attitude controller is able to track the reference trajectory. Consequently the control input is expressed as

$$\boldsymbol{u} = -r^{-1}\mathbf{B}^T\mathbf{P}\left(\mathbf{x} - \mathbf{x}_{ref}\right) + r^{-1}\mathbf{B}^T\boldsymbol{\lambda}_{ref}, \boldsymbol{\lambda}_{ref} = \mathbf{P}\mathbf{x}_{ref}. \tag{42}$$

The second of Eq. (42) is used to obtain the matrix, P. Like in Eq. (34), the choice of the sign in the first of Eq. (42) will depend on the direction of the desired torque, forward or reverse torque.

5. Simplified formulation of the optimal angular rate trajectory synthesis problem

Some authors (for example, see Aghili [4]) have formulated the attitude rate acquisition problem without including the gravity gradient torques. Thus the quaternion kinematics could be ignored. Thus in this case one could set $\boldsymbol{\lambda}_{\mathbf{q}}(t) \equiv 0$, which results in considerable simplification of the trajectory synthesis problem. The downside of the approach is that the quaternion kinematics is ignored and consequently the quaternion attitude could be quite arbitrary. Quite often after a detumbling manoeuver, a precise orientation must be acquired. The required attitude could be acquired in an independent manoeuver and the methodology for this is developed in the next section. The associated tracking problem, which involves tracking the complete state vector set point, must then be separately addressed. Typically this is done by using a barrier Lyapunov function as illustrated by Vepa [17].

6. Optimal attitude orientation acquisition trajectory synthesis

To begin with the quaternion kinematics is given by Eqs. (9) and (11), and can be expressed in one of two alternate forms as.

$$\frac{d\mathbf{q}}{dt}=\frac{1}{2}\mathbf{A}_{\omega}(\boldsymbol{\omega})\mathbf{q}=\frac{1}{2}\boldsymbol{\Gamma}(\mathbf{q})\boldsymbol{\omega}. \tag{43}$$

In Eq. (43), the angular velocity vector is treated as a control variable and expressed as

$$\boldsymbol{\omega}=|\boldsymbol{\omega}|_{\max}\boldsymbol{u}, \tag{44}$$

where the direction vector $\boldsymbol{u}$ is parametrized by an equation similar to Eq. (21). Thus,

$$\boldsymbol{u}=\begin{bmatrix}\sin\alpha\cos\beta & \cos\alpha\cos\beta & \sin\beta\end{bmatrix}^T. \tag{45}$$

When one is interested in the problem of finding the directional control

$$\boldsymbol{u}=\boldsymbol{u}(t),\, t_0\le t\le t_f, \tag{46}$$

the angular velocity direction time history is sought, such that it minimizes the cost functional:

$$J=0.5(\mathbf{q}(t)-\mathbf{q}_d)^T\mathbf{Q}_f(\mathbf{q}(t)-\mathbf{q}_d)\Big|_{t=t_f}=\Phi\{\mathbf{q}(t)\}|_{t=t_f}, \tag{47}$$

subject to, Eqs. (43), (44) and (45). The corresponding Hamiltonian function is

$$H=\boldsymbol{\lambda}_{\mathbf{q}}^T\left(\frac{1}{2}\mathbf{A}_{\omega}(\boldsymbol{\omega})\mathbf{q}\right)=\boldsymbol{\lambda}_{\mathbf{q}}^T\left(\frac{1}{2}\boldsymbol{\Gamma}(\mathbf{q})\boldsymbol{\omega}\right)=\frac{|\boldsymbol{\omega}|_{\max}}{2}\left(\boldsymbol{\lambda}_{\mathbf{q}}^T\boldsymbol{\Gamma}(\mathbf{q})\right)\boldsymbol{u}. \tag{48}$$

The corresponding co-state differential equations are

$$\frac{d}{dt}\boldsymbol{\lambda}_{\mathbf{q}}^T(t)=-\frac{\partial H}{\partial\mathbf{q}}=-\frac{1}{2}\boldsymbol{\lambda}_{\mathbf{q}}^T\mathbf{A}_{\omega}(\boldsymbol{\omega}). \tag{49}$$

By using an argument similar to the one used in developing Eqs. (31)–(34), the optimal control is given by

$$\boldsymbol{u}=-\boldsymbol{\Gamma}^T(\mathbf{q})\boldsymbol{\lambda}_{\mathbf{q}}/\left|\boldsymbol{\Gamma}^T(\mathbf{q})\boldsymbol{\lambda}_{\mathbf{q}}\right|=\boldsymbol{\Gamma}(\mathbf{q})\boldsymbol{\lambda}_{\mathbf{q}}/\left|\boldsymbol{\Gamma}(\mathbf{q})\boldsymbol{\lambda}_{\mathbf{q}}\right|. \tag{50}$$

For the co-state boundary conditions one has

$$\lambda_{\mathbf{q}}(t_f) = \left.\frac{\partial \Phi\{\mathbf{q}(t)\}}{\partial \mathbf{q}}\right|_{t=t_f} = \mathbf{Q}_f(\mathbf{q}(t_f) - \mathbf{q}_d). \tag{51}$$

Once the control is found from Eqs. (50) and (44) is used to define the angular velocity vector and Eqs. (19) and (20) to define the optimal input control torque.

7. Shape based optimal trajectory synthesis

An alternative approach to the optimization based on the integration of co-states is to use a shape based approach as outlined by Caubet and Biggs [18, 19]. For purposes of comparison the shape based approach serves as a useful alternative. In a shape based approach, each of the quaternion components are expressed as a summation of polynomials in terms of a time variable, multiplied by coefficients which may be determined by applying the relevant boundary conditions at the initial and final values of the time variable over a finite time frame. Thus, for example, the quaternion components are expressed as

$$\begin{aligned} q_i = {} & q_{i0} + \dot{q}_{i0} t_f \left(\frac{t}{t_f}\right) + \left(\dot{q}_{i0} - \dot{q}_i\big|_{t=t_f}\right) t_f \left(\frac{t}{t_f}\right)^2 \left(1 - \frac{t}{t_f}\right) \\ & + \left(q_{if} - q_{i0} - \dot{q}_{i0} t_f\right) \frac{t^2}{t_f^2} \left(3 - 2\frac{t}{t_f}\right) + \sum_{j=5}^{N+4} \bar{e}_{ij-4} \left((j-4)\left(\frac{t}{t_f}\right)^2 - (j-3)\left(\frac{t}{t_f}\right)^3 + \left(\frac{t}{t_f}\right)^{j-1}\right), \end{aligned} \tag{52}$$

where the coefficients $\bar{e}_{ij-4}$ are yet to be determined. They are determined by minimizing the cost function

$$J = \left(\int_0^1 (q_1^2 + q_2^2 + q_3^2 + q_4^2 - 1) d\left(\frac{t}{t_f}\right)\right)^2. \tag{53}$$

Once all the coefficients of the quaternion components q_i are determined, the angular velocity vector is defined by the inverse of the relation given by Eq. (43) which is

$$\boldsymbol{\omega} = \begin{bmatrix} q_4 & q_3 & -q_2 & -q_1 \\ -q_3 & q_4 & q_1 & -q_2 \\ q_2 & -q_1 & q_4 & -q_3 \end{bmatrix} \begin{bmatrix} \dot{q}_1 \\ \dot{q}_2 \\ \dot{q}_3 \\ \dot{q}_4 \end{bmatrix}. \tag{54}$$

The angular velocity vector $\boldsymbol{\omega}$ is evaluated for a range of non-dimensional time values between 0 and 1. From the ratio of the maximum of this set, defining the angular velocity time history in terms of the non-dimensional time variable, and the maximum allowable angular velocity magnitude, the length of the time frame t_f over which the control torques must be applied may be found. From the angular velocity vector $\boldsymbol{\omega}$ the torques that must be applied to the satellite including the gravity gradient torques may be found. From a range of choices for N

(say $1 \leq N \leq 6$) in the Eq. (52) defining the quaternion, the one that gives the lowest value for t_f is selected.

8. Typical simulation examples

The first example considered the attitude dynamics is defined by Eqs. (18) and (19). Thus the gravity gradient torques acting on the spacecraft are included in the dynamic model and they are responsible for coupling the attitude quaternion kinematics and the angular velocity dynamics. The objective is to spin the spacecraft so the final angular velocity vector is given by $\boldsymbol{\omega} = [1 \quad 1 \quad 1]^T rads/s$. The initial angular velocity vector is $\boldsymbol{\omega} = [0 \quad 0 \quad 0]^T$. The spacecraft is fitted with magnetic torque actuators and the maximum three axis torques are limited to $\mathbf{T}_c = [0.62 \quad 1 \quad 1]^T Nm$. The diagonal non-zero elements of the weighting matrix $\mathbf{W}$ in Eq. (34) and the principal moment of inertia of the spacecraft are respectively given by

$$\mathbf{W}_{diag} = [0.24 \quad 0.9 \quad 1], \mathbf{I} = [2.27293 \quad 3.27331 \quad 0.3989] kgm^2. \qquad (55)$$

In all examples the solution of the TPBVP is done using MATLAB's function, *bvp4c.m*. Whenever there was a need to solve an initial value problem, the equations were integrated using MATLAB's *ode45.m*.

In all cases, the time variable was made non-dimensional so it raged from [0–1]. The integration time step was chosen to be relatively small initially ($\Delta t = 0.0001$), and automatically and iteratively reduced linearly as the final time was approached. The iterations were terminated when no further improvement in the accuracy of the predicted final time was feasible.

Figure 1 shows the time history of the reference quaternion components and **Figure 2** shows the corresponding angular velocities (p, q, r). **Figures 3** and **4** show the corresponding, actual, quaternion components and the corresponding angular velocity components, where an approximate optimal linear feedback law based on Eq. (42) is used to track the reference trajectory.

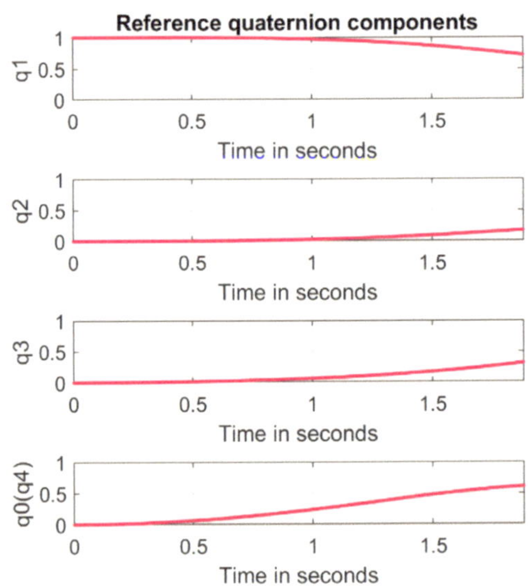

Figure 1.
Time history of the reference quaternion components for the first example.

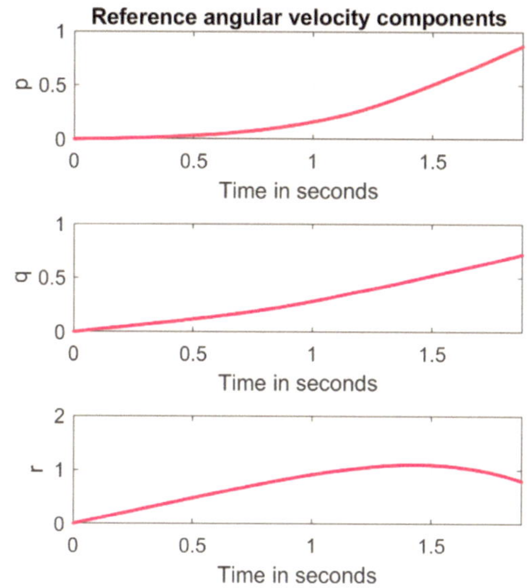

Figure 2.
Time history of the corresponding reference velocity components for the first example.

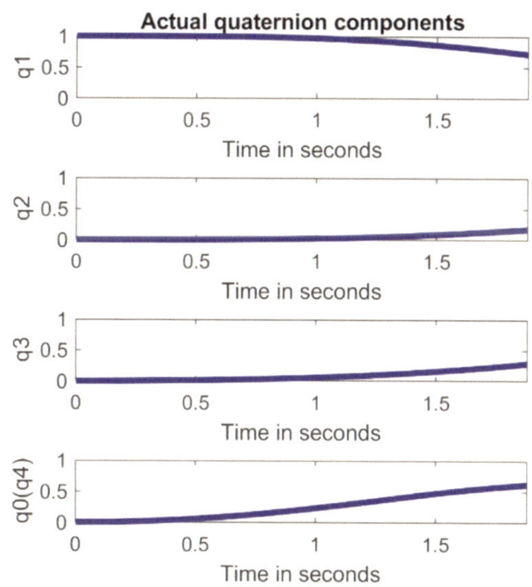

Figure 3.
Time history of the actual quaternion components tracked by the feedback controller for the first example.

In the next example, the simplified attitude dynamics is used with the gravity gradient torques neglected. This decouples the angular rate dynamics from the attitude quaternion dynamics, which need to be considered for synthesizing the reference trajectories. The reference angular velocities are then integrated to obtain the spacecraft's quaternion attitude time history. **Figure 5** shows the reference angular velocity components.

Figure 6 shows the errors in the actual angular velocity components when compared with corresponding reference values and **Figure 7** the corresponding quaternion components. **Figure 8** shows the attitude in terms of the Euler axis and the Euler principal angle components.

In the final example it is desired to alter the attitude quaternion of the spacecraft, so as to point the spacecraft in a desired direction. In this case on the

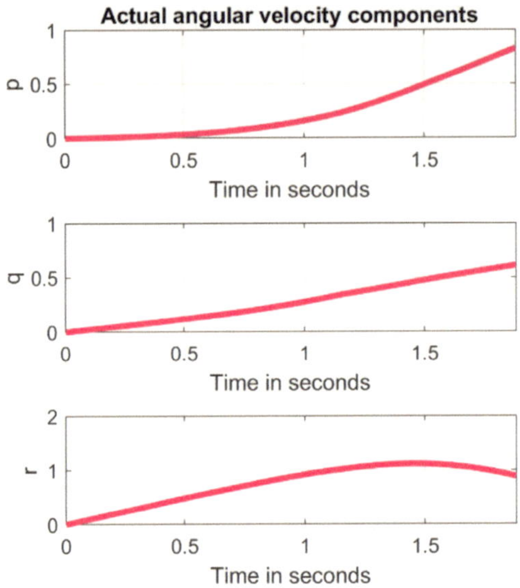

Figure 4.
Time history of the actual velocity components tracked by the feedback controller for the first example.

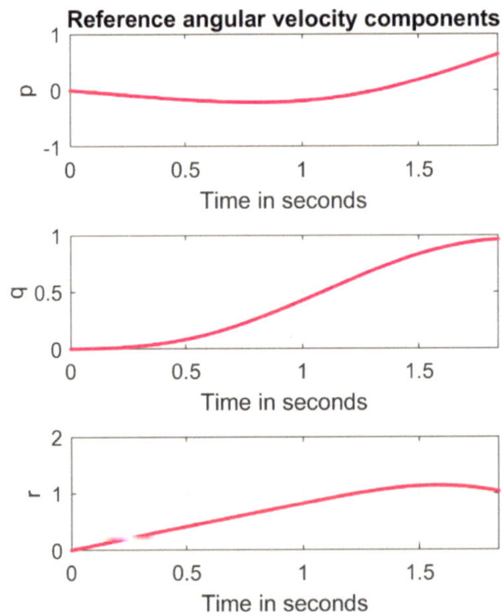

Figure 5.
Time history of the reference velocity components for the second example.

quaternion kinematics defined by Eq. (44) are used. The maximum angular velocity of the spacecraft is assumed to be limited to 0.001 rad/s. The desired pointing direction is specified as a yaw, roll pitch Euler angle sequence given by $[\psi \quad \theta \quad \phi] = [22° \quad 25° \quad 30°]$ corresponding to the components of the quaternion $\mathbf{q}_d = [0.2068 \quad 0.2518 \quad 0.1682 \quad 0.9303]$.

In **Figure 9** are shown the reference optimal quaternion components and in **Figure 10** are shown the corresponding angular velocity components. The optimum torque components required to affect the attitude change are shown in **Figure 11**.

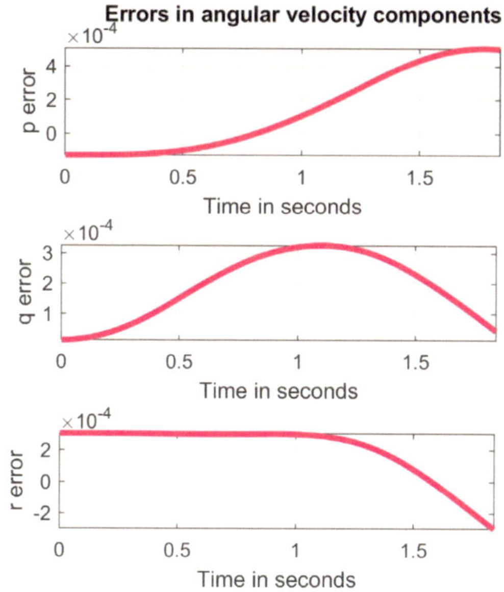

Figure 6.
Time history of the errors in the actual velocity components corresponding to ***Figure*** *5, tracked by the feedback controller.*

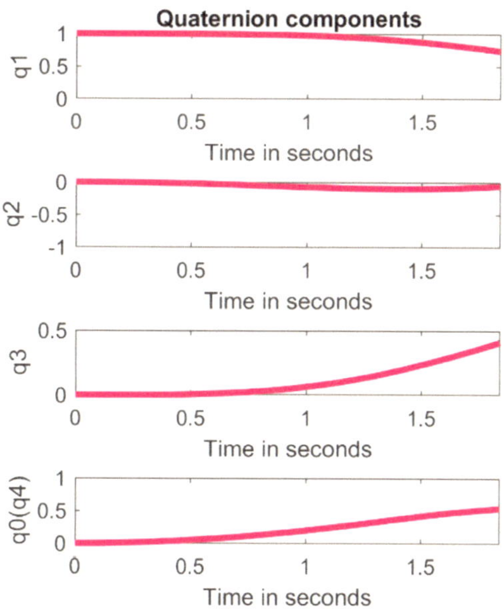

Figure 7.
Time history of the corresponding quaternion components tracked by the feedback controller for the second example.

These include compensation for the gravity gradient torques. They show that they could be easily achieved by low thrust electric actuators such as electro-spray thrusters.

When small reaction wheels are used Eq. (19) may be modified to include the momentum of the wheels and the control inputs to the wheels could also be

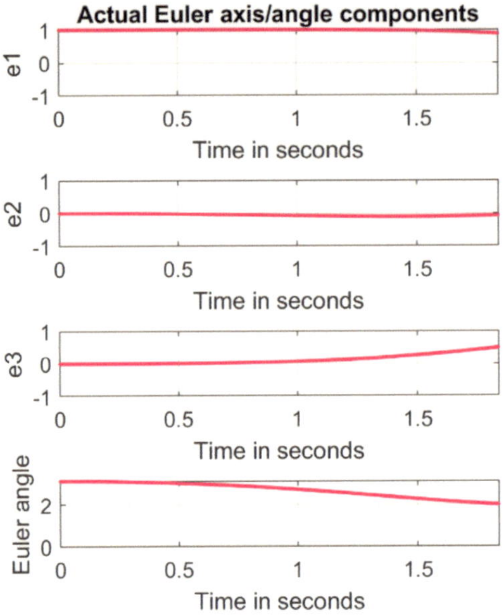

Figure 8.
Time history of the attitude components in terms of the Euler axis and the Euler principal angle components for the second example.

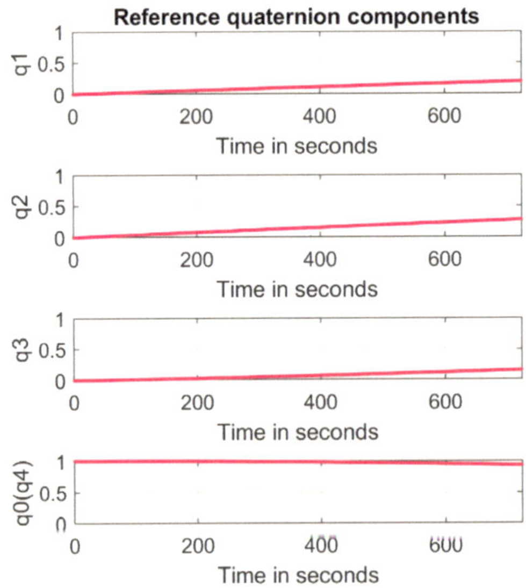

Figure 9.
Time history of the reference quaternion components for the final example.

estimated. If reaction wheels are used much larger torques are possible and the time over which they are used could be shortened. In **Figure 12** are shown the attitude time history components in terms of the Euler axis and Euler principal angle components.

This example is also solved using the shape based approach discussed briefly in Section 7. In **Figure 13** are shown the required applied torque components obtained by the shape based approach with $N = 1$ in Eq. (52). The time frame over which the control must be applied is $t_f = 733.3s$, which is the lowest for all N considered and is about the same as the time required by the approach based on the integration

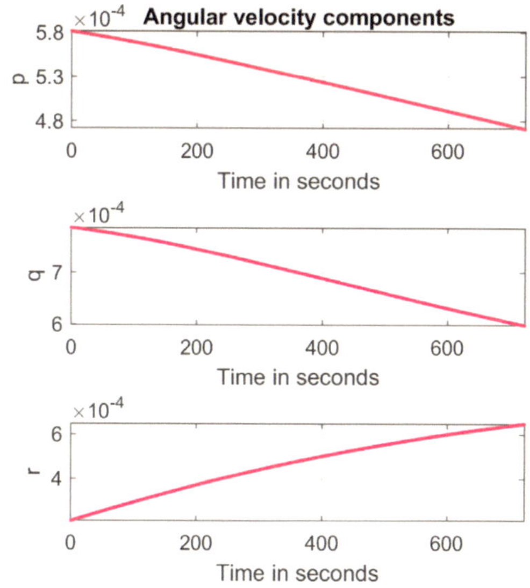

Figure 10.
Time history of the corresponding reference velocity components for the final example.

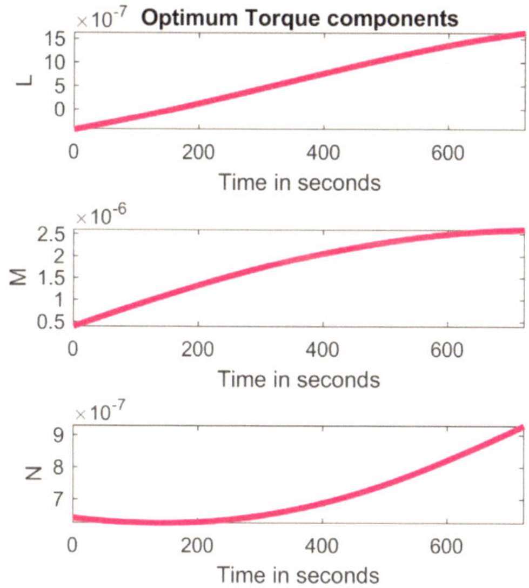

Figure 11.
Time history of the corresponding reference torque components for the final example.

of co-states. However the reference torque components shown in **Figure 13**, are much larger than those plotted in **Figure 11**, and for this reason, they are not referred to as 'optimum torques.' The corresponding gravity gradient torques, acting on a satellite orbiting the Earth at the geostationary orbit radius, are also shown in **Figure 14**. Although the gravity gradient torques are of the same orders of magnitude as the reference optimum torque components in **Figure 11**, they are much smaller than the corresponding torque components obtained by the shape based approach and shown in **Figure 13**. It must be recognized that the gravity gradient torques become much larger as the spacecraft orbits the Earth at a much closer orbit radius.

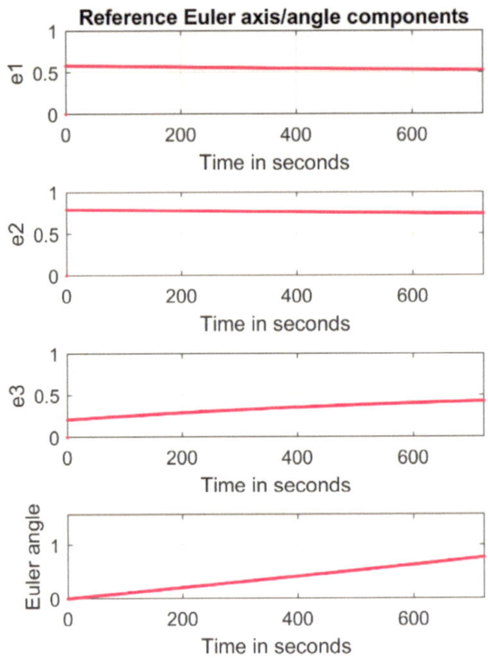

Figure 12.
Time history of the attitude components in terms of the Euler axis and the Euler principal angle components for the final example.

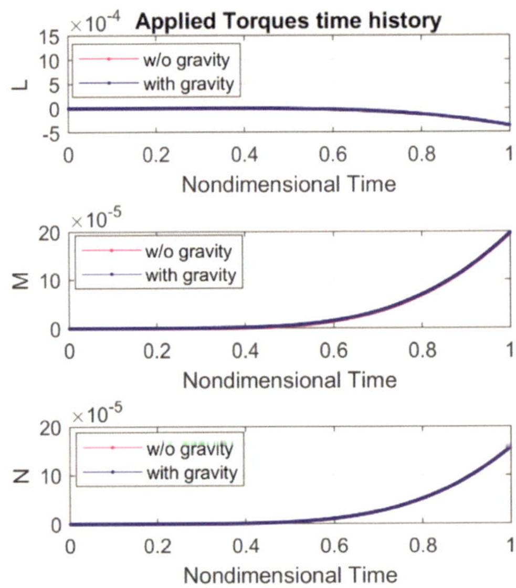

Figure 13.
Time history of the applied torque components for the final example, obtained by the shape based approach.

9. Discussion and conclusions

A close examination of the results in **Figures 9** and **12** shows that while the quaternion component time histories are not linearly varying, the Euler axis and Euler principal angle components are almost linear. This allows for linear extrapolation of the trajectories if when desired. It is also observed that the acquisition of

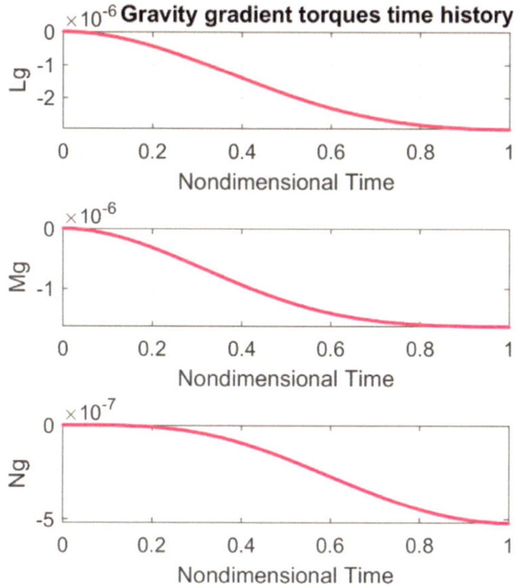

Figure 14.
Time history of the gravity gradient torque components for the final example, obtained by the shape based approach.

the Euler axis is relative fast in comparison with growth rate of the Euler angle which is relatively slower. The kinematics of the Euler axis seems to represent a fast sub-system while the growth of the Euler angle represents the slow sub-system. This observation, facilitates the construction of approximate sub-optimal trajectories where in the Euler axis is acquired instantly and the Euler angle increases or decreases linearly with time. Once a sub-optimal solution can be defined in terms of interpolating polynomial it can also be further optimized by using Eq. (13) and shape based methods as those proposed by Caubet and Biggs [18, 19], quite rapidly and if need be, by the pseudo spectral method, or other direct collocation methods. The advantage of further optimization using shape based methods is that the precise shape of the desired output could be achieved avoiding overshoot. However depending on the choice of the output shape function, the control could be restrictive and so the magnitudes of the torques required could be much larger in comparison with the co-states approach.

For the preceding example, where a set angular velocity components were desired, the velocity components and the corresponding quaternion components are shown in **Figures 5–7** as they vary with time. **Figure 8** also shows that the Euler axis and Euler principal angle components are varying as quadratic functions of time. A similar conclusion cannot be drawn as far as the angular velocity components and the components of the quaternion. This again is extremely useful in applying low order polynomials for developing formulae for extrapolating the optimal trajectories, by converting the quaternion components to the domain of the Euler axis and Euler principal angle components. It also facilitates the integration of various optimal segments into a single trajectory over an extended time frame.

From the first example, comparing **Figures 1** and **2** with **Figures 3** and **4**, it is seen that the optimum tracking feedback control law obtained by linearly approximating the relationship between the states and co-states by Eq. (40), performs well. The errors between these two sets of trajectories, the reference trajectory and the actual tracked trajectory, are always within 5% of the corresponding reference value, over the time frame of the plots.

In this paper, either the simplest form of the attitude dynamics or the basic kinematic equations alone are used to construct the optimal trajectories. The required control torques are obtained from the inverse dynamic relations. The usefulness of transforming the attitude representation to the Euler axis and Euler principal angle components, as it facilitates the application of low order polynomials for the construction of approximate sub-optimal trajectories, is demonstrated. Furthermore it is shown how optimal feedback control laws may be constructed from the solution for the optimal trajectories, for tracking the reference trajectories.

Conflict of interest

There are no conflicts of interest to declare.

Author details

Ranjan Vepa
School of Engineering and Material Science, Queen Mary University of London, London, UK

*Address all correspondence to: r.vepa@qmul.ac.uk

IntechOpen

© 2020 The Author(s). Licensee IntechOpen. Distributed under the terms of the Creative Commons Attribution - NonCommercial 4.0 License (https://creativecommons.org/licenses/by-nc/4.0/), which permits use, distribution and reproduction for non-commercial purposes, provided the original is properly cited.

References

[1] Lee B-H, Lee B-U, Oh H-S, Lee S-H, Rhee S-E. Time optimal attitude maneuver strategies for the agile spacecraft with reaction wheels and thrusters. Journal of Mechanical Science and Technology. 2005;**19**(9):1695-1705

[2] Yoshida K, Dimitrov D, Nakanishi H. On the capture of tumbling satellite by a space robot. In: IEE/RSJ International Conference on Intelligent Robots and Systems; 2006. pp. 4127-4132

[3] Aghili F. Optimal control for robotic capturing and passivation of a tumbling satellite, AIAA Guidance, Navigation and Control Conference. In: AIAA Guidance, Navigation and Control Conference and Exhibit; Aug., 2008; Honolulu, Hawaii; pp. 18–21 AIAA 2008-7274, DOI: 10.2514/6.2008-7274

[4] Aghili F. Time-optimal de-tumbling control of spacecraft. Journal of Guidance, Control, and Dynamics. 2009;**32**(5):1671-1675

[5] Yang CC, Wu CJ. Time-optimal de-tumbling control of a rigid spacecraft. Journal of Vibration and Control. 2008; **14**(4):553-570. DOI: 10.1177/1077546307080034

[6] Liu HY, Wang HN, Chen ZM. De-tumbling controller and attitude acquisition for micro-satellite based on magnetic torque. Journal of Astronautics. 2007;**28**(2):333-337

[7] Zhang L, Yu C, Zhang S, Cai H. Optimal attitude trajectory planning method for CMG actuated spacecraft. Proceedings of the Institution of Mechanical Engineers, Part G: Journal of Aerospace Engineering. 2016;**232**(1): 131-142

[8] Sharma R, Tewari A. Optimal nonlinear tracking of spacecraft attitude maneuvers. IEEE Transactions on Control Systems Technology. 2004; **12**(5):677-682. DOI: 10.1109/TCST.2004.825060

[9] Hegrenas O, Gravdahl JT, Tondel P. Attitude control by means of explicit model predictive control, via multi-parametric quadratic programming. In: Proceedings of American Control Conference; Piscataway, NJ: IEEE; 2005. pp. 901-906

[10] Lobo K, Lang J, Starks A, Sands T. Analysis of deterministic artificial intelligence for inertia modifications and orbital disturbance. International Journal of Control Science and Engineering. 2018;**8**(3):53-62

[11] Sands T, Kim JJ, Agrawal BN. Improved hamiltonian adaptive control of spacecraft. In: Proceedings of the IEEE Aerospace Conference; MT, USA: Big Sky; 2009. pp. 7-14

[12] Nakatani S, Sands T. Simulation of spacecraft damage tolerance and adaptive controls. In: Proceedings of the IEEE Aerospace Conference; MT, USA: Big Sky; 2014. pp. 1-8

[13] Baker K, Cooper M, Heidlauf P, Sands T. Autonomous trajectory generation for deterministic artificial intelligence. Electrical and Electronic Engineering. 2018;**8**(3):59-68

[14] Smeresky B, Rizzo A, Sands T. Kinematics in the information age. Mathematics. 2018;**6**(9):148

[15] Bryson AE Jr, Ho YC. Applied Optimal Control. Waltham, MA, USA: Ginn and Company; 1969

[16] Conway BA, editor. Spacecraft Trajectory Optimization. 1st ed. New York, NY, USA: Cambridge University Press; 2010

[17] Vepa R. Dynamics and Control of Autonomous Space Vehicles and

Robotics. New York, NY, USA: Cambridge University Press; 2019

[18] Caubet A, Biggs JD. Optimal attitude motion planner for large slew maneuvers using a shape-based method. In: AIAA Guidance Navigation and Control Conference; Boston, USA; 2013. pp. 1-9. DOI 10.2514/6.2013–4719

[19] Caubet A, Biggs JD. An efficient sub-optimal motion planning method for attitude maneuvers. In: Proceedings AstroNet-II International Final Conference; Germany: Springer; 2016. pp. 31-46

Chapter 8

Parameter Optimization for Spacecraft Attitude Stabilization Using Magnetorquers

Renato Bruni and Fabio Celani

Abstract

The attitude stabilization of a spacecraft that uses magnetorquers as torque actuators is a very important task. Depending on the availability of angular rate sensors on the spacecraft, control laws can be designed either by using measurements of both attitude and attitude rate or by using measurements of attitude only. The parameters of both types of control laws are usually determined by means of a simple trial-and-error approach. Evidently, such an approach has several drawbacks. This chapter describes recently developed systematic approaches for determining the parameters using derivative-free optimization techniques. These approaches allow to find the parameter values that minimize the settling time of the attitude error or an indirect measure of this error. However, such cost indices depend also on initial conditions of the spacecraft, which are not known in advance. Thus, a min-max optimization problem is formulated, whose solution provides values of the parameters minimizing the chosen cost index under the worst initial conditions. The chapter also provides numerical results showing the effectiveness of the described approaches.

Keywords: derivative-free optimization, spacecraft attitude control, robust optimization, min-max problems, magnetic actuators

1. Introduction

A *magnetorquer* (or magnetic torquer) is a torque actuator widely used for attitude control in satellites, especially those flying in low Earth orbits. The magnetorquer generates a magnetic dipole that interacts with the Earth magnetic field, thus generating a torque used to control the spacecraft attitude. The control torque generated by magnetorquers is constrained to belong to a plane orthogonal to the Earth magnetic field; hence, magnetorquers may be supported by additional torque actuators to achieve full three-axis control (see [1], Chapter 7). However, attitude control systems using only magnetorquers represent a feasible option especially for low-cost satellites or for satellites with a fault in the main attitude control system. Therefore, attitude control of spacecraft using magnetorquers is a very important topic in aerospace engineering.

Many control laws have been designed in such a setting, and an overview can be found in [2]. In particular, Celani [3] shows that attitude stabilization using only magnetorquers can be achieved by proportional derivative-like (PD-like) control which requires measurements of both attitude and attitude rate. The same work shows that attitude stabilization can also be obtained by using attitude-only feedback, which has the advantage of not requiring the installation of angular rate sensors (rate gyros), thus saving in cost, volume, and weight. Those control laws contain parameters. Thus, numerical values must be assigned to those parameters to practically implement the laws. A common way to find those values is by using a so-called trial-and-error approach, consisting in trying several values and rejecting those which are not acceptable, and eventually choosing the best among the acceptable ones. This approach is affected by two important drawbacks: (a) it is often very time-consuming, and (b) it is not systematic. Thus, even if an obtained solution is satisfactory, it is not known if protracting the search could lead to better solutions and which would be the magnitude of the possible improvement.

On the other hand, a systematic approach for determining the parameters should aim at finding the values which minimize the settling time of the attitude error. Such an approach has been recently proposed in [4–6]. However, this is not an easy objective to pursue. The settling time depends not only on the parameters but also on the initial conditions of the spacecraft. To overcome this issue, the above works propose to compute the values of the parameters that minimize the settling time obtained under the worst initial conditions, so as to provide averagely good results and to set as bound the worst-case behavior. Hence, in the above works, the problem is modeled as a min-max problem, and the obtained parameters' values are called *robust optimal* values.

This min-max problem is considerably challenging, since solving the main minimization problem (upper-level) requires solving a maximization problem (lower-level) at every evaluation of its objective function. A decomposition is not possible because the worst initial conditions are not determined in general: they, in turn, depend on the adopted parameters. Optimizing this min-max problem involves other two mathematical issues:

1. Settling time cannot be expressed as an analytical function of parameters and initial conditions; therefore the specification of an explicit optimization model is not possible.

2. Settling time is not even continuous with respect to both parameters and initial conditions, in the sense that small variations of the latter may result in a substantial gap in the variation of the settling time.

To overcome these issues, the approach proposed in [4–6] relies on the use of derivative-free optimization algorithms as building blocks. These techniques do not need first-order information on the objective function nor do they need its analytical expression. They only need to compute the objective function over a number of points by means of simulations.

For the design of the specific optimization algorithm, a distinction must be done between spacecraft equipped with angular rate sensors, studied in [5], and spacecraft not equipped with those sensors, studied in [6]. This because, in the first case, the control is based on the availability of measurements of both attitude and attitude rates and is realized through a *PD-like control*, while in the second case, the control can rely only on attitude feedback, and it is called *attitude-only feedback control*.

The first case, though presenting the mentioned difficulties, turns out to be the easier among the two, since the PD-like control law has only two parameters, and their robust optimal values can be successfully determined by means of a global search optimization procedure of the type of DIRECT algorithm [7], as proposed in [5]. In the second case, the control law contains four parameters, and their range of variation is wider than the first case. Thus, the determination of the robust optimal parameters becomes even harder, and a very complex derivative-free optimization algorithm based on a combination of both local and global search had to be developed. Moreover, numerical experience has shown that dealing with a discontinuous objective function makes the determination of the optimal solution very challenging even when using derivative-free algorithms. Thus, in Ref. [4, 6], the objective function is changed to the so-called integral time absolute error (ITAE) which is continuous with respect to both the parameters and the initial conditions. Such a change is acceptable since it has been shown that minimizing the ITAE is approximately equivalent to minimizing the settling time (see [8]).

This chapter describes in detail the abovementioned approaches to the determination of robust optimal values for the parameters in order to minimize the settling time or the ITAE obtained under the worst initial conditions. The exposition will highlight in particular the following main contributions: (i) the definition of a new systematic approach for the determination of the parameters for the PD-like control algorithm and for the attitude-only feedback, (ii) the formulation of a min-max optimization model to find a robust optimal solution to both problems, and (iii) the development of derivative-free optimization strategies to tackle the min-max problems.

The chapter is organized as follows: Section 2 defines the spacecraft model and the control algorithms for the two cases mentioned above; Section 3 describes the optimization model and the solution algorithm for the case of PD-like control; Section 4 provides some computational experience for this first case; Section 5 describes the optimization model and the solution algorithm for the case of attitude-only feedback control; Section 6 provides again computational experience for this second case.

2. Spacecraft model and control algorithms

The following coordinate frames are introduced to describe the attitude dynamics of an Earth-orbiting spacecraft and the Earth magnetic field:

Earth-centered inertial frame $\mathcal{F}_i$. The origin of the frame is the center of the Earth, the x_i axis coincides with the vernal equinox direction, the z_i axis is the axis of rotation of the Earth and points northward, and the y_i axis completes the frame (see [1], Chapter 2.6.1).

Spacecraft body frame $\mathcal{F}_b$. Its origin is in the spacecraft mass center. The axes are attached to the spacecraft and are selected so that the (inertial) pointing objective is having $\mathcal{F}_b$ aligned with $\mathcal{F}_i$.

Since the objective is having $\mathcal{F}_b$ aligned to $\mathcal{F}_i$, consider the relative kinematics and dynamics of the satellite with respect to the inertial frame. Let the attitude of $\mathcal{F}_b$ with respect to $\mathcal{F}_i$ be represented by quaternion $q = \begin{bmatrix} q_1 & q_2 & q_3 & q_4 \end{bmatrix}^T = \begin{bmatrix} q_v^T & q_4 \end{bmatrix}^T$. The corresponding direction cosine matrix is equal to

$$C(q) = (q_4^2 - q_v^T q_v)I + 2q_v q_v^T - 2q_4 q_v^\times, \tag{1}$$

where I is the identity matrix (see [9], Section 5.4). Moreover, given $a \in \mathbb{R}^3$ symbol $a^\times$ denotes the skew symmetric matrix

$$a^\times := \begin{bmatrix} 0 & -a_3 & a_2 \\ a_3 & 0 & -a_1 \\ -a_2 & a_1 & 0 \end{bmatrix} \tag{2}$$

so that the cross product $a \times b$ can be expressed as the matrix multiplication $a^{\times}b$. The attitude kinematics equation is equal to $\dot{q} = W(q)\omega$ (see [9], Section 5.5.3), where $\omega \in \mathbb{R}^3$ is the spacecraft angular velocity resolved in $\mathcal{F}_b$ and

$$W(q) := \frac{1}{2}\begin{bmatrix} q_4 I + q_v^{\times} \\ -q_v^T \end{bmatrix}. \tag{3}$$

The attitude dynamics equation resolved in body frame is equal to $J\dot{\omega} = -\omega^{\times}J\omega + T$, where $J \in \mathbb{R}^{3\times3}$ is the inertia matrix of the spacecraft and T is the control torque resolved in $\mathcal{F}_b$ (see [9]). Three magnetic coils aligned with the $\mathcal{F}_b$ axes equip the spacecraft. Thus, the following magnetic attitude control torque is created

$$T = m_{coils} \times B^b = -B^{b\times}\ m_{coils}. \tag{4}$$

In the previous expression, $m_{coils} \in \mathbb{R}^3$ is the column matrix of the magnetic dipole moments for the three coils, and B^b is the Earth magnetic field at spacecraft resolved in body frame $\mathcal{F}_b$ (see [1], Section 7.4.1). Let B^i be the Earth magnetic field at spacecraft expressed in inertial frame $\mathcal{F}_i$. Observe that B^i changes in time, at least because of the motion of the spacecraft along the orbit. Hence,

$$B^b(q,t) = C(q)B^i(t) \tag{5}$$

showing the explicit dependence of B^b on both q and t. The previous equations grouped together lead to the following system

$$\begin{aligned} \dot{q} &= W(q)\omega \\ J\dot{\omega} &= -\omega^{\times}J\omega - B^b(q,t)^{\times}\ m_{coils} \end{aligned} \tag{6}$$

where m_{coils} is the control variable. Let us characterize the dependence on time of $B^b(q,t)$, which is equivalent to characterize the time dependence of $B^i(t)$. Assume a circular orbit with radius R. Through the adoption of the so-called dipole model of the Earth magnetic field (see [10], Appendix H), obtain:

$$B^i(t) = \frac{\mu_m}{R^3}\left[3\left(\left(\hat{m}^i(t)\right)^T \hat{r}^i(t)\right)\hat{r}^i(t) - \hat{m}^i(t)\right]. \tag{7}$$

In the previous equation, μ_m is the total dipole strength, $r^i(t)$ is the position of the spacecraft resolved in $\mathcal{F}_i$, and $\hat{r}^i(t)$ is the column matrix of the direction cosines of $r^i(t)$. The coordinates of vector $\hat{m}^i(t)$ are the direction cosines of the Earth magnetic dipole expressed in $\mathcal{F}_i$ which are written as follows:

$$\hat{m}^i(t) = \begin{bmatrix} \sin(\theta_m)\cos(\omega_e t + \alpha_0) \\ \sin(\theta_m)\sin(\omega_e t + \alpha_0) \\ \cos(\theta_m) \end{bmatrix} \tag{8}$$

where θ_m is the coelevation of the dipole, $\omega_e = 360.99$ deg/day is the average angular velocity of the Earth, and α_0 is the dipole right ascension at $t = 0$. Set $\mu_m = 7.746\ 10^{15}$ Wb m and $\theta_m = 170.0°$ as reported in [11].

Eq. (7) shows that to characterize the time dependence of $B^i(t)$, one needs to determine an expression for $r^i(t)$ which is the spacecraft position vector resolved in $\mathcal{F}_i$.

Define the orbital plane coordinate system a_p, b_p whose origin is at the Earth center and with a_p axis coinciding with the line of nodes. Then, the center of mass of the satellite is positioned in

$$\begin{aligned} a^p(t) &= R\cos(nt+\psi) \\ b^p(t) &= R\sin(nt+\psi) \end{aligned} \tag{9}$$

where n is the orbital rate and ψ is spacecraft argument at time $t = 0$. It is possible to determine the mass center coordinates in the inertial frame $\mathcal{F}_i$ from (9) by means of a rotation matrix which depends on the inclination *incl* of the orbit and on the right ascension of the ascending node (RAAN) Ω (see [1], Section 2.6.2). By inserting the expressions of the latter coordinates into (7), we obtain an explicit expression for $B^i(t)$.

Since $C(q) = I$ for $q = \begin{bmatrix} q_v^T & q_4 \end{bmatrix}^T = \pm\overline{q}$ where $\overline{q} = \begin{bmatrix} 0 & 0 & 0 & 1 \end{bmatrix}^T$ (see (1)), then the goal is designing control strategies for m_{coils} so that $q_v \to 0$ and $\omega \to 0$. Reference [3] proposes the following stabilizing proportional derivative (PD)-like control law, which is a modification of those in [12, 13]:

$$m_{coils} = -m^\star_{coils}\ \mathrm{sat}\left(\frac{1}{m^\star_{coils}}B^b(q,t)\times\left(\kappa_p q_v + \kappa_d\omega\right)\right). \tag{10}$$

In the previous equation, $m^\star_{coils}$ is the saturation limit on each magnetic dipole moment, "sat" denotes the standard saturation function, and κ_p and κ_d are parameters. As shown in [3], if the orbit's inclination *incl* is not too low, there are large ranges for the values of $\kappa_p > 0$ and $\kappa_d > 0$ which lead to local exponentially stability of equilibrium $(q,\omega) = (\overline{q}, 0)$ for the closed-loop systems (6) and (10).

The following attitude-only feedback, obtained as modification of one described in [12], is also proposed in [3]

$$\begin{aligned} \dot{\delta} &= \alpha(q-\beta\delta) \\ m_{coils} &= -m^*_{coils}\mathrm{sat}\left(\frac{1}{m^*_{coils}}B^b(q,t)\times\left(\kappa_1 q_v + \kappa_2\alpha\beta W(q)^T(q-\beta\delta)\right)\right). \end{aligned} \tag{11}$$

In (11), $\delta\in\mathbb{R}^4$ is an internal state of the controller, $\kappa_1\ \kappa_2\ \alpha\ \beta$ are all parameters, and $W(q)$ was introduced in (3). Note that the previous equation describes an attitude feedback, since it requires only the measure of attitude q and not of attitude rate ω. It has the advantage of not requiring the installation of rate gyros obtaining savings in cost, volume, and weight. As shown in [3], if the orbit's inclination *incl* is not too low, there are large ranges for the values of $\kappa_1 > 0$, $\kappa_2 > 0$ $\alpha > 0$, $\beta > 0$ which lead to local exponential stability of equilibrium $(q,\omega,\delta) = \left(\overline{q}, 0, \frac{1}{\epsilon\lambda}\overline{q}\right)$ achieved for the closed-loop system (6) and (11).

For both control laws, there are no precise indications for choosing the parameters. In practice, they are usually computed recurring to a trial-and-error approach. This causes the two main limitations: (1) the computation is quite time-consuming, and (2) it is not systematic. This means that when a satisfactory set of parameter values is finally obtained, it is unknown whether continuing the search could allow to discover new parameter values producing a better performance of the closed-loop system. And, if the search is continued, it is unknown when it should be stopped and what could be the possible improvements obtained with this additional search. In any case, it can easily happen to neglect values providing an overall better performance, unless all possible values are tried. However, such an exhaustive search is generally impracticable, because the search space is way too vast to be completely explored. Therefore, we describe here a different approach to find the values of the feedback parameters. Since the desired attitude is obtained when $q_v = 0$, we define the settling time t_{si} for each component q_i, with $i \in \{1, 2, 3\}$, as:

$$t_{si} := \min \ t \quad \text{s.t.} |q_i(t)| \leq \nu \quad \forall t \geq t_{si}. \tag{12}$$

In other words, this is the time needed for $|q_i|$ to permanently remain under ν. Value $0 < \nu < 1$ can be set depending on the desired final value of q_i. Finally, we can define the settling time t_s for the whole quaternion q as the time required by the slowest component of q_v, hence

$$t_s := \max_{i=1,2,3} t_{si} \tag{13}$$

Now, having set the spacecraft initial conditions to specific values, one can determine the values of the parameters that minimize the settling time t_s for each control law (10) and (11).

3. Determination of optimal parameters for the case of PD-like control

In the case of PD-like control (10), the minimization of the settling time can be formulated as follows

$$\min_{\kappa_p \geq 0, \ \ \kappa_d \geq 0} t_s. \tag{14}$$

In order to practically solve the problem, the feasible set should be reasonably bounded and not infinite. Hence, we define two upper bounds $\widehat{\kappa_p}$ and $\widehat{\kappa_d}$ for the gains κ_p and κ_d. These values can usually be determined for the specific problem; we do this for our case study in Section 3.3. Thus, the problem becomes as follows, where the dependence of t_s on the initial conditions $q_0 := q(0)$, $\omega_0 := \omega(0)$, and ψ (see (9)) is explicitly indicated,

$$\min_{(\kappa_p, \kappa_d) \in K} t_s(\kappa_p, \kappa_d, q_0, \omega_0, \psi), \tag{15}$$

and the feasible set is $K = \{(\kappa_p, \kappa_d) : \ 0 \leq \kappa_p \leq \hat{\kappa_p}, \ \ 0 < \kappa_d \leq \hat{\kappa_d}\}$. Even though t_s obviously depends on κ_p and κ_d and on initial conditions, it is not possible in practice to express this relation in analytical form. Moreover, t_s is discontinuous with respect to κ_p, κ_d and initial conditions See for example in ([14], p. 233) the prove that the settling time of the step response is not continuous with respect to the system parameters.

Now, problem (15) has the following features: (*i*) an analytic expression of the objective function cannot be given, and (*ii*) the objective function is not continuous with respect to the decision variables. Therefore, we cannot use standard optimization techniques. Instead, we need to use derivative-free optimization.

3.1 From the simple min to the min-max problem

Given the initial conditions (q_0, ω_0, ψ), problem (15) can be solved to optimality by using the global optimization derivative-free technique described in Section 3.2.

However, when the initial conditions change, then that solution may be no longer optimal. An example of this is given in Section 4.1. Therefore, a robust approach consists in searching for the optimal solution to (15) under the *worst* initial condition. Such a worst-case approach is commonly used in similar cases. In other words, we accept to pay something, in terms of objective function values, in the easy cases, but we obtain in return advantages in the more difficult situations. However, the worst initial condition is not a priori known, since it depends on the chosen values of κ_p and κ_d. Therefore, the problem cannot be decomposed and should be solved as a whole.

The set of values describing the initial conditions is given by:

$$S = \left\{ \left(q_0, \omega_0, \psi\right) : \|q_{0v}\| \leq 1, q_{04} = \left(1 - q_{0v}^T q_{0v}\right)^{1/2}, \right. \\ \left. |\omega_{01}| \leq \hat{\omega_{01}}, |\omega_{02}| \leq \hat{\omega_{02}}, |\omega_{03}| \leq \hat{\omega_{03}}, 0 \leq \psi < 2\pi \right\} \tag{16}$$

Set S includes any possible initial attitude and any possible initial argument ψ for the spacecraft; only the magnitude of the initial angular rate is limited. The worst-case minimization of t_s can be modeled as follows:

$$\min_{\left(\kappa_p, \kappa_d\right) \in K} \quad \max_{\left(q_0, \omega_0, \psi\right) \in S} \quad t_s\left(\kappa_p, \kappa_d, q_0, \omega_0, \psi\right). \tag{17}$$

To use the derivative-free algorithm explained in Section 3.2, the feasible set of each optimization problem must be now converted into a hyperrectangle. This can be obtained by expressing $\|q_{0v}\| \leq 1$ in spherical coordinates (ρ, ϕ, θ):

$$S = \left\{ \left(q_0, \omega_0, \psi\right) : q_{01} = \rho \sin\theta \cos\phi, \;\; q_{02} = \rho \sin\theta \sin\phi, \;\; q_{03} = \rho \cos\theta, \right. \\ q_{40} = \left(1 - q_{0v}^T q_{0v}\right)^{1/2}, 0 \leq \rho \leq 1, \;\; 0 \leq \phi < 2\pi, \;\; 0 \leq \theta \leq \pi, \\ \left. |\omega_{01}| \leq \hat{\omega_{01}}, \;\; \omega_{02}| \leq \hat{\omega_{02}}| \;\; \omega_{03}, \leq \hat{\omega_{03}}| \;\; 0 \leq \psi < 2\pi| \right\}. \tag{18}$$

The dependence of t_s on q_0 can be expressed as dependence on the variables (ρ, ϕ, θ), as follows. We introduce the hyperrectangle:

$$H = \{(\rho, \phi, \theta, \omega_0, \psi) : \;\; 0 \leq \rho \leq 1, \;\; 0 \leq \phi < 2\pi, \;\; 0 \leq \theta \leq \pi, \\ |\omega_{01}| \leq \hat{\omega_{01}}, \;\; \omega_{02}| \leq \hat{\omega_{02}}| \;\; \omega_{03}, \leq \hat{\omega_{03}}| \;\; 0 \leq \psi < 2\pi| \}. \tag{19}$$

Now, the min-max problem (17) can be equivalently reformulated as follows:

$$\min_{\left(\kappa_p, \kappa_d\right) \in K} \quad \max_{(\rho, \phi, \theta, \omega_0, \psi) \in H} \quad t_s\left(\kappa_p, \kappa_d, \rho, \phi, \theta, \omega_0, \psi\right). \tag{20}$$

3.2 The whole derivative-free optimization approach

This section explains the min-max procedure to solve problem (20). Its building blocks are given by the following derivative-free algorithms:

3.2.1 The global strategy

Lipschitzian methods constitute a main approach in non-differentiable optimization. However, they are limited by their requirement of knowing the value of the Lipschitz constant. On the other hand, the search may be conducted without knowing the Lipschitz constant if we use DIRECT-type algorithms [15]. These techniques are based on the partitioning of the feasible region into hyperrectangles and on their examination in a specific order. The feasible region starts as a single hyperrectangle that is internally normalized to a unit hyperrectangle. At each iteration k, the algorithm partitions the hyperrectangles identified in the previous $(k-1)$th iteration, obtaining a collection of smaller hyperrectangles $\mathcal{H}^{(k)} = \{H_1, \ldots, H_k\}$, and evaluates the objective function in their central points. The computation of the objective value allows to identify potentially optimal hyperrectangles within $\mathcal{H}^{(k)}$. These hyperrectangles are further partitioned and investigated in the next $(k+1)$th iteration of the algorithm, while the rest of them is simply discarded. The algorithm stops when the hyperrectangle size becomes very small or when the maximum number of iterations is reached. This algorithm guarantees to converge to the global optimum of the function if sampling is dense enough. However, a large number of function evaluations might be needed to obtain dense sampling, and consequently large computational times could be required.

3.2.2 The local strategy

When the function under optimization is sufficiently regular, it can be optimized by using gradient-based methods, which require the computation of the first-order derivatives of the objective function. However, the basic strategy underlying these methods could be replicated even if no information on the derivatives is available. This is the basic idea of the SDBOX algorithm [16], which can be described as follows. After the selection of a starting point, it cyclically determines a feasible and *good* descent direction for the objective function, and then it performs a *sufficiently* large step along such direction. To find the good feasible descent direction, at each iteration k, the algorithm computes the local behavior of the objective function at the variation of the i-th variable. If a move of length α along d_i gives a feasible point where the function is reduced enough, the algorithm applies a linesearch technique d_i to find the stepsize α_k. Otherwise, the algorithm tries the same operations in the opposite direction $-d_i$. If both d_i and $-d_i$ are not able to obtain a sufficient decrease, the stepsize α is decreased. The linesearch technique does not require information on the slope of the objective function. Note that the convergence of the algorithm was proved in [16] for minimization of a continuously differentiable function; however, the same technique is often used to optimize different types of functions. This approach is considerably faster than the global strategy; however it is well known that a poor choice of the initial point leads to poor solutions.

3.2.3 The whole procedure

To simplify the description of the robust optimization approach, we now rename the set of variables (κ_p, κ_d) as x belonging to a feasible set $F_x = \{lb_i \leq x_i \leq ub_i, \;\; i = 1, \ldots, n\} \subset \mathbb{R}^n$ (in our case $F_x = K$ and $n = 2$) and the set of initial conditions $(\rho, \phi, \theta, \omega_0, \psi)$ as y belonging to a feasible set $F_y = \left\{lb_j \leq y_j \leq ub_j, \;\; j = 1, \ldots, m\right\} \subset \mathbb{R}^m$ (in our case $F_y = H$ and $m = 7$, since ω_0 has three components). Moreover, we now use $g(x, y)$ to denote the objective function (in our case t_s). Its analytical expression is not known; however its value can be computed via a software simulation.

The upper-level minimization problem $\min_{x \in F_x} g(x, y)$ must be solved by means of a global strategy applied one time: no starting points can be considered here; hence, there is no ground for a local strategy. Thus, the procedure will contain an external loop implementing the global strategy. This loop essentially computes the value of g

obtained for a number *maxeval_ext* of points x. Let $\overline{x}$ be one of them. The evaluation of $\overline{x}$ requires however to solve one lower-lever maximization problem $\max_{y \in F_y} g(\overline{x}, y)$. Hence, the lower-lever problem should be solved many times, and its corresponding computation time must be reduced to a few seconds. We can solve it in three ways: (1) with the global strategy but using a necessarily small maximum number of function evaluations *maxeval_int*; (2) with the local strategy, allowing the same number of function evaluations from a single start; and (3) with the local strategy, using a multistart from different starting points, still allowing *maxeval_int* total function evaluations. This last option allows to use information that happens to be available in the choice of the starting points. In our case, from the physics of the problem, we suppose that good solutions are in the vicinity of extremal values of angular velocity, so we take them as starting points. We report the whole procedure:

Procedure 1: Solve min-max

Input: A $g(x, y)$ computable by means of a software simulation for $x_i \in [lb_i, ub_i]$ with $i = 1, \ldots, n$ and $y_j \in [lb_j, ub_j]$ with $j = 1, \ldots, m$.

Output: A robust optimum value $x^* = arg \min_{x \in F_x} \left(\max_{y \in F_y} g(x, y) \right)$.

External loop:
Solve the upper-level problem $\min_{x \in F_x} f(x)$, with $f(x) = \max_{y \in F_y} g(x, y)$ by using DIRECT for *maxeval_ext* evaluations of f, and return x^*.
The k-th evaluation works with point $x^{(k)}$

Internal loop:
Solve the lower-level problem $\max_{y \in F_y} g\left(x^{(k)}, y\right)$

by using

$$\begin{cases} \text{DIRECT performing } maxeval\_int \text{ evaluations } of\ g \\ \text{SDBOX single start performing } maxeval\_int \text{ evaluations } of\ g \\ \text{SDBOX multistart with } s \text{ starting points, performing} \\ \dfrac{maxeval\_int}{s} \text{ evaluations of } g \text{ for each of them} \end{cases}$$

4. Computational results for the case of PD-like control

We apply the above approach to solve the numerical example studied in [3]. The inertia matrix of the spacecraft is equal to $J = diag[27,\ 17,\ 25]\ \text{kgm}^2$; the saturation level for each magnetic dipole moment is given by $m^*_{coils} = 10\ \text{Am}^2$. The orbit has an inclination of $incl = 87°$ and an altitude of 450 km. The orbital period in these conditions is about 5609 s. The value Ω of RAAN is 0. Given these values, we can compute an upper bound $\hat{\kappa_p}$ for κ_p. Simulation shows that $\|B^i(t)\| \geq B_{min} = 2.4 \times 10^{-5}$ T for the considered orbit; since $\|B^b(q, t)\| = \|B^i(t)\|$, it follows that $\|B^b(q, t)\| \geq B_{min}$. Assume that the resolution of the attitude sensor is $q_s = 0.04$ in terms of quaternion component. In this case, an upper bound for κ_p can be computed by enforcing that each component of the term $B^b(q, t) \times \kappa_p q_v$ (see (10)) does not exceed the saturation limit $m^*_{coils} = 10\ \text{Am}^2$ when $B^b(q,t)$ and q_v are orthogonal, $\|B^b(q,t)\| = B_{min}$ and $|q_i| = q_r$ for one index $i \in 1, 2, 3$, and $q_i = 0$ for the other ones. This is achieved if κ_p is smaller than $m^*_{coils}/(B_{min}q_r) = 1.0417 \times 10^8$; thus, we set $\hat{\kappa_p} = 10^8$. Assuming that the attitude rate sensor has a resolution of $\omega_r = 4 \times 10^{-4}$ rad/s, by a parallel argument it follows that an upper bound for κ_d is given by $m^*_{coils}/(B_{min}\omega_r) = 1.0417 \times 10^9$; thus, we set $\hat{\kappa_d} = 10^9$.

4.1 Fixed initial conditions

Consider an initial state characterized by attitude equal to the target attitude $q_0 = \bar{q}$ (which corresponds to having $\rho = 0$ and any value for ϕ and θ) and by the following initial angular rate:

$$\omega_0 = [0.02 \quad 0.02 \quad -0.03]^\text{T} \ \text{rad/s}. \tag{21}$$

This example corresponds to a spacecraft possessing the desired attitude and no angular momentum, which received an impact from a small object leading to an instantaneous change in the spacecraft angular rate. We assign to ψ a random value over the interval $[0,\ 2\pi[$ by setting $\psi = 0.332$ rad.

Gains κ_p and κ_d have been found by trial-and-error in [3] as $\kappa_p = 2 \times 10^5$ and $\kappa_d = 3 \times 10^8$. The settling time corresponding to these values of the gains is t_s = 4.002 orbital periods, while the vast majority of possible gain values would produce settling times larger than 10 orbital periods. On the contrary, if we use the optimization algorithm described in Section 3.2, we obtain the following fixed initial condition optimal gains after only 32,821 iterations (i.e., 251 s of computation):

$$\begin{aligned} \kappa_p &= 205761.316872 \\ \kappa_d &= 99382716.049383 \end{aligned} \tag{22}$$

The corresponding optimal settling time is $t_s = 1.775$ orbital periods. Hence, the optimization algorithm can provide a significant improvement in convergence speed.

However, the above gains might become no longer optimal if we assume different initial conditions. Consider, for example, the case of initial conditions with the same attitude $q_0 = \bar{q}$ (i.e., $\rho = 0$ and any ϕ and θ), the same argument $\psi = 0.332$ rad, but a different initial angular rate

$$\omega_0 = [0.1 \quad 0.1 \quad 0.1]^\text{T} \ \text{rad/s}. \tag{23}$$

The above gains yield a settling time t_s = 3.562, and they are no longer optimal, as shown in Section 4.2. Indeed, the optimal values of the gains would need to be determined for every possible initial condition, which is clearly impossible in practice. As seen in Section 3.1, we can instead search for the gain values providing the best performance under the worst initial condition. It has been found that the worst initial conditions corresponding to gains (22) are

$$\begin{aligned} &\rho = 0.5, \quad \phi = 0.0, \quad \theta = 2.356194490192345, \quad \psi = 1.570796326794897, \\ &\omega_0 = [-0.1 \quad 0.1 \quad 0.1]^\text{T}. \end{aligned} \tag{24}$$

	DIRECT + DIRECT	DIRECT + SDBOX single start	DIRECT + SDBOX multistart
κ_p	205493.82716049382	203930.04115226338	209581.61865569273
κ_d	117283950.61728397	154320987.65432101	117283950.61728397
t_s	3.359	3.142	3.742
ext eval	10,641	11,127	10,011
Time (s)	172,242.6	57,255.1	89,640.8

Table 1.
Results of Solve min-max with maxiter_ext = 10,000 and maxiter_int = 2200.

Gains	DIRECT	SDBOX multistart	Grid search
	eval = 40,000	*eval* = 40,000 (5000 × 8)	*eval* = 78,125 (5^7)
$\kappa_p = 205493.82716049382$	3.876	3.872	3.990
$\kappa_d = 117283950.61728397$			
$\kappa_p = 203930.04115226338$	4.223	4.149	4.362
$\kappa_d = 154320987.65432101$			
$\kappa_p = 209581.61865569273$	3.895	3.744	4.016
$\kappa_d = 117283950.61728397$			

Table 2.
Accurate evaluation of the previously obtained gains.

Under conditions (24), the above values of κ_p and κ_d yield a settling time $t_s = 4.338$.

4.2 Variable initial conditions

We now search for κ_p and κ_d using the robust optimization approach. We allow initial conditions to vary as follows: $0 \le \rho \le 1$, $0 \le \phi \le 2\pi$, $0 \le \theta \le \pi$, $|\omega_{01}| \le 0.1$, $|\omega_{02}| \le 0.1$, $|\omega_{03}| \le 0.1$, and $0 \le \psi \le 2\pi$. We allow a maximum of 2,200 function evaluations in each internal loop, in order to practically solve the problem. **Table 1** reports the results of Procedure Solve min-max with *maxeval_ext* = 10,000 and *maxeval_int* = 2200.

When using SDBOX multistart, the starting points are $\rho = 0.5$, $\phi = \pi$, $\theta = \pi/2$, $\psi = \pi$, and all the eight combinations of extreme angular velocities

$$[\ 0.1 \quad 0.1 \quad 0.1]^{\mathrm{T}},\ [\ 0.1 \quad 0.1 \quad -0.1]^{\mathrm{T}}, \ldots, [-0.1 \quad -0.1 \quad -0.1]^{\mathrm{T}}. \tag{25}$$

These results show that (i) SDBOX multistart is able to reach the highest value of the objective t_s even with the small number of computations allowed in the internal loop and (ii) the different results in the solution of the lower-level problems cause a different evolution of the upper-level search. To understand which among the three points above is the best choice, we continue the analysis in **Table 2**. We fix parameters κ_p and κ_d, and we solve the maximization problem with increased accuracy by allowing more function evaluations. We do this by means of global strategy, local multistart strategy (which proved to dominate the single start one), and an exhaustive grid search, which is much slower but used here as reference.

This analysis shows that the first point has the smallest maximum settling time. In conclusion, the robust optimal solution is:

$$\kappa_p = 205493.82716049382, \quad \kappa_d = 117283950.61728397. \tag{26}$$

To evaluate the performance of the above solution (26), we try it with the three different initial conditions reported above. If the initial state is (21), the settling time becomes $t_s = 1.945$ (instead of 1.775). Hence, there is a small worsening. Clearly, no improvement was possible because gains (22) are optimal for state (21). If the initial state is given by (23), we obtain $t_s = 2.958$ (instead of 3.562). Hence, there is an improvement. This could be expected, though not guaranteed, since gains (22) have a higher worst-case result than gains (26).

Finally, if the initial state is given by (24), then we obtain $t_s = 3.370$ (instead of 4.338). Hence, there is a substantial improvement. Indeed, this was certain, since gains (26) have a worst-case result considerably better than gains (22). In conclusion, the gains computed with the proposed robust optimization approach offer improvements in the difficult situations. Moreover, even if they may be suboptimal in the easier cases, the balance appears beneficial.

5. Determination of optimal parameters for the case of attitude-only feedback control

As observed, the objective function t_s is not continuous with respect to the parameters, and this introduces numerical difficulties in solving the optimization problem. Thus, one can consider an alternative objective function named integral time absolute error (ITAE) [8], indicated by Γ:

$$\Gamma = \int_0^{T_f} t\|q_v(t)\|dt \tag{27}$$

where $\|\cdot\|$ represents the Euclidean norm and T_f is a time which is selected large enough. The ITAE carries the benefit of being continuous with respect to the design parameters. Continuous differentiability cannot be guaranteed. However, since $(q, \omega, \delta) = \left(\pm\overline{q}, 0, \frac{1}{\beta}\overline{q}\right)$ are equilibria of the closed-loop system (6) and (11), it is very unlikely that q_v becomes $[0\ \ 0\ \ 0]^T$ at some finite time. Note that this would make Γ not continuously differentiable with respect to the design parameters because $\|q_v\|$ is not continuously differentiable at $q_v = [0\ \ 0\ \ 0]^T$.

Minimizing the ITAE is known to lead to nearly optimal solutions with regard to the settling time. In fact, in very simple situations, it is shown analytically that minimizing the ITAE leads to solutions that minimize the settling time. In more complex scenarios, it was numerically shown that minimizing the ITAE gives solutions that are very close to the optimal ones in terms of settling time (see [8]).

By introducing physically reasonable upper bounds $\hat{\kappa_1}, \hat{\kappa_2}, \hat{\alpha}, \hat{\beta}$ for the design parameters, we obtain the feasible set $K = \{(\kappa_1, \kappa_2, \alpha, \beta) : \ 0 \leq \kappa_1 \leq \hat{\kappa_1}, \ 0 \leq \kappa_2 \leq \hat{\kappa_2}, \ 0 \leq \alpha \leq \hat{\alpha}, \ 0 \leq \beta \leq \hat{\beta}\}$. Now, our optimization problem is:

$$\min_{(\kappa_1, \kappa_2, \alpha, \beta) \in K} \Gamma. \tag{28}$$

Given specific initial conditions of the spacecraft, problem (28) can be solved by a suitable use of derivative-free techniques. However, if the initial conditions of the spacecraft change, that solution might be no longer optimal. Since several different initial conditions for the spacecraft are possible in practical situations, a robust solution is an optimal solution to problem (28) under the *worst* spacecraft initial conditions. Such a worst-case optimization is widely employed in such scenarios, because by adopting the latter approach, we can give an efficient bound on the objective value in spite of the uncertainty on the spacecraft initial conditions. However, the worst initial conditions for the spacecraft cannot be determined a priori, since they depend on the selected values of $\kappa_1, \kappa_2, \alpha, \beta$. The initial conditions of the spacecraft are given by $q_0 = q(0)$, $\omega_0 = \omega(0)$, $0 \leq \psi < 2\pi$, and $0 \leq \alpha_0 < 2\pi$. We chose the set of their possible values as:

$$S = \Big\{ \left(q_0, \omega_0, \psi, \alpha_0\right) : \ \|q_{0v}\| \leq 1, \ q_{04} = \left(1 - q_{0v}^T q_{0v}\right)^{1/2}, \\ |\omega_{01}| \leq \hat{\omega_{01}}, \ \ \omega_{02}| \leq \hat{\omega_{02}}| \ \ \omega_{03}, \leq \hat{\omega_{03}}| \ \ 0 \leq \psi < 2\pi| \ \ 0 \leq \alpha_0 < 2\pi, \Big\}. \tag{29}$$

Note that S includes all possible initial attitudes, all possible right ascensions of the Earth magnetic dipole at time $t = 0$, and all possible initial arguments ψ for the spacecraft. It only constrains the amplitude of the initial angular velocity. The minimization of Γ under the worst spacecraft initial conditions is equivalent to formulating the following min-max problem:

$$\min_{(\kappa_1, \kappa_2, \alpha, \beta) \in K} \ \ \max_{\left(q_0, \omega_0, \psi, \alpha_0\right) \in S} \Gamma. \tag{30}$$

To apply the optimization techniques, we convert the feasible set of each optimization problem into a hyperrectangle by expressing the set $\|q_{0v}\| \leq 1$ in spherical coordinates (ρ, ϕ, θ):

$$S = \left\{ \left(q_0, \omega_0, \psi, \alpha_0\right) : q_{01} = \rho \sin\theta \cos\phi, \ \ q_{02} = \rho \sin\theta \sin\phi, \ \ q_{03} = \rho \cos\theta, \\ q_{40} = \left(1 - q_{0v}^T q_{0v}\right)^{1/2}, 0 \leq \rho \leq 1, \ \ 0 \leq \phi < 2\pi, \ \ 0 \leq \theta \leq \pi, \\ \omega_{01}| \leq \hat{\omega_{01}}| \ \ \omega_{02}, \leq \hat{\omega_{02}}| \ \ \omega_{03}| \leq \hat{\omega_{03}}, \ \ 0 \leq \psi < 2\pi| \ \ 0 \leq \alpha_0 < 2\pi| \right\}. \tag{31}$$

The dependence of Γ on q_0 can now be expressed as dependence on the variables (ρ, ϕ, θ). Consequently, after having introduced the hyperrectangle

$$H = \{(\rho, \phi, \theta, \omega_0, \psi, \alpha_0) : \ \ 0 \leq \rho \leq 1, \ \ 0 \leq \phi < 2\pi, \ \ 0 \leq \theta \leq \pi, \\ |\omega_{01}| \leq \hat{\omega_{01}}, \ \ \omega_{02}| \leq \hat{\omega_{02}}| \ \ \omega_{03}, \leq \hat{\omega_{03}}| \ \ 0 \leq \psi < 2\pi| \ \ 0 \leq \alpha_0 < 2\pi, \}, \tag{32}$$

the min-max problem (30) can be equivalently reformulated as follows:

$$\min_{(\kappa_1, \kappa_2, \alpha, \beta) \in K} \ \ \max_{(\rho, \phi, \theta, \omega_0, \psi, \alpha_0) \in H} \Gamma. \tag{33}$$

5.1 Combining global and local search

To simplify the description of the proposed approach, we now rename the set of design parameters $(\kappa_1, \kappa_2, \alpha, \beta)$ as x belonging to a feasible set $F_x = \{lbx_i \leq x_i \leq ubx_i, \ \ i = 1, \ldots, n\} \subset \mathbb{R}^n$ (in our case $F_x = K$ and $n = 4$) and the set of initial conditions $(\rho, \phi, \theta, \omega_0, \psi, \alpha_0)$ as y belonging to a feasible set $F_y = \left\{lby_j \leq y_j \leq uby_j, \ \ j = 1, \ldots, m\right\} \subset \mathbb{R}^m$ (in our case $F_y = H$ and $m = 8$, since ω_0

possesses three components). Indicate by $f(x,y)$ the function giving the objective value (in our case Γ). When y is fixed (the initial conditions are assigned), we simply write $\bar{y}$ in it; when x is fixed (the design parameters are assigned), we write $\bar{x}$ in it. Problem (28) can be written as

$$\min_{x \in F_x} f(x,\bar{y}) \tag{34}$$

and may be tackled by a global derivative-free optimization algorithm of the type of DIRECT [7], already described in Section 3. Those methods work without the need for analytically writing the objective function; they only need to compute it in a number of points by using simulations. Due to the so-called *everywhere dense* property, such an algorithm reaches a global optimum if the sampling is dense enough. However, a dense search may require that the function is evaluated many times.

In our case, all $lbx_i = 0$, since design parameters have to be greater than or equal to zero. Values ubx_i can typically be set to very large values based on some physical considerations. For instance, $\hat{\kappa_1}$ can easily become equal to 10^9, since its maximum feasible value can be determined knowing the saturation level of the coil moments, the minimum amplitude of the geomagnetic field, and the attitude sensor resolution. However, when considering these values, a sufficiently dense exploration of the feasible set F_x requires a number of function evaluations such that the corresponding run time is impracticable. On the other hand, a sampling that uses a practically sustainable number of function evaluations does not produce solutions substantially better than random solutions. The presence of a large number of local minima makes the optimization task particularly difficult. In such conditions, the evaluation of the generic H_h using only one point may be very inaccurate at the first iterations of the algorithm, because the dimension of the hyperrectangles is too large. By proceeding with the iterations, the hyperrectangles becomes smaller, but their number, and consequently the run time, increases exceedingly.

On the other hand, if a fast but effective probing technique to early identify the "promising" H_h would be available, then one could explore densely only such promising regions in reasonable times. Hence, a probing technique based on the use of the local derivative-free optimization algorithm SDBOX was proposed in [6]. This algorithm was originally presented in [16] as a globally convergent algorithm for the minimization of a continuously differentiable function, but in practice it can be employed to optimize different types of functions as a good trade-off between efficiency and convergence properties. Most interestingly for our case, given an initial guess $(\kappa_1', \kappa_2', \alpha', \beta') = x'$, this algorithm should be able to find good solutions in short times in the neighborhood of x'. We describe below a solution approach combining these global and local strategies to solve problem (28).

Procedure 2: Solve min combining global and local search

Input: A vector $\bar{y} \in F_y$ and a function $f(x,\bar{y})$ computable by means of a software simulation for any $x \in F_x$. Values for the parameters p, *maxeval*, *maxsubsets*, *maxiter*, *maxpost*.

Output: A solution x^{**} approximating one vector in *arg* $\min_{x \in F_x} f(x,\bar{y})$.

1. Normalize and grid partition the whole feasible set F_x into a collection of hyperrectangles $\mathcal{H}^{(1)} = \{H_1, \ldots, H_p\}$ similarly to the initial phase of DIRECT.

2. For each $H_h \in \mathcal{H}^{(1)}$, compute the value f_h of the solution obtained by *maxeval* iterations of SDBOX in H_h starting from its central point. This is an upper bound on the value of the best solution in H_h and constitutes our "evaluation" of H_h.

3. Take a number *maxsubsets* of hyperrectangles corresponding to the smallest of the above f_h values.

4. Take the region given by the union of those subsets, and "convexify" it by including also the additional subsets required to convert it into an hyperrectangle $F_x{}^*$.

5. Switch to DIRECT algorithm to continue the search in $F_x{}^*$ allowing *maxiter* function evaluations. This search can now be dense using reasonable time, and it gives a solution x^*.

6. Try to improve x^* by using *maxpost* iterations of a local search method, finally obtaining a solution x^{**} to problem (28).

Depending on the practical case, the number of hyperrectangles p in the partitioning phase (step 1) and the number of iterations *maxeval* in the evaluation phase (step 2) must be selected in order to allow a fixed computational time to the evaluation phase. Indeed, we search for a compromise between speed and effectiveness at this stage. Those parameters must be tuned by also considering that the accuracy of the evaluation of each H_h depends not only on *maxeval* but also on the span of each H_h, which in turn depends on p. However, there is no need that all the hyperrectangles in the initial partition $H^{(1)}$ have the same size. Their size could be growing with the absolute values of the coordinates so that p is kept smaller. The number *maxsubsets* in the selection phase (step 3) must be chosen so that $F_x{}^*$ remains much smaller than F_x; otherwise the benefits of the proposed procedure is reduced. The number of function evaluations *maxiter* in the standard DIRECT phase (step 5) is chosen so that the search in $F_x{}^*$ is dense enough. This is now possible because of the size reduction in F_x. The post-optimization phase (step 6) can be executed with SDBOX or with CS-DFN [17], which is another linesearch-based method which uses a dense set of search directions and not only the coordinates ones. CS-DFN does not require f to be continuously differentiable; however, it could be more computationally expensive on smooth problems. For example, in our case, Γ is at least continuous, and even if there are no theoretical arguments to ensure its continuous differentiability, in practice this property may often occur. In conclusion, we can choose between SDBOX and CS-DFN depending on the presence or absence of the continuous differentiability of f, and if we set *maxpost* to be sufficiently large, we obtain a solution x^{**} which satisfies necessary conditions for a local optimum. Moreover, x^{**} should approximate one of the global optima, because, given the dense search in F^*, it should provide one of the global minima of F^*, and the regions of $F\{F^*$, which were less "promising," should not contain better solutions. Note that the accuracy and the properties of the evaluation phase can be modified, depending on the computational request of the practical case, either by modifying the number of iterations or even by employing an alternative evaluation algorithm, still keeping the same algorithmic framework.

On the other hand, problem (33) is made of an upper-level minimization problem and a lower-level maximization one. By changing the names of the set of variables as explained at the beginning of the section, the problem becomes

$$\min_{x \in F_x} \left(\max_{y \in F_y} f(x,y) \right) = \min_{x \in F_x} g(x) \tag{35}$$

with function g such that its value at the generic point $\overline{x}$ is given by the solution of the lower-level problem:

$$g(\overline{x}) = \max_{y \in F_y} f(\overline{x},y). \tag{36}$$

The upper-level problem is solved through an external loop applying the combination of local and global search described as Procedure 2. This loop computes, using a parameter *maxeval_ext*, to define the overall maximum number of function evaluations and the value of g corresponding to different points of F_x. Let $\overline{x}$ be one of them; then the evaluation of $g(\overline{x})$ requires the solution of one lower-level maximization problem $\max_{y \in F_y} f(\overline{x},y)$. Thus, the lower-level problem must be solved up to *maxeval_ext* times. Consequently, solving it in a few seconds is crucial.

For this problem, the global strategy either would perform a very poor search or would need excessive time. Therefore, the local strategy appears to be the only feasible choice for the lower-level problem. However, taking as initial guess the center of the feasible set F_y does not lead to good solutions of the maximization problem within the limited available time.

In this case, based on the problem physics, it can be assumed that good solutions of the maximization problem are in the neighborhood of extreme values of angular velocity. Therefore, we solve the lower-level problem in a nested loop by using a local search with multistart. We take as initial guesses the eight combinations of extreme values for the three components of the angular velocity $\pm\omega\hat{_{01}}$, $\pm\omega\hat{_{02}}$, and $\pm\omega\hat{_{03}}$. This can be done with SDBOX or CS-DFN, depending on the presence or absence of the continuous differentiability of f. In our case, Γ is at least continuous also with regard to the initial conditions of the spacecraft. Again, there are no theoretical arguments to ensure its continuous differentiability; however in practice this may often happen. For each solution of the lower-level problem, we allow a necessarily small maximum number of function evaluations *maxeval_int*. In conclusion, this produces a solution x_R^* for problem (33). We report here the whole procedure:

Procedure 3: Solve min-max combining global and local search

Input: A $f(x,y)$ computable by means of a software simulation for any $x \in F_x$ and $y \in F_y$. Values for the parameters *maxeval_ext*, *maxeval_int*, s.

Output: A robust solution x_R^* approximating one vector in
$arg \min_{x \in F_x} \left(\max_{y \in F_y} f(x,y) \right)$.

External loop:
Solve the upper-level problem $\min_{x \in F_x} g(x)$, with $g(\overline{x}) = \max_{y \in F_y} f(\overline{x},y)\ \forall \overline{x} \in F_x$, by using Procedure 2 with *maxeval_ext* total evaluations of g, and return x_R^*. Given $\overline{x}$, the evaluation of $g(\overline{x})$ is performed by the internal loop.

Internal loop:
Take $\overline{x}$, and solve the lower-level problem $\max_{y \in F_y} f(\overline{x},y)$ by using multistart local search with s starting points, performing $\frac{maxeval\_int}{s}$ evaluations of f for each of them.

6. Computational results for the case of attitude-only feedback control

We apply the above method to solve the numerical example presented in [3]. The spacecraft inertia matrix is $J = diag[27,\ 17,\ 25]\ \text{kg m}^2$, and the saturation level for each dipole moment is $m^*_{coils} = 10\text{A m}^2$. The orbit inclination is $incl = 87°$, and the orbit altitude is 450 km; the right ascension of the ascending node Ω is equal to 0. Upper bounds $\hat{\kappa_1}, \hat{\kappa_2}, \hat{\alpha}, \hat{\beta}$ are selected as $\left(10^9, 10^9, 10^4, 10^{-3}\right)$.

At the beginning we consider the easier situation of known spacecraft initial conditions, and we present, in Section 6.1, the results of Procedure 2 in solving this problem. We report also a comparison with the classical DIRECT method. Subsequently, in Section 6.2, the more realistic case of a spacecraft having variable initial conditions is considered.

6.1 Fixed initial conditions

We consider here the case of the above described spacecraft with known fixed initial conditions; thus we deal with problem (28) using the following values:

$$(\rho, \phi, \theta, \omega_0, \psi, \alpha_0) = (0, 0, 0, 0.02, 0.02, -0.03, 0.9416, 4.5392). \qquad (37)$$

The best solution obtained in [3] by trial and error search achieves a value of ITAE=3.7×10^7, while the vast majority of the solutions have the ITAE limit value of about 1.2×10^9. This upper limit is related to the value of T_f in the definition of ITAE (27), which is chosen equal to 56,009 s corresponding to 10 orbital periods. Practically this means that when we reach the limit value of ITAE, the corresponding settling time would be roughly larger than 10 orbital periods. Then, that solution is not an attractive one, and we are not interested in determining it with further precision. Even if this choice causes a flattening in the values of ITAE, the use of such a finite T_f is necessary to run the simulations that compute the ITAE in practice.

Table 3 reports two solution attempts carried on with the standard DIRECT algorithm on the whole feasible set K with, respectively, 50,000 and 100,000 iterations, followed by 1000 iterations of local search refinement using CS-DFN. In spite of the significant computational effort (the running times of these experiments, respectively, correspond to about 3 days and 1 week), the obtained solutions have values of ITAE greater than 1.1×10^9 that is not very different from the ITAE value of a random solution. Evidently, the search was not dense enough to explore

Algorithm	Solution	Obj. value	Time
DIRECT 50,000 + CS-DFN 1000	κ_1 = 913405022.139 κ_2 = 195426826.870 α = 9794.170752422 β = 0.000000000000	1,142,470,478.101	262,000 + 5460 s
DIRECT 100,000 + CS-DFN 1000	κ_1 = 500798437.500 κ_2 = 159788790.177 α = 9996.679486501 β = 0.000000000003	1,129,234,873.703	565,200 + 5040 s
Procedure 2: Combining global and local strategies	κ_1 = 246494.579020 κ_2 = 233333315.349 α = 92.5925925927 β = 0.000129629629	**8,021,573.4077**	201,500 + 15 + 4 s

Table 3.
Comparison of Procedure 2 and standard DIRECT with 50,000 or 100,000 iterations.

the feasible set as it would be needed. Other similar attempts with standard DIRECT do not achieve better results. The same table also reports Procedure 2 of Section 5.1. As observable, this procedure is able to reach a much better solution, with a value of ITAE of about 8×10^6. Note that this solution is also considerably better than the best solution obtained by trial and error in days of work.

The promising region of the feasible set K has been identified with the previously described probing technique. The values of *maxelav* and p are selected so that this step is executed in reasonable time, according to the following considerations. One single function evaluation takes a time which is very variable and goes from fractions of seconds to several tenths of seconds; but a very rough average function evaluation time can be assessed as equal to 1 s. By selecting *maxeval* = 10, the evaluation of each hyperrectangle would require roughly 10 s. Thus, to finish within 2 or 3 days of computation, we select $p = 40,960$, which is obtained by making 64 partitions on the domain of κ_1, 64 partitions on the domain of κ_2, and 10 partitions on the domain of α. The intervals corresponding to these partitions do not have the same size; they increase with the absolute value of the coordinates. The following hyperrectangle is obtained as convexification of the collection of the most promising regions, as in Step 4 of Procedure 2:

$$\begin{aligned} K^* &= \{(\kappa_1, \kappa_2, \alpha, \beta) : 200000 \le \kappa_1 \le 260000, 200000000 \\ &\le \kappa_2 \le 310000000, 0 \le \alpha \le 100, 0 \le \beta \le 0.001\}. \end{aligned} \tag{38}$$

The determination of K^* actually needs 409,600 iterations of SDBOX and 201,500 s of computations (about 56 hours). Note that the time necessary for each function evaluation is generally much faster in K^*, where it can be less than 0.01 s than in the rest of K. Now, by applying standard DIRECT strategy over the feasible set K^*, as in Step 5 of Procedure 2, after 3007 iterations and only 15 s, we obtain the solution:

$$\kappa_1 = 200781.893004, \quad \kappa_2 = 226268861.454, \quad \alpha = 39.3004115226, \quad \beta = 0.0005 \tag{39}$$

whose value is ITAE = 9.072×10^6. Note that the volume of the set K^* is considerably smaller than that of K: it is only 1/15151515.15 of the volume of K. To better appreciate the difference, consider that an exploration of K with the same degree of density used on K^* would require $15 \times 15,151,515.15 = 227,272,727.25$ s that roughly corresponds to more than 7 years, if the simulation times on K were the same as on K^*. Since they are often much slower, the time needed would be even more.

Solution (39) can be further refined by using the local search strategy, as in Step 6 of Procedure 2. By performing 1000 iterations of the local search CS-DFN, which can move along a dense set of directions [17], the solution (39) is improved in 4 s to ITAE = 8.021×10^6 with the solution:

$$\begin{aligned} &\kappa_1 = 246494.579020, \quad \kappa_2 = 233333315.349, \quad \alpha = 92.5925925927, \\ &\beta = 0.000129629629. \end{aligned} \tag{40}$$

We also tested in this post-optimization phase the local search SDBOX, which moves only along the coordinate directions [16]. With 5000 additional iterations of SDBOX that requires 11 s, solution (39) is improved to the following new solution, which has ITAE = 9.062×10^6

$$\begin{aligned}\kappa_1 &= 200781.893004, \;\; \kappa_2 = 226268812.597, \;\; \alpha = 39.3004115226, \\ \beta &= 0.000500000024.\end{aligned} \tag{41}$$

Solution (40) corresponds to x^{**} of Procedure 2. Note that this is actually an approximation of a theoretical optimal solution. Since there is probably no optimal solutions available for comparison, we evaluate it by using the following considerations.

The settling time (see (12) and (13)) corresponding to (40) is computed as $t_s = 6280$ s. Then, we search for a lower bound $\underline{t_s}$ on the minimum settling time of the considered case, through physical considerations. The initial conditions (37) correspond to having the spacecraft with the desired attitude but with a nonzero initial angular velocity $\omega(0) = [0.02 \;\; 0.02 \;\; -0.03]^T$. Consider now simple rotations about each single body axis, and for each simple rotation, compute lower bounds $\underline{t_{sx}}$, $\underline{t_{sy}}$, $\underline{t_{sz}}$ of the times necessary to move to the desired attitude with final zero angular rate. Then, a rough lower bound for the settling time is given by $\underline{t_s} = \max\left\{\underline{t_{sx}}, \underline{t_{sy}}, \underline{t_{sz}}\right\}$. Value $\underline{t_{sx}}$ can be computed using the equation which describes rotation about the x body axis, which is given by

$$\ddot{\phi} = J_x T_x \tag{42}$$

where ϕ is the roll angle. The amplitude of torque T_x is limited by an upper limit T^*, which can be found using (4). Indeed, numerical simulations show that $\|B^i\| \le B^* = 5 \cdot 10^{-5}$ T. Since $\|B^b\| = \|B^i\|$ (see (5)), then $\|B^b\| \le B^*$. Moreover, each component of m_{coils} is bounded by $m^*_{coils} = 10 \;\; \text{A} \;\; m^2$ and then $\|m_{coils}\| \le \sqrt{3} \;\; m^*_{coils}$. Thus, $T^* = \sqrt{3} \;\; B^* \;\; m^*_{coils} = 5\sqrt{3} \cdot 10^{-4}$ N m. Next, the minimum time to bring the state of system (42) subject to the constraint $|T_x| \le T^*$, from the initial state $\phi = 0 \;\; \dot{\phi} = \omega_x(0)$ to the final state $\phi = 0 \;\; \dot{\phi} = 0$, is given by (see [18], Section 7.2)

$$\underline{t_{sx}} = \frac{J_x}{T^*}\left(1 + \sqrt{2}\right)|\omega_x(0)| = 1505 \;\; \text{s}. \tag{43}$$

Similar considerations hold for rotations about y and z axes leading to

$$\begin{aligned}\underline{t_{sy}} &= \frac{J_y}{T^*}\left(1 + \sqrt{2}\right)|\omega_y(0)| = 948 \;\; \text{s}. \\ \underline{t_{sz}} &= \frac{J_z}{T^*}\left(1 + \sqrt{2}\right)|\omega_z(0)| = 2091 \;\; \text{s}.\end{aligned} \tag{44}$$

Then, $\underline{t_s} = \max\left\{\underline{t_{sx}}, \underline{t_{sy}}, \underline{t_{sz}}\right\} = 2091$ s. Therefore, solution (40) takes just about 70 min more than the minimum time necessary to rotate the spacecraft about a single body axis at the maximum speed allowed by the available magnetorquers so that it reaches the desired rest position. Thus, solution (40) does not appear to be too far from an optimal solution. Note also that the above computed bound is very conservative, in the sense that the minimum spacecraft evolution time surely cannot require less, though it could very easily require more.

6.2 Variable initial conditions

Now, problem (33) is solved by employing the above described Procedure 3. We select 20 partitions on the domain of κ_1, 20 on that of κ_2, and 10 on that of α. We therefore perform 40,000 evaluations of $g(x)$ for the identification of the new K^* and then another 3000 + 1000 evaluations of $g(x)$ to solve the problem on this K^*, for a total of $maxeval\_ext = 44,000$.

For the internal loop, as described in Procedure 3, we employ as initial guesses the eight combinations of extreme values for the three components of the angular velocity $\pm\hat{\omega_{01}}$, $\pm\hat{\omega_{02}}$, and $\pm\hat{\omega_{03}}$. We allow 64 iterations per point, for a total of 512 iterations per single lower-level problem, which requires slightly less than 3 s in average within K^*. Run times are greater in the rest of K; however, we try to keep them under control by allowing to exit the internal loop when the value of $f(\overline{x}, y)$ is large enough to reach the limit value for ITAE of about 1.2×10^9. Then, the whole nested loop procedure provides the following solution in about 342,000 s (about 95 h)

$$\begin{aligned} \kappa_1 &= 227777.777778, \quad \kappa_2 = 294444444.444, \\ \alpha &= 83.3333333333, \beta = 0.00061095869532. \end{aligned} \tag{45}$$

The ITAE value of this solution is 2.176×10^7 for the initial conditions (37) instead of ITAE $= 8.021 \times 10^6$ of solution (40). However, this solution is a robust solution: by varying the initial conditions in H, the worst value that can be obtained is ITAE $= 1.357 \times 10^8$ that is still considerably better than average solutions, whose vast majority has the limit value for ITAE of 1.2×10^9. As a comparison, the worst value achievable by varying the initial conditions in H for solution (40) is ITAE $=$ 1.103×10^9 that is very close to the limit value for ITAE. Indeed, the above limit value for ITAE is very easily obtainable for almost any generic tuple $(\overline{\kappa}_1, \overline{\kappa}_2, \overline{\alpha}, \overline{\beta})$ by simply searching for difficult initial conditions. Note also that this value of ITAE is obtained in some attempts in solving problem (33) by using standard DIRECT algorithm on the whole feasible set K allowing no more than 1 day of computation. More time-consuming attempts in solving problem (33) using standard DIRECT algorithm cannot be practically accomplished. This holds for the following motivations. In the case of fixed initial conditions (Section 6.1), 1 week of computation was not enough to explore the search space. In this case, there is an internal loop requiring 512 function evaluations instead of one single function evaluation. Hence, any serious attempt would need to allocate months of computation and would obtain results probably similar to those obtained in Section 6.1.

7. Conclusions

The attitude of a spacecraft can be controlled using only magnetorquers by means of a PD-like control or an attitude feedback. However, for both control laws, design parameters must be assigned. These parameters may be conveniently selected so that they minimize the spacecraft settling time, or an indirect measure of it, either for fixed initial conditions of the spacecraft or under the worst initial conditions. This latter choice gives an upper bound on the minimum value of the objective obtainable by varying the initial conditions. This chapter has described solution approaches based on an innovative use of global and local derivative-free optimization techniques to practically solve these computationally demanding problems. This approach is able to provide robust solutions to the considered application in reasonable times.

Author details

Renato Bruni[1*†] and Fabio Celani[2†]

1 Department of Computer Control and Management Engineering "Antonio Ruberti", Sapienza University of Rome, Roma, Italy

2 School of Aerospace Engineering, Sapienza University of Rome, Roma, Italy

*Address all correspondence to: bruni@diag.uniroma1.it

† These authors contributed equally.

IntechOpen

© 2020 The Author(s). Licensee IntechOpen. Distributed under the terms of the Creative Commons Attribution - NonCommercial 4.0 License (https://creativecommons.org/licenses/by-nc/4.0/), which permits use, distribution and reproduction for non-commercial purposes, provided the original is properly cited. BY-NC

References

[1] Sidi MJ. Spacecraft Dynamics and control. New York: Cambridge University Press; 1997

[2] Silani E, Lovera M. Magnetic spacecraft attitude control: A survey and some new results. Control Engineering Practice. 2005;**13**(3):357-371

[3] Celani F. Robust three-axis attitude stabilization for inertial pointing spacecraft using magnetorquers. Acta Astronautica. 2015;**107**:87-96

[4] Bruni R, Celani F. Determining optimal parameters in magnetic spacecraft stabilization via attitude feedback. AIP Proceedings. 2016;**1776**: 090032

[5] Bruni R, Celani F. A robust optimization approach for magnetic spacecraft attitude stabilization. Journal of Optimization Theory and Applications. 2017;**173**:994-1012

[6] Bruni R, Celani F. Combining global and local strategies to optimize parameters in magnetic spacecraft control via attitude feedback. Journal of Optimization Theory and Applications. 2019;**181**:997-1014

[7] Jones DR, Perttunen CD, Stuckman BE. Lipschitzian optimization without the Lipschitz constant. Journal of Optimization Theory and Application. 1993;**79**(1):157-181

[8] Graham D, Lawthrop RC. The synthesis of optimum response: Criteria and standard forms, part 2. Transactions of the AIEE. 1953;**72**:273-288

[9] Wie B. Space Vehicle Dynamics and Control. Reston, VA: AIAA; 2008

[10] Wertz JR, editor. Spacecraft Attitude Determination and Control. Norwell, MA: Kluwer Academic; 1978

[11] Rodriguez-Vazquez AL, Martin-Prats MA, Bernelli-Zazzera F. Full magnetic satellite attitude control using ASRE method. In: 1st IAA Conference on Dynamics and Control of Space Systems. 2012

[12] Lovera M, Astolfi A. Spacecraft attitude control using magnetic actuators. Automatica. 2004;**40**(8): 1405-1414

[13] Lovera M, Astolfi A. Global magnetic attitude control of inertially pointing spacecraft. Journal of Guidance, Control, and Dynamics. 2005;**28**(5):1065-1067

[14] Ogata K. Modern Control Engineering. 4th ed. Upper Saddle River, NJ: Prentice Hall; 2002

[15] Jones DR. DIRECT global optimization. In: Floudas CA, Pardalos PM, editors. Encyclopedia of Optimization. Berlin: Springer; 2009. pp. 725-735

[16] Lucidi S, Sciandrone M. A derivative-free algorithm for bound constrained optimization. Computational Optimization and Applications. 2002;**21**:119-142

[17] Fasano G, Liuzzi G, Lucidi S, Rinaldi F. A linesearch-based derivative-free approach for nonsmooth constrained optimization. SIAM Journal on Optimization. 2014;**24**(3):959-992

[18] Athans M, Falb PL. Optimal Control: An Introduction to the Theory and Its Applications. New York: Dover; 2007

Chapter 9

Precise Attitude Control Techniques: Performance Analysis From Classical to Variable Structure Control

Elisa Capello and Matteo Dentis

Abstract

Small satellites have begun to play an important role in space research, especially about new technology development and attitude control. The main objective of this research is the design of a *robust* flight software, in which the key feature is suitably designed control laws to guarantee the robustness to uncertainties and external disturbances. To accomplish the desired mission task and to design the robust software, a classical Proportional Integrative Derivative (PID) method and two robust control system technologies are provided, focusing on applications related to small satellites and on the real-time implementability. Starting from PID approach, simulations are performed to prove the effectiveness of the proposed control systems in different scenarios and in terms of pointing stability and accuracy, including uncertainties, measurement errors, and hardware constraints. Different control techniques are analyzed: (i) a tube-based robust model predictive control (MPC) and (ii) a variable gain continuous twisting (CT) sliding mode controller. Both controllers are compared with loop shaping PID controller.

Keywords: precise attitude control, continuous twisting controller, model predictive control, pointing stability, variable structure control

1. Introduction

The attitude tracking of rigid bodies, i.e., spacecraft systems, is an active research area [1, 2], since every system works in a harsh environment far from direct human control. In detail, a space system requires a subsystem that can autonomously handle and control the attitude dynamics. This subsystem has the main purpose of controlling the orientation of the spacecraft with respect to an inertial reference frame. This subsystem includes sensors and actuators to measure the orientation and to apply the torques needed to change the orientation.

This research will focus on small satellites, i.e., object with mass lower than 500 kg. The common features of these objects are the small volume and mass, and these features allow small satellites to be launched as cargo and later being deployed by an other spacecraft or as payload of a single launch vehicle. Since the cost of launch is heavily affected by the payload mass, small satellites offer a relatively low-cost solution to space access. However, reduced dimensions bring new difficulties, since small satellites are more sensitive to disturbances and perturbations than larger satellites. The attitude control problem of a spacecraft in the presence of disturbance and/or uncertainties has been extensively studied. Many different control strategies have been proposed in literature, including adaptive controllers [3–5], robust control methods [6, 7] or H_2/H_∞ controller [8]. The importance of robust controllers for attitude tracking and of the definition of the mathematical model is pointed out by Dasdemir [9], in which a quaternion-based control is proposed. Even if external disturbances are included, only sinusoidal variations of them are considered, and zero-tracking error is proposed. In [10] adaptive gains of a sliding mode controller are designed to counteract the presence of failures. Moreover, in [10] actuator limitations and dynamical constraints are also included.

Due to the presence of uncertainties and dynamical constraints, the main objective of this research is to design and compare different *robust* control systems for attitude tracking. The two proposed control methodologies are based on model predictive control theory [11, 12] and on variable structure theory [13, 14]. A first definition of robustness, although not so rigorous, can be the capability of the control system to work well under sets of parameters different from the nominal one. For example, these parameters can be uncertainties within the system, not known but bounded.

Both of the proposed methods have advantages and drawbacks, briefly described in the following. As explained in [15], a model predictive control approach is able to effectively handle constraints on torque magnitude and attitude angles and can be more effective than other classical methods. However, a high computational effort is required to solve online the optimization problem. Focusing on robust approach, in this chapter, a tube-based robust MPC (TRMPC) is proposed, which focuses on two main goals: (i) the robustness to additive disturbances and (ii) the computational efficiency of a classical MPC, due to an offline evaluation of the constraints. Thanks to this control strategy, the uncertain future trajectories lie in sequence of sets, known as *tubes*, and the online MPC scheme is applied *only* to the nominal trajectories, representing the center of the tube itself as in [16].

The second proposed methodology is based on variable structure strategy [13], in which the control law is a function of the system state and changes among the possible structures according to some suitably defined rules. In particular, in sliding mode controller (SMC) systems, a switching function is designed, which implicitly defines a sliding surface corresponding to the points in the state space in which the switching function is zero. At any time, the structure applied by the control law depends on the position of the state with respect to the sliding surface. When a sliding motion is established, the closed-loop system is in sliding mode, and its trajectories are constrained on the sliding surface. Sliding mode methods provide controllers which can counteract uncertainties and disturbances, if the perturbations affecting the system are matched and bounded (first-order SMC) [17] or smooth matched disturbances with bounded gradient (second-order SMC) [14, 18]. One of the main drawbacks of SMC methods is the chattering phenomenon, which excites the high-frequency unmodeled dynamics in practical applications. Moreover, the performance of the control system is affected by the quality of the measurements and of the computation frequency of the system [19]. For this reason, our idea is to design a Continuous Twisting Sliding Mode Controller (CTSMC) with adaptive gains [20, 21]. The peculiarities of this controller are (i) adaptation of the gains, (ii) continuous control inputs, and (iii) the external disturbances which are included in the definition of the control gains.

For both the proposed controllers onboard, hardware limitations are included in the design and implementation. As clearly explained in [22], limited computational resources and reduced sampling time can be a problem in the design of robust controller. Moreover, the sample frequency reduction in the SMC design implies a residual chattering, which can be reduced with the introduction of a hyperbolic tangent [23]. A comparison with a Proportional Integrative Derivative (PD) controller, in which the gains are defined from loop shaping theory [24], is also proposed.

The chapter is structured as follows. In Section 2 mathematical models of the spacecraft are described, with an emphasis on the external disturbance definition. Section 3 is focused on the definition of mathematics and implementation of the two proposed control laws: (i) TRMPC and (ii) CTSMC. In this section the real-time implementability of both controllers is included. Simulation results are detailed in Section 4, in which a *nadir* pointing nominal attitude with an additional *roll* pointing is performed. Finally, concluding remarks are proposed in Section 5.

2. Mathematical models

Section 2 introduces in detail all the mathematical models, designed and implemented for the design of an orbital simulator. Attitude dynamics and quaternion kinematics are introduced to identify the system. Moreover, a detailed description of all the external disturbances is provided, with a focus on LEO maneuvers. Finally, a complete description of the different actuation configurations is included, starting from NASA configuration to tetrahedral.

2.1 Spacecraft attitude dynamics

In the present section, the mathematical model of the rigid body attitude dynamics is described. The rigid body attitude dynamics are described by Euler's equation:

$$\dot{\omega}_B = J^{-1}[-\omega_B \times (J \cdot \omega_B + H_{rws}) - T_{rws} + T_{ext}] \tag{1}$$

where $\dot{\omega}_B \in \mathbb{R}^3$ is the angular acceleration in body frame, $\omega_B \in \mathbb{R}^3$ is the spacecraft angular velocity in body frame, $J = \left[J_x, J_y, J_z\right]^T \in \mathbb{R}^{3\times3}$ is the inertia matrix of the spacecraft, $H_{rws} \in \mathbb{R}^3$ is the total angular momentum of the reaction wheel system expressed in body frame, $T_{rws} \in \mathbb{R}^3$ is the torque provided by reaction wheels to the spacecraft in body frame, and $T_{ext} \in \mathbb{R}^3$ is the external torque applied to the spacecraft. This torque may be the sum of the external disturbance torque plus the torque provided by the thrusters of the reaction control system (RCS) and/or the torque provided by magnetic torquers (MGT). The torque provided by RCS or MGT is usually applied in order to slow down the reaction wheels (the main actuation system for attitude control) once the maximum angular momentum of the wheels is reached. This so-called momentum desaturation will be deeply discussed in Section 2.3. Note the "-" sign applied to T_{rws}: since the torque applied to the spacecraft body by the reaction wheels is a *reaction* torque, i.e., the torque applied to the spacecraft structure has opposite sign with respect to the torque generated by the electric motors of the wheels, this torque has to be considered *internal*, and hence the *minus* sign is considered. Torque is achieved by momentum exchange between the reaction wheels and the rest of the spacecraft body. Eventually, the spacecraft angular velocity, ω_B, is obtained by time-integrating Eq. (1).

The attitude of the spacecraft is expressed in *quaternion* form. The use of quaternions is very common in space applications. The advantages of using the quaternion notation are that singularities that may occur using the classical Euler angles[1] are completely eliminated, since the quaternion expresses a specific attitude unambiguously. In the following, some fundamentals required to understand the use of quaternions will be provided, while for a more detailed discussion, please refer to [25].

A quaternion is basically a vector of four elements:

$$q = [q_1, q_2, q_3, q_4]^T \tag{2}$$

where the vectorial part, $q_v = [q_1, q_2, q_3]^T$, represents a scaled form of the eigenvector, i.e., the axis of rotation, and the scalar component, q_4, represents the magnitude of the rotation. A fundamental property for an attitude quaternion is that

$$q_1^2 + q_2^2 + q_3^2 + q_4^2 = 1 \tag{3}$$

A quaternion of the form

$$q = [0, 0, 0, 1]^T \tag{4}$$

is the *unitary* quaternion, and it represents an attitude which is perfectly aligned with the selected reference frame.

The quaternion kinematics is described by

$$\dot{q} = \frac{1}{2}\Sigma(q)\cdot\omega_B = \frac{1}{2}\begin{bmatrix} q_4 & -q_3 & q_2 \\ q_3 & q_4 & -q_1 \\ -q_2 & q_1 & q_4 \\ -q_1 & -q_2 & -q_3 \end{bmatrix}\cdot\omega_B \tag{5}$$

where $q \in \mathbb{R}^4$ is the current attitude quaternion and ω_B is the spacecraft angular velocity. Time-integrating Eq. (5) allows the determination of the true attitude of the spacecraft in terms of quaternion. For attitude control purposes, there are two main quaternion mathematical operations that must be known: the quaternion product and the quaternion inverse. The quaternion product is defined as

$$q \otimes p = \begin{bmatrix} p_4 \cdot q_v + q_4 \cdot p_v - q_v \times p_v \\ q_4 \cdot p_4 - q_v \cdot p_v \end{bmatrix} \tag{6}$$

The quaternion inverse is defined as the normalization of the conjugate quaternion, i.e.,

$$q^{-1} = \frac{q^*}{\|q\|^2} = \frac{[-q_1, -q_2, -q_3, q_4]^T}{\|q\|^2} \tag{7}$$

2.2 External disturbances

In the low Earth orbit (LEO) environment, four main sources of orbital disturbances which affect the attitude and orbit dynamics of a spacecraft can be found:

[1] For example, in the rotation sequence 3-2-1, known as Yaw-Pitch-Roll sequence, the singularity occurs for a pitch angle of 90 deg.

residual atmospheric drag, gravitational disturbances, solar radiation, and electromagnetic disturbances. The last two sources of disturbances present a lower magnitude with respect to the other sources, so they are described but not detailed in the following. A brief summary of LEO disturbances is presented in [26].

2.2.1 Atmospheric drag

The drag disturbance is due to the residual atmospheric gases which impact on the surface of the spacecraft. At high altitude, above 100 km, the residual atmosphere cannot be considered a continuum medium anymore, and a discrete medium model must be considered: the free molecular flow model. Indeed, the residual atmospheric drag is computed by considering the momentum exchange between the gas particles and the spacecraft. When hitting the surface of the spacecraft, some particles may be completely or partially reflected, depending on the number of factors, such as surface and incident flow temperature, molecular weight of particles, speed of the spacecraft, and more. To consider such effects, the accommodation coefficient has been introduced in order to compute the spacecraft drag coefficient by Cook [27]. The analysis of Cook shows that in some cases it may be useful to accurately compute the drag coefficient of a spacecraft; however, there are other uncertainties in the atmospheric models (determination of the exact temperature, density, magnetic flux, and more) such that the computation of the total drag force is affected by uncertainties higher than uncertainties which affect the computation of the drag coefficient. The work of Koppenwallner [28] analyze the real data of three spacecraft (CHAMP, GRACE, and GOCE) in order to derive a method to determine the drag coefficient of slender spacecraft. In this method, the effect of the accommodation coefficient is evaluated and compared with the results of CFD analysis. This method allows an easy computation of the drag coefficient depending on the ratio between the frontal and side areas. As already mentioned, this method can be applied only to spacecraft with simple shape. The work of Huges [29] proposes the computation of the drag force and torque integrating the incident flow on the surface of the spacecraft, without computing the drag coefficient. In this case, the accommodation coefficient is always required to compute the normal and tangential drag force applied to the surface. To compute the drag torque, the computation of the center of pressure is required. For simple shape, e.g., a plate with constant mass density, the center of pressure is located in the center of the plate area. For complex shapes, the total center of pressure can be computed by summing the contribution of the number of plates used to discretize the whole spacecraft surface.

For the purpose of this work, the computation of the residual drag force is carried out using the easiest method:

$$F_d = \frac{1}{2}\rho V_{SC}^2 S_{front} C_D \tag{8}$$

where F_d is the total drag force acting on the center of pressure of the spacecraft and aligned with the spacecraft inertial velocity V_{SC}, ρ is the atmospheric density, S_{front} is the exposed frontal area perpendicular to the direction of motion, and C_D is the drag coefficient. According to the analysis of Cook, the value of C_D is set equal to 2.2. In order to consider uncertainties in both computation of the drag coefficient and estimation of the center of pressure, the value of C_D may be increased up to 10–20%, considering a worst-case scenario.

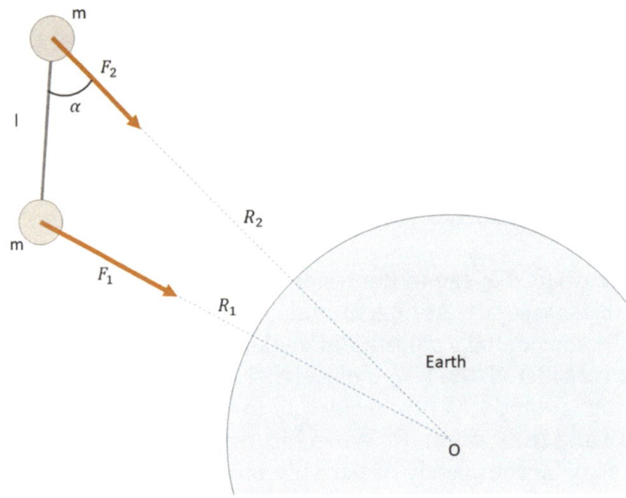

Figure 1.
Simple gravity gradient torque example.

The drag torque is eventually computed:

$$M_d = r_{cp} \times F_d \tag{9}$$

where r_{cp} is the distance between the center of pressure and the center of mass of the spacecraft. As already mentioned, the center of pressure may be estimated by discretizing the spacecraft in a number of simple plates and summing the contribution of each plate. It has to be remarked that the estimation of the center of pressure is subjected to uncertainties.

2.2.2 Gravity gradient

Gravitational torque is due to the differential gravitational force which acts on the spacecraft. As shown in **Figure 1**, two equal masses m orbiting the Earth linked together by a mass-less rigid beam of length l are subjected to two forces F_1 and F_2 equal to

$$F_1 = -G\frac{m \cdot M_E}{R_1^2} \qquad F_2 = -G\frac{m \cdot M_E}{R_2^2} \tag{10}$$

where $G = 6.67 \cdot 10^{-11}$ Nm^2/kg^2 is the universal gravitational constant, M_E is the mass of the Earth, and R_1 and R_2 are the distances from the center of the Earth of the two masses, respectively. The *minus* sign in Eq. (10) is due to the fact that the force is directed to the center of the central body. Since $R_1 < R_2$, it follows that $F_1 > F_2$; hence a torque M_{gg} is generated:

$$M_{gg} = (F_2 - F_1)\frac{l}{2}\sin\alpha \tag{11}$$

Extending this behavior to the whole body of a spacecraft, the following result is obtained:

$$M_{gg} = 3\frac{GM_E}{R^3}\hat{o}_3 \times (J\hat{o}_3) \tag{12}$$

where R is the distance of the spacecraft from the center of the Earth, J is the inertia matrix of the spacecraft, and o_3 is the third column of the direction cosine matrix from the body to local vertical frame. As an alternative, $\hat{o}_3 = -r_{ECI}/\|r_{ECI}\|$, considering $r_{ECI} \in \mathbb{R}^3$ as the position of the spacecraft with respect to the inertial frame.

Equation (12) has been obtained applying Eq. (10) to the infinitesimal mass element dm of the spacecraft, computing the cross product between the infinitesimal force df_{gg} and the distance dl of dm from the center of mass of the spacecraft and integrating over all the spacecraft volume. The volume integration results in the computation of the inertia matrix. Since the gravity gradient torque tends to align the minimum inertia axis of the spacecraft with respect to the local vertical frame, gravity gradient stabilization has also been used as passive stabilization method. This method ensures a two-axis stabilization, since only roll and pitch axes with respect to the local vertical frame are affected by gravity gradient torque, while the yaw axis is torque-free. The equilibrium stability of gravity gradient stabilization is ensured only if the nadir inertia axis of the spacecraft is the smallest inertia axis: if this condition is not satisfied, the equilibrium condition may be unstable, and an active stabilization of the spacecraft is required. For a deeper analysis, the reader can refer to [25].

2.2.3 Other disturbance torques

As mentioned at the beginning of Section 2.2, there are other sources of torque disturbances which affect the attitude dynamics of a spacecraft. The first is a magnetic torque due to the residual magnetic dipole that can affect the spacecraft. This torque is due to the interaction between the residual magnetic dipole d and the magnetic field B generated by the Earth such that $M_{mag} = d \times B$. This is the same phenomenon that aligns the needle of a compass with the magnetic North of the Earth. Spacecraft are usually designed in order to minimize, up to neutralize, its residual magnetic dipole. However, the use of magnetic torquers (as described in Section 2.3) generates a desired magnetic dipole in order to obtain a desired torque, which can be used to control the attitude of very small spacecraft (usually CubeSats), or to desaturate momentum actuators (reaction wheels and gyroscopes). In general, since the spacecraft designer is able to properly cope with the magnetic dipole, this disturbance can be neglected.

An additional disturbance torque is due to solar radiation. This torque is produced by photons emitted by the Sun which exchange momentum with the surface of the spacecraft. Clearly, this disturbance is present only if the orbit of the spacecraft is in the sunlight. When the orbit of the spacecraft is shadowed by the eclipse of the Earth, this disturbance is not present. The magnitude of this torque depends on the activity of the Sun, which presents seasonal variations according to the Sun's cycles. For spacecraft with a compact shape and without long appendages, the magnitude of this torque is usually lower than the effect of drag and gravity; hence, this disturbance may be neglected.

Other disturbances are due to internal torques, due to fuel sloshing, and/or due to flexible modes which are present for spacecraft with flexible solar arrays or appendages [30]. In particular, the attitude control of a flexible spacecraft must be properly designed, since a bad-designed closed-loop attitude control may excite flexible modes of the spacecraft instead of damping them, causing a catastrophic failure of the mission.

2.3 Actuators

To control and stabilize the attitude of a spacecraft, it is necessary to apply torques to the spacecraft structure in order to reorient it and/or to maintain the desired attitude counteracting external disturbances. There exist a number of actuators which are able to accomplish this task, with different working principles and application scope:

- Reaction control thrusters (RCT)
- Magnetic torquers (MGT)
- Control moment gyroscopes (CGM)
- Reaction wheels (RWS) and/or momentum wheels

Reaction control thrusters are basically small rocket motors (chemical thrusters). The simplest RCT is composed of a combustion chamber and an expansion nozzle. Both fuel and oxidizer are introduced in the combustion chamber, and they are ignited (with a capaciitive discharge, torch ignition, or using hypergolic propellants), and then the exhaust flows in the nozzle[2] accelerating and exchanging linear momentum, producing thrust. To produce torque, RCT are grouped in a RCS, and they are commanded coupled: two thrusters, installed at a certain distance from each other and firing in opposite direction, are activated simultaneously, producing a defined torque. If thrusters are throttable, i.e., the amount of thrust can be regulated, also the produced torque can be regulated; differently, if they are not throttable, only a fixed amount of thrust, and consequently torque, is generated, then other modulation techniques should be used to obtain a throttable torque (e.g., pulse-width modulation (PWM) modulation). For large spacecraft, e.g., Space Shuttle, chemical thrusters are used, since they produce a high amount of thrust. For smaller spacecraft, cold gas thrusters may also also be used: in this actuators the combustion chamber is eliminated, and the thrust is produced by the gas flowing from the storage tank to the nozzle by opening (or closing) the flow control valve (FCV). Such type of thrusters is not throttable.

RCTs may be used for two different purposes. The main use is to desaturate reaction wheels and/or control moment gyros: they are activated in order to provide the external torque which contrasts the torque due to the wheel deceleration or gyros reorientation. An additional method is that they are used combined with a different main actuator system (RWS or CGM) when the required torque is too high to be provided by the main system. RCT are also used to generate the force required to accomplish the orbit control task (execution of orbital maneuvers or station-keeping).

Magnetic torquers produce force, and torque, by generating an electromagnetic field which interact with the magnetic field of the Earth. Since their performance is linked to the electromagnetic field of the Earth, their effectiveness decreases with the orbit altitude, and they cannot be used during interplanetary missions. For these reasons, they are mainly used in small or all-electric spacecraft for desaturation of momentum accumulation devices (RWS or CGM). Examples of application are found in the Hubble Space Telescope [31].

Control moment gyroscopes consist in an assembly of gyroscopes spinning about an axis which can be tilted and consequently change the angular momentum vector of the devices. Differently from RCT and MGT, which are mostly on/off actuators,

---

[2] Mainly convergent-divergent nozzles are used: the flow accelerates up to supersonic velocity in the convergent part, and then it continues to expand in the divergent segment.

CMG allows to control the attitude of a spacecraft reaching a very high accuracy, since they can generate a continuous torque by changing the CMG angular momentum. Compared to reaction wheels, CMG are more effective and efficient, since they can produce a higher torque: indeed, they are used to control the attitude of the International Space Station [32]. For mechanical reasons, they are subjected to saturation: when they reach the *gimbal lock*, they are stuck, and they cannot provide further angular momentum change (i.e., torque). When this condition is reached, an external torque must be applied in order to allow CMG to reorient themselves in the original configuration, making them able to provide torque again and restore the attitude control capabilities of the system.

Reaction wheels are basically disk masses which rotate about a fixed spin axis driven by an electric motor. By varying the rotation speed, i.e., applying a torque to the wheel, by reaction the same amount of torque is transmitted to the spacecraft structure, and it starts moving. These devices are the most commonly used actuators for fine attitude control due to its simplicity and lower complexity than the CMG, even though they are able to provide less torque and less accuracy than CMGs. RWS are also subjected to two different types of saturation: torque saturation, due to the maximum current that flows into the wires of the electric motor, and momentum saturation, due to the maximum spin velocity before breaking bearings. Once the momentum saturation is reached, RWS must be desaturated in a similar way as for CMG, and external torque, usually provided by RCT or MGT, is applied to counteract the deceleration of the wheels. Once the original condition is reached, RWS starts to provide torque again.

To fully control the three-axis attitude, there are necessarily three reaction wheels, but for redundancy issues, there are usually installed four reaction wheels: in this way, if a wheel failure occurs, a three-axis attitude control is always ensured, even though pointing accuracy may be degraded. The reaction wheel system configuration can be installed using a number of different wheel accommodation, mostly depending on the specific mission since a main control axis may be required. The mostly used configurations are:

- NASA 3 + 1: three wheels are aligned with the spacecraft main inertia axis, while the fourth wheel is inclined by 45 deg. with respect to all planes. During nominal operations, only the wheels aligned with the body axis are used, while the fourth wheel is used only during wheel failure.

- Pyramidal: the wheels' spin axes are pointing toward the faces of a pyramid with square base. This means that wheels are inclined with a fixed angle with respect to the base of the pyramid, generating torque along the height of the pyramid, and the projections of the wheel momentum lie in the pyramid base, generating torque along the last two directions.

- Tetrahedral: as for the pyramid, in this configuration the spin axes of the wheels are pointing to the faces of a tetrahedron. Hence, a wheel generates torque only along the height of the tetrahedron, while the other three wheels generate torque both in the tetrahedron base and along the height.

Note that, except for the NASA configuration, it is necessary to allocate the three-axis control torque to the four wheel assembly. To do this, it is common to use the Moore-Penrose pseudo-inverse matrix, A^+, since the non-square control allocation matrix $A \in \mathbb{R}^{3\times 4}$ cannot be inverted. The three reaction wheel configuration yet presented is depicted in **Figure 2**. For the NASA 3 + 1 configuration, angles α and β are both equal to 45 deg.; for the pyramidal configuration, β is selected by the

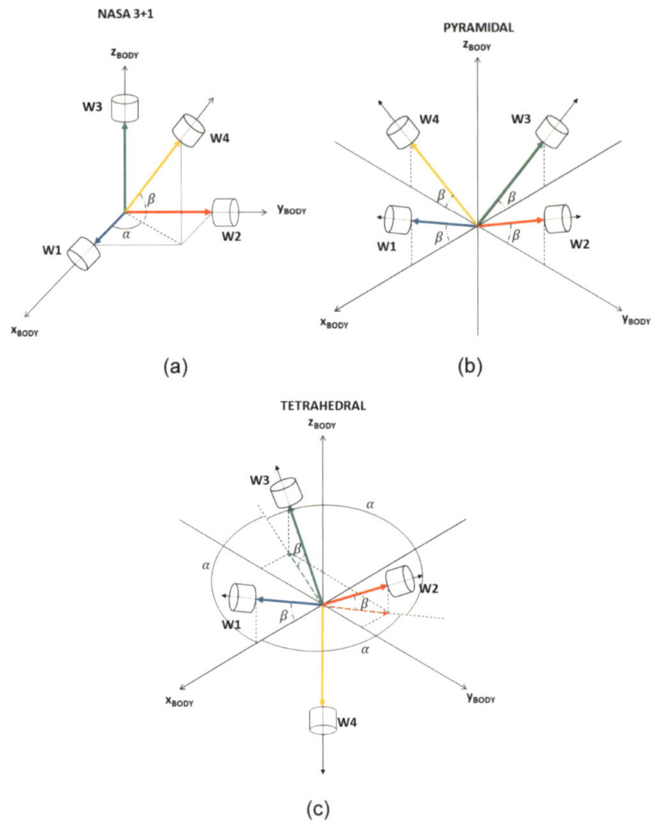

Figure 2.
Reaction wheel configurations. (a) NASA 3+1. (b) Pyramidal. (c) Tetrahedral.

designer (and angle of 35.26 deg. maximizes the volume of the control envelope), while α, not depicted in **Figure 2**, is equal to 90 deg.; for the tetrahedral configuration, angle α is equal to 120 deg., in order to obtain a symmetric tetrahedron, while angle β is set by the designer (a value of 19.47 deg. ensure the most regular envelope).

In the following of this chapter, reaction wheels are considered as the main actuation system for attitude control accommodated according to the pyramidal configuration.

3. Attitude control laws

The objective of Section 3 is related to mathematics and design of two *robust* control strategies for attitude tracking. As detailed in Introduction, two control systems are considered in this chapter: (i) a robust MPC strategy, based on the concept of *tube*, and (ii) a continous SMC strategy, based on the method of twisting algorithm. Both controllers are compared with a PID controller, in which the gains are based on the theory proposed by [24].

3.1 Tube-based robust model predictive control

Tube-based model predictive control (TRMPC) is a class of robust controllers, i.e., controllers which are able to cope external disturbances and uncertainties which affect the system. The concept of *tube* has been introduced in classical MPC in order to improve robustness of such controllers and in order to force the perturbed system dynamics to converge to the center of it. This *tube* is generated by propagating the unperturbed system dynamics. The outer-bounding tube is generated in order to take into account all possible realization of the disturbances w, which are assumed to be included in a set $\mathbb{W}$ (i.e., $\mathrm{w} \in \mathbb{W}$). Starting from a classical MPC problem, implemented with respect to the nominal unperturbed system dynamics, the TRMPC is subject to *tighten constraints*, which ensure robustness of the TRMPC to external disturbances w. For the control algorithm design, let us consider the following discrete time-invariant state-space system in which persistent disturbances w_k are included:

$$\mathrm{x}_{k+1} = \mathrm{A}_d\mathrm{x}_k + \mathrm{B}_d\mathrm{u}_k + \mathrm{w}_k, \tag{13}$$

where x_k and u_k represent the discrete-time state vector and the control signal at time k, respectively.

Let us assume that the system is required to satisfy hard constraints on both state and input:

$$\mathrm{x}_k \in \mathbb{X}, \quad \mathrm{u}_k \in \mathbb{U}, \tag{14}$$

where $\mathbb{X} \subset \mathbb{R}^n$ and $\mathbb{U} \subset \mathbb{R}^m$ are compact and convex polytopes [citaz]. For the definition of the disturbance, w_k is considered as a realization of a stochastic process, an independent and identically distributed (i.i.d.) zero-mean random variable, with a convex and bounded support $\mathbb{W} \subset \mathbb{R}^n$ containing the origin.

As already introduced before, the TRMPC approach is based on the concept of *tube*. The center of this *tube* corresponds to the nominal undisturbed trajectory, which dynamics is defined as

$$\mathrm{z}_{k+1} = \mathrm{A}_d\mathrm{z}_k + \mathrm{B}_d\mathrm{v}_k, \tag{15}$$

where z_k and v_k are the discrete-time nominal state and input, respectively. **Figure 3** provides a representation of the outer-bounding tube at the k-th time step centered on the nominal trajectory at each i-th step over a N prediction horizon [33].

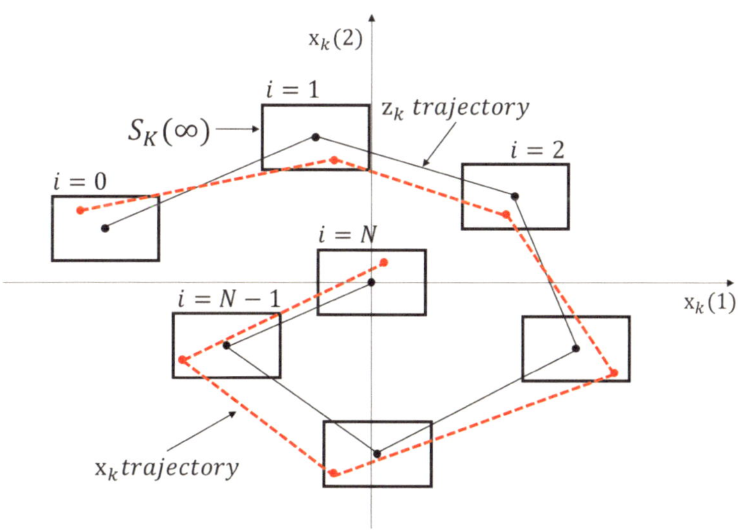

Figure 3.
Outer-bounding tube representation at the k-th time step over a prediction horizon of N.

Two features of this controller are (i) the TRMPC which allows to steer the uncertain trajectories to the nominal one via a classical MPC approach and (ii) the robustness which ensured tightening the constraints with respect to the initial ones in Eq. (14), as detailed in [12]. Then, the following time-varying feedback control law related to the i-th step ahead k is defined:

$$\mathrm{u}_{i|k} = \mathrm{v}_{i|k} + K\left(\mathrm{x}_{i|k} - \mathrm{z}_{i|k}\right), \tag{16}$$

in which K is defined such that $\mathrm{A}_K = \mathrm{A}_d + \mathrm{B}_d K$ is robustly stable and K quadratically stabilizes the system (17) with respect to the parametric uncertainty q. Finally, the closed-loop dynamics can be rewritten as follows:

$$\mathrm{x}_{i+1|k} = (\mathrm{A}_d + \mathrm{B}_d K)\mathrm{x}_{i|k} + \mathrm{B}_d \mathrm{v}_{i|k} + \mathrm{w}_{i|k}. \tag{17}$$

Moreover, to stabilize the system with respect to parametric uncertainty q, due to, for example, neglected nonlinearities, a linear matrix inequality (LMI) approach is applied.

In the next paragraph, the definition of the feedback gain matrix (summarized in Algorithm 1 is analyzed in detail. First, the terminal state constraint set $\mathbb{X}_N$ (given any $\mathrm{x}_k \in \mathbb{X}_N$), a matrix $P \in \mathbb{R}^{n \times n}$, $P \succ 0$, exists such that

$$(\mathrm{A}_d + \mathrm{B}_d K)^T P (\mathrm{A}_d + \mathrm{B}_d K) + Q + K^T R K - P \preccurlyeq 0. \tag{18}$$

$Q \in \mathbb{R}^{n \times n}$, $Q \succ 0$, and $R \in \mathbb{R}^{m \times m}$, and $R \succ 0$ are diagonal positive definite matrices. As proposed in [33], solving the following LMI system allows to obtain the feedback gain matrix K:

$$\begin{cases} Q + K^T R K + (\mathrm{A}_d^+ + \mathrm{B}_d^+ K)^T \tilde{P} (\mathrm{A}_d^+ + \mathrm{B}_d^+ K) - \tilde{P} \preccurlyeq 0, \\ Q + K^T R K + (\mathrm{A}_d^+ + \mathrm{B}_d^- K)^T \tilde{P} (\mathrm{A}_d^+ + \mathrm{B}_d^- K) - \tilde{P} \preccurlyeq 0, \\ Q + K^T R K + (\mathrm{A}_d^- + \mathrm{B}_d^+ K)^T \tilde{P} (\mathrm{A}_d^- + \mathrm{B}_d^+ K) - \tilde{P} \preccurlyeq 0, \\ Q + K^T R K + (\mathrm{A}_d^- + \mathrm{B}_d^- K)^T \tilde{P} (\mathrm{A}_d^- + \mathrm{B}_d^- K) - \tilde{P} \preccurlyeq 0, \end{cases} \tag{19}$$

in which the following edge-uncertain system matrices are defined: $\mathrm{A}_d^- = \mathrm{A}_d(\mathrm{q}^-)$, $\mathrm{A}_d^+ = \mathrm{A}_d(\mathrm{q}^+)$, $\mathrm{B}_d^- = \mathrm{B}_d(\mathrm{q}^-)$, and $\mathrm{B}_d^+ = \mathrm{B}_d(\mathrm{q}^+)$. The proposed LMIs are based on the well-known edge theorem, which is an extension of Kharitonov theorem, as explained in [34, 35].

Algorithm 1 Feedback Gain Evaluation

1: **procedure**
2: **for** each i-th vertex $\left(\boldsymbol{A}^i(\mathrm{q}), \boldsymbol{B}^i(\mathrm{q})\right)$ **do**
3: $\mathrm{sys}_i = \mathbf{Q} + \boldsymbol{X}\boldsymbol{A}_i^T + \boldsymbol{A}_i^T \boldsymbol{X} - \boldsymbol{\Upsilon}^T \boldsymbol{B}_i^T - \boldsymbol{B}_i^T \boldsymbol{\Upsilon}$
4: **end for**
5: Solve $\left[X \succ 0, \mathrm{sys}_i \prec 0\right]$
6: Get $\boldsymbol{X}$ and $\boldsymbol{\Upsilon}$
7: Get $\boldsymbol{K} = \boldsymbol{\Upsilon}\boldsymbol{X}^{-1}$
8: **end procedure**

Since bounded disturbance $\mathrm{w}_{i|k}$ is included in the system dynamics, the deviation of the actual state $\mathrm{x}_{i|k}$ with respect to the nominal one $\mathrm{z}_{i|k}$ i step ahead time k can be defined as

$$\mathrm{e}_{i+1|k} = (\mathrm{A}_d + \mathrm{B}_d K)\mathrm{e}_{i|k} + \mathrm{w}_{i|k}. \tag{20}$$

As anticipated before, tightened constraint sets shall be considered for the nominal system, properly designed starting from an outer approximation of the minimal robust positively invariant (mRPI) set [36] for (20):

$$S_K(\infty) \doteq \sum_{j=0}^{\infty} \mathrm{A}_K^j \mathbb{W}, \tag{21}$$

in compliance with the guidelines provided in [12]. The set is mRPI for (20) because *only* additive disturbance has been considered affecting the system dynamics as in Eq. (13). It is important to highlight that the parametric uncertainty is included only for evaluating the feedback gain matrix K that quadratically stabilizes the closed-loop disturbed dynamics, but no uncertainty has been included in the control design.

Hence, if the time-invariant control law (16) is employed and the nominal system (15) satisfies the tightened constraint sets,

$$\begin{aligned} \mathrm{z}_{i|k} &\in \mathbb{Z} \subseteq \mathbb{X} \ominus S_K(\infty), \\ \mathrm{v}_{i|k} &\in \mathbb{V} \subseteq \mathbb{U} \ominus K S_K(\infty). \end{aligned} \tag{22}$$

the initial constraints $\mathrm{x}_{i|k} \in \mathbb{X}$ and $\mathrm{u}_{i|k} \in \mathbb{U}$ are robustly satisfied at each time step k, only for *restricted disturbances* [12].

Since $x_{i|k} = z_{i|k} + e_{i|k}$, where $e_{i|k} \in S_K(\infty)$, it follows that $H_x x_{i|k} \leq h_x$ if

$$H_x z_{i|k} \leq h_x - \Phi_\infty, \tag{23}$$

with $\Phi_\infty = \max\limits_{e_{i|k}} \left\{ H_x e_{i|k} \mid e_{i|k} \in S_K(\infty) \right\}$. Thus,

$$\hat{\mathbb{Z}} = \left\{ z_{i|k} \in \mathbb{R}^n \mid H_x z_{i|k} \leq h_x - \Phi_\infty \right\} \tag{24}$$

represents a suitable constraint set for the nominal state $z_{i|k}$ in order to obtain an inner approximation $\mathbb{Z}$ of $\hat{\mathbb{Z}}$, where $\hat{\mathbb{Z}} = \mathbb{X} \ominus S_K(\infty)$. To evaluate Φ_∞, it is possible to compute an upper bound of this set solving a simple linear programming [33]. We should define $\Phi_\infty \leq (1-\beta)^{-1}\Phi_T$, with $\beta \in (0, 1)$, where

$$\Phi_T = \max_{w_{i|k}} \left\{ H_x \sum_{j=0}^{T-1} A_K^j \mathbb{W}, w_{i|k} \in \mathbb{W} \right\}, \tag{25}$$

is the solution of a linear programming problem. Hence, it is possible to obtain an upper bound of Φ_∞ properly selecting β as close as desired to 1. Then, the constraint set $\mathbb{Z}$ can be defined by

$$\mathbb{Z} \doteq \left\{ z_{\ell|k} \in \mathbb{R}^n \mid H_x z_{\ell|k} \leq h_x - (1-\beta)^{-1}\Phi_T \right\} \subseteq \hat{\mathbb{Z}}. \tag{26}$$

In a similar way, the constraint set on the control input $\mathbb{V}$ can be approximated as

$$\mathbb{V} \doteq \left\{ v_{i|k} \in \mathbb{R}^m \mid H_u v_{i|k} \leq h_u - (1-\beta)^{-1} K \Phi_T \right\} \subseteq \hat{\mathbb{V}}. \tag{27}$$

starting from the initial control input constraint set $\mathbb{U}$ and being $\hat{\mathbb{V}} = \mathbb{U} \ominus KS_K(\infty)$. More details can be found in [12, 33].

Then, the finite horizon optimal quadratic cost can be defined for the nominal dynamics in terms of nominal state $z_{i|k}$ and nominal control input $v_{i|k}$ as

$$J_N(z_k, \mathbf{v}_k) = \sum_{i=0}^{N-1} \left(z_{i|k}^T Q z_{i|k} + v_{i|k}^T R v_{i|k} \right) + z_{N|k}^T P z_{N|k}, \tag{28}$$

where $\mathbf{v}_k$ represents the control sequence over a N-step prediction horizon. $P \in \mathbb{R}^{n \times n}$ is the solution of the discrete Algebraic Riccati Equation [37]. Thus, the nominal finite horizon optimal control problem can be stated as follows:

$$\min_{\mathbf{v}_k} \; J_N(z_k, \mathbf{v}_k) \tag{29a}$$

$$\begin{aligned} \text{s.t.} \quad & z_{i+1|k} = A_d z_{i|k} + B_d v_{i|k}, \quad z_{0|k} = x_k, \\ & z_{i|k} \in \mathbb{Z}, \quad i \in [1, N-1], \\ & v_{i|k} \in \mathbb{V}, \quad i \in [0, N-1], \\ & z_{N|k} \in \mathbb{Z}_N, \end{aligned} \tag{29b}$$

with $\mathbb{Z}_N \subseteq \mathbb{X}_N \ominus S_K(\infty)$. The first control action $v_{0|k}^*$ of the optimal sequence $\mathbf{v}_{k'}^*$ solution of Eq. (29a) (29b) represents the optimal control applied to the nominal system, while the correspondent control on the uncertain system is defined according to Eq. (16). The final TRMPC algorithm can be summarized as shown in Algorithm 2.

Algorithm 2 TRMPC Algorithm

1: **procedure**
2: *Offline*: Evaluate the feedback gain matrix K and the nominal constraint sets $\mathbb{Z}$ (Eq. (26) and $\mathbb{V}$ from Eq. (27).
3: *Online*: Initialization: for $k = 0$, set $z_{0|k} = x_k = x_0$.
4: At current time k, evaluate x_k, z_k.
5: **for** $i = 0 : N - 1$ **do**
6: Solve Eq. (29) (30)
7: **end for**
8: Get $\mathbf{v}_0^*$ and extract the first control action v_0^*.
9: Evaluate u_k according to (16).
10: Evaluate z_{k+1} applying $\mathbf{v}_0^*$ on (15) and x_{k+1} applying u_k on Eq. (13).
11: **end procedure**

Summarizing, the TRMPC control scheme consists in a classical MPC controller applied to the nominal unperturbed dynamics z subjected to the tightened constraints $\mathbb{Z}$ and $\mathbb{V}$. Then, to the computed control v, it has been added a second control $K(x - z)$ which steers the perturbed dynamics x to converge to the nominal one. The feedback matrix K should be designed properly in order to ensure that the perturbed dynamics satisfy the original constraints $\mathbb{X}$ and $\mathbb{U}$.

3.1.1 Comments on real-time implementability

For the implementation of the TRMPC controller on a real hardware, it is necessary to linearize the equation of motion in order to obtain a linear time-invariant system in the classical form:

$$\begin{aligned} x &= Ax + Bu + B_w w \\ y &= Cx + Du. \end{aligned} \tag{31}$$

In the case of attitude control of a spacecraft, the state and control vectors are assumed to be $x = \left[q_{e_1}, q_{e_2}, q_{e_3}, \omega_{e_1}, \omega_{e_2}, \omega_{e_3}\right]^T \in \mathbb{R}^6$, where q_{e_1}, q_{e_2}, and q_{e_3} are the vectorial components of the quaternion error $q_e = q \otimes q_{des}^{-1} \in \mathbb{R}^3$, where the symbol $\otimes$ is referred to the quaternion multiplication, while ω_i is the body angular velocity. Matrix A has been obtained by linearizing kinematics and Euler's equations of motion, which results in

$$A = \begin{bmatrix} A_{11} & A_{12} \\ A_{21} & A_{22} \end{bmatrix}, \tag{32}$$

with

$$A_{11} = \begin{bmatrix} 0 & \omega_{d_3} & -\omega_{d_2} \\ -\omega_{d_3} & 0 & \omega_{d_1} \\ \omega_{d_2} & -\omega_{d_1} & 0 \end{bmatrix}, A_{12} = \frac{1}{2}\begin{bmatrix} 1 & 0 & 0 \\ 0 & 1 & 0 \\ 0 & 0 & 1 \end{bmatrix}, A_{21} = \varnothing^{(3,3)},$$
$$A_{22} = \begin{bmatrix} 0 & k_1\omega_3 & k_1\omega_2 \\ k_2\omega_3 & 0 & k_2\omega_1 \\ k_3\omega_2 & k_3\omega_1 & 0 \end{bmatrix},$$

with $k_1 = \frac{J_y - J_z}{J_x}$, $k_2 = \frac{J_z - J_x}{J_y}$, and $k_3 = \frac{J_x - J_y}{J_z}$. All the parameters are computed in an equilibrium point x_0. The other matrices of the state-space formulation in Eq. (31) are

$$B = \begin{bmatrix} \varnothing^{(3,3)} \\ J^{-1} \end{bmatrix}, C = \mathbb{I}^{(6,6)}, D = \varnothing^{(6,3)},$$

where $\mathbb{I}$ and $\varnothing$ are, respectively, the identity and null matrices of proper dimensions. The unperturbed MPC optimization problem is

$$\begin{aligned} J_N(\mathbf{z}_k, \mathbf{v}_k) &= \sum_{i=0}^{N-1}\left(\mathbf{z}_{i|k}^T Q \mathbf{z}_{i|k} + \mathbf{v}_{i|k}^T R \mathbf{v}_{i|k}\right) + \mathbf{z}_{N|k}^T P \mathbf{z}_{N|k} = \\ &= \frac{1}{2}\mathbf{z}^T(0)\Upsilon \mathbf{z}(0) + \min_{v}\left(\frac{1}{2}V^T H V + \mathbf{z}^T(0) F V\right), \\ &\text{s.t.} \quad LV \leq W \end{aligned} \tag{33}$$

where $W = \begin{bmatrix} W_v \\ W_z \end{bmatrix}$ are the system constraints and $V \in \mathbb{R}^{mN}$ is the optimal control vector sequence. Finally, the other matrices are defined as

$$\begin{aligned} \Upsilon &= 2\left(Q + \tilde{A}^T \tilde{Q} \tilde{A}\right), \\ H &= 2\left(\tilde{R} + \tilde{B}^T \tilde{Q} \tilde{B}\right), \\ F &= 2\left(\tilde{A}^T \tilde{Q} \tilde{B}\right), \end{aligned}$$

in which

$$\tilde{A} = \begin{bmatrix} A_d \\ A_d^2 \\ \vdots \\ A_d^N \end{bmatrix}, \tilde{B} = \begin{bmatrix} B_d & 0 & \dots & 0 \\ A_d B_d & B_d & & \vdots \\ \vdots & \vdots & & \vdots \\ A_d^{N-1} B_d & A_d^{N-2} B_d & \dots & B_d \end{bmatrix}, \tilde{Q} = \begin{bmatrix} Q & 0 & \dots & 0 \\ 0 & \ddots & & \vdots \\ \vdots & \dots & Q & 0 \\ 0 & \dots & 0 & P \end{bmatrix}$$

$$\tilde{R} = \begin{bmatrix} R & 0 & 0 \\ 0 & \ddots & 0 \\ 0 & 0 & R \end{bmatrix},$$

where R is the weighing matrix related to the control and P is the terminal weighing matrix.

The online optimization problem is solved with the *Quad-Wright solver*, which is an improved version of the *quadprog* Matlab solver. The solver returns the optimal control sequence V according to the inputs H, F, L, and W. The first m components of the control vector sequence are then translated into the control v to compute the final control u and to propagate the nominal unperturbed dynamics z. The feedback gain matrix K is computed offline. Solving the LMI is possible to compute the matrix K which stabilizes the uncertain system and the terminal weighing matrix P which ensures satisfaction of the terminal constraints. This is a derivation of the edge theorem, as discussed in [38]. If no uncertainties are considered, the gain K can be computed using LQR design techniques.

3.2 Variable gain continuous twisting sliding mode controller

Sliding mode control (SMC) found its origin as discontinuous nonlinear controller for nonlinear variable structure systems (VSS). The basic principle of SMC is to steer the system state to lie on a *sliding manifold*: once the system state reaches the sliding manifold, the sliding mode condition holds by applying a discontinuous high-frequency control, despite the external disturbance affecting the system dynamics. One of the advantages of using SMC is that such control law is robust with respect to the so-called matched uncertainties, i.e., uncertainties which affect the control channel of the system. The basic assumption is that such uncertainties must be bounded. The classical first-order SMC is the following:

$$u = -\rho \cdot \text{sign}(\sigma) \tag{34}$$

The gain ρ is selected such that the uncertainty d is bounded by the value of ρ, i.e., $|d| < \rho$. σ is the *sliding variable* and it is designed considering the system state. For an SMC as Eq. (34), the sliding variable is selected as $\sigma = x_2 + K \cdot x_1$ when a two-state system is considered and $\dot{x}_1 = x_2$ is considered. The working principle of SMC is to steer to zero the variable σ in finite time. A critical issue is the proper design of the sliding variable: for different and more advanced controllers, for example, the twisting SMC (TSMC), the sliding variable can be defined in a different way. In [20] a summary of different sliding mode controllers can be found.

Since the classical SMC stabilize a dynamic system by applying a high-frequency—theoretically infinite—discontinuous control, such type of controller cannot be used with actuators that must be controlled with a continuous signal, such as reaction wheels. To adapt SMC to continuous actuators, some smoothing techniques can be introduced in order to compute a continuous signal: for example, the "sign" function can be smoothed in the neighborhood of the origin by using the hyperbolic tangent or other similar functions. A more advanced technique is to use a higher-order SMC, as the case of the super-twisting SMC (STWSMC) or the continuous twisting SMC (CTSMC): using such techniques, a discontinuous control is applied as *virtual* control, while the *physical* control will result in as continuous. In the present work, the CTSMC will be further investigated. Let us consider the system:

$$\begin{cases} \dot{q}_e = \frac{\omega}{2} \\ \dot{\omega}_e = J^{-1}u + \Delta(t) \end{cases} \tag{35}$$

which represents a linearized attitude dynamics. q_e and ω_e are the quaternion error and the angular velocity error, respectively, u is the control torque, J is the spacecraft inertia, and $\Delta(t)$ is the external disturbance. The external disturbance is assumed to be *Lipschitz*, i.e., it is differentiable with bounded derivative: $\dot{\Delta}(t) \leq \mu$. According to the system (35), a continuous twisting SMC can be designed as

$$\begin{cases} u = -k_1|q_e|^{\frac{1}{3}}\text{sign}(q_e) - k_2|\omega_e|^{\frac{1}{2}}\text{sign}(\omega_e) + \eta \\ \dot{\eta} = -k_3|q_e|\text{sign}(q_e) - k_4|\omega_e|\text{sign}(\omega_e) \end{cases} \tag{36}$$

According to this control design, it results in $\sigma = q_e$ and $\dot{\sigma} = \omega_e$. Control gains k_i can be designed as [21]

$$k_1 = 7 \qquad k_2 = 5 \qquad k_3 = 2.3 \qquad k_4 = 1.1 \tag{37}$$

Such gains ensure robustness against Lipschitz disturbances $\Delta(t)$ bounded by $\mu = 1$. Since in practical cases the Lipschitz disturbances are usually bounded by $\dot{\Delta}(t) \leq \mu^* = L\mu \neq 1$, control gains can be scaled as

$$k_{p1} = k_1 L^{\frac{2}{3}} \qquad k_{p2} = k_2 L^{\frac{1}{2}} \qquad k_{p3} = k_3 L \qquad k_{p4} = k_4 L \tag{38}$$

The controller of Eq. (36) can be redesigned by substituting gains k_i with gains k_{pi}. The adaptation mechanism of the continuous twisting SMC consists in varying the scaling gain L such as

$$\dot{L}(t) = \begin{cases} l, & \text{if } T_e \neq 0 \text{ or } |(q_e, \omega_e)| > \epsilon \\ 0, & \text{if } T_e = 0 \text{ or } |(q_e, \omega_e)| \leq \epsilon \end{cases} \tag{39}$$

where ϵ is a threshold value, introduced since it is not possible to consider the exact *zero* value of q_e and ω_e (in practical applications), and $T_e(t)$ is a timer which is reset while the value of q_e and ω_e is within ϵ for at least a defined time span. The value of l must be chosen in order to reject the increase of $\dot{\Delta}$. In addition, a similar discharge strategy, i.e., considering $l < 0$, can be introduced if the values of control gains are higher than required and an excessive chattering is induced.

3.2.1 Comments on real-time implementability

A brief discussion about finite-time controllers is provided. In general, SMC controllers, if properly designed, ensure the convergence $\sigma = 0$ if the controller frequency is infinite. In practical applications, this is obviously not feasible, and only a high but finite controller frequency can be realized. Considering the controller sample time ΔT_{con}, as stated in [21], with the current set of gains (37), it is possible to achieve an accuracy of $|q_e| \le 19\Delta T_{con}^3$ and $|\omega_e| \le 19\Delta T_{con}^2$. A consequence of using a finite-time frequency controller is that, even though the CTSMC control computes theoretically a continuous command, a residual chattering is always observed. As mentioned before, the use of smoothing techniques may mitigate, or even eliminate, residual chattering. It has to be highlighted that for very low controller frequency (<10 Hz), gains computation (37) and (38) may fail and a more empirical tuning may be required, as well as the accuracy considerations may not be applicable anymore.

4. Simulation results

Both control algorithms, described in Section 3, have been applied in an Earth observation mission scenario. The considered spacecraft has a total mass of 500 kg and principal tensor of inertia (i.e., "*diagonal*" inertia matrix) with $I_{xx} = I_{yy} = I_{zz} =$ 100 kgm^2. The spacecraft has a cubic shape with a frontal area of 2 m^2. For the drag torque computation, a drag coefficient of $C_D = 2.2$ is considered, and the center of pressure is supposed to be located in $[0, -0.05, 0.01]^T$ m with respect to the center of mass of the spacecraft. The reference orbit is an equatorial orbit of 600 km of altitude. The desired quaternion and angular velocity with respect to the inertial reference frame are depicted in **Figure 4**: the mission scenario consists in a *nadir* pointing nominal attitude with an additional *roll* pointing of 10 deg. after 1000 s. The *roll* pointing phase lasts about 500 s, and then the reference attitude returns to *nadir* pointing. The maneuver is propagated for half an orbital period.

The pointing errors for the three proposed controllers (PID, CTSMC, and TRMPC controllers) are depicted in **Figures 5–7**. From these figures, the maximum angular error obtained with the CTSMC (**Figure 7**) is more than one order of magnitude lower than the maximum error obtained with the other two controllers. In a similar way, the angular error obtained during the pointing phase (zoomed

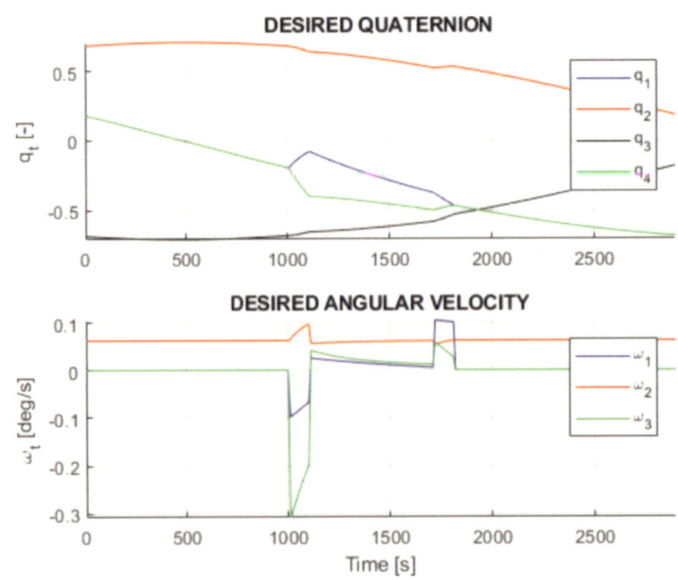

Figure 4.
Desired inertial quaternion and angular velocity.

figures) is close to 10^{-5} deg. for the SMC controller. This is due to the fact that the CTSMC tries to regulate exactly the state variables to zero, while PID and TRMPC, in general, allow a small but finite pointing error, unless a proper integral action is introduced in the controllers: in the presented PID controller, the integral action is negligible with respect to the proportional and derivative actions.

In **Figure 8** the computed inputs are depicted, in terms of reaction wheel torque and momentum for each control law. For the TRMPC, the computed input (i.e., torque in body frame) has been derived as $T_{cmd} = A_{rws} \cdot T_{MPC,rws}$, where A_{rws} is the torque allocation matrix and $T_{MPC,rws}$ is the optimal control already allocated to each wheel and directly computed by the control system. Indeed, one advantage of using

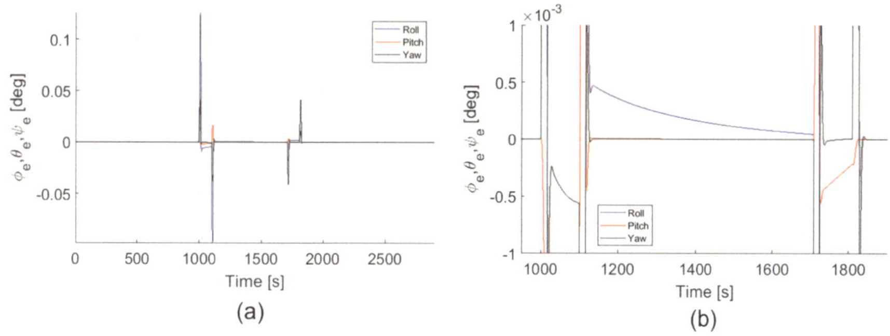

Figure 5.
Attitude tracking in terms of Euler angles with PID controller. (a) Euler Angle error. (b) Zoom of Euler Angle error.

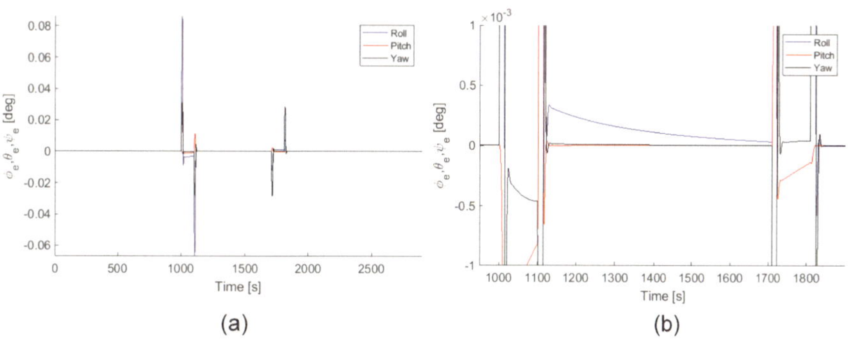

Figure 6.
Attitude tracking in terms of Euler angles with TRMPC controller. (a) Euler Angle error. (b) Zoom of Euler Angle error.

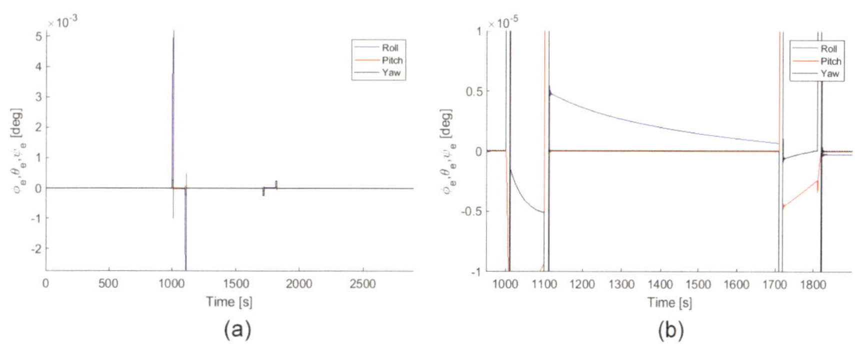

Figure 7.
Attitude tracking in terms of Euler angles with CTSMC controller. (a) Euler Angle error. (b) Zoom of Euler Angle error.

an MPC controller is that the reaction wheel configuration can be taken into account in the controller design, computing directly the required reaction wheel torque. In a different way, for PID and CTSMC, the output of the control system is the required torque in body axis, T_{cmd}, and then the required torque is allocated to each reaction wheels by the pseudo-inverse matrix $A^{+}_{rws} = A^{T}_{rws}\left(A_{rws}A_{rws^T}\right)^{-1}$, i.e., $T_{rws} = A^{+}_{rws} \cdot T_{cmd}$. As highlighted in **Figure 8**, PID and TRMPC show a similar torque profile, while CTSMC presents higher peaks. As already explained in Section 3, the CTSMC control system regulates exactly the state variables to zero.

In **Figure 9**, the chattering effect, observed in the sliding mode controller, is highlighted. Even if some chattering mitigation strategy is proposed [23], the chattering can be observed due to the low controller update frequency (10 Hz).

Eventually, the pointing stability performance index has been evaluated for all the three proposed control algorithms. Pointing stability is defined as *the time-varying effects on science instrument pointing caused by the dynamic interactions between spacecraft structure and mechanisms, attitude control and determination errors, structural thermal distortion, and other environmental disturbances* [39]. For the

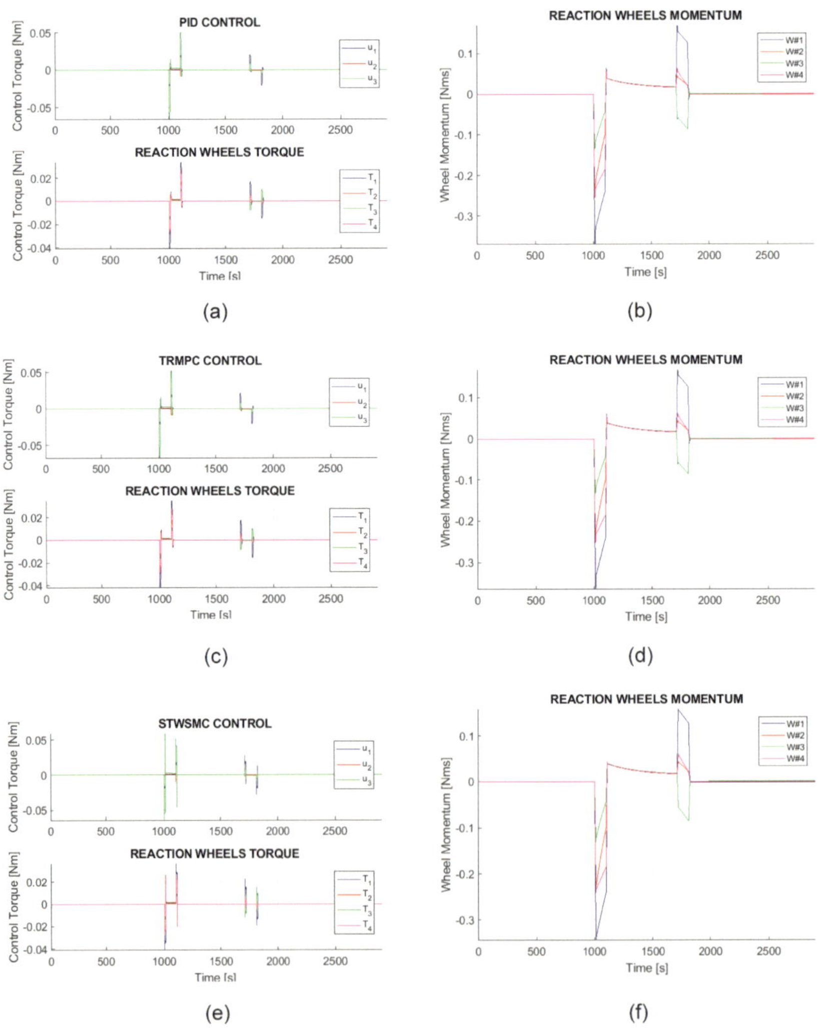

Figure 8.
Computed torque, reaction wheel torque, and momentum. (a) PID torque. (b) PID momentum.(c) TRMPC torque. (d) TRMPC momentum.(e) CTSMC torque. (f) CTSMC momentum.

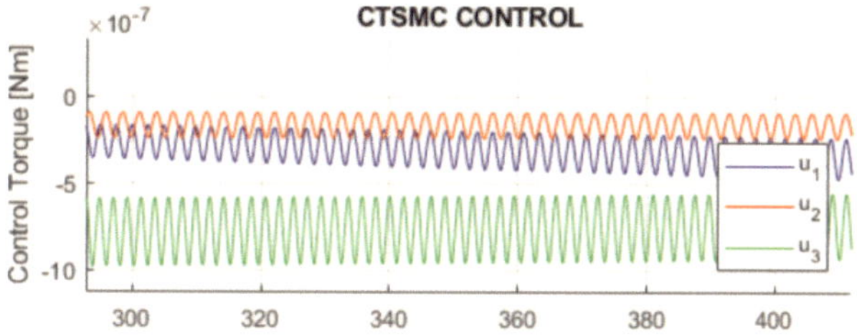

Figure 9.
Chattering effects of CTSMC control system.

Controller	Roll [deg]	Pitch [deg]	Yaw [deg]
PID	$5.31 \cdot 10^{-2}$	$5.75 \cdot 10^{-3}$	$1.30 \cdot 10^{-2}$
TRMPC	$2.42 \cdot 10^{-2}$	$1.24 \cdot 10^{-3}$	$4.63 \cdot 10^{-3}$
CTSMC	$3.23 \cdot 10^{-4}$	$1.47 \cdot 10^{-6}$	$1.61 \cdot 10^{-6}$

Table 1.
Pointing stability results: 3σ values.

mathematical evaluation of the pointing stability, the guidelines expressed in [40] have been followed:

$$\sigma^2_{STA}(\Delta t_s) = \mathrm{E}\left[(e(t) - e(t - \Delta t_s))^2\right]$$

where $\mathrm{E}[\cdot]$ is the expected value, Δt_s is the stability time, and $e(t)$ is the instantaneous angular error. The pointing stability has been evaluated only during the roll pointing maneuver, since it is the most critical mission phase. The pointing stability has been evaluated over a time span of 10 s. The pointing stability is defined as degrees over a time period, which represents the variation of the angular error over the defined time period. A summary of the pointing stability results is reported in **Table 1**. It can be observed that for each controller the pointing stability never exceed to $5.5 \cdot 10^{-2}$ deg. over 10 s. If a drawback should be evaluated, the lower is the pointing error, the lower is the value of pointing stability, which implies a higher stability. Indeed, the CTSMC shows the higher pointing stability performance, since it causes the lower pointing error. In addition, the oscillating pointing error induced by the chattering torque command is mostly canceled, considering the relatively long stability time of 10 s.

5. Conclusions

In the proposed research, small spacecraft attitude tracking with two advanced robust control laws has been developed: (i) a tube-based robust model predictive control and (ii) a continuous twisting sliding mode controller. A detailed mathematical model, with an emphasis on the external disturbances and the actuation systems, is included. Quaternion-based kinematics is proposed, even if Euler angle kinematics is also described, as more intuitive. Mathematics and some implementation details are also described, including the real-time implementability of both controllers. The performance of both controllers are compared with a PID controller. The advantages of using SMC have been presented, highlighting the lower pointing error over the whole considered mission. Moreover, chattering effects of using SMC have been observed, even though they are strongly reduced with respect to classical SMC. It has been also observed that the use of an optimal and robust control, TRMPC, causes an increment in the pointing performance as well, including the advantage of computing directly the optimal control torque for a complex reaction wheel configuration, instead of allocating the reaction wheel torque with the classical pseudo-inverse allocation.

Conflict of interest

The authors declare no conflict of interest.

Author details

Elisa Capello[1,2*†] and Matteo Dentis[1†]

1 Department of Mechanical and Aerospace Engineering, Politecnico di Torino, Torino, Italy

2 Institute of Electronics, Computer and Telecommunication Engineering, National Research Council of Italy (CNR-IEIIT), Torino, Italy

*Address all correspondence to: elisa.capello@polito.it

† These authors contributed equally.

IntechOpen

© 2020 The Author(s). Licensee IntechOpen. Distributed under the terms of the Creative Commons Attribution - NonCommercial 4.0 License (https://creativecommons.org/licenses/by-nc/4.0/), which permits use, distribution and reproduction for non-commercial purposes, provided the original is properly cited. (cc) BY-NC

References

[1] Wen JY, Kreutz-Delgado K. The attitude control problem. IEEE Transactions on Automatic Control. 1991;**36**(10):1148-1162

[2] Chaturverdi NA, Sanyal AK, McClamroch NH. Rigid-body attitude control. IEEE Control Systems Magazine. 2011;**31**(3):30-51

[3] Bai Y, Biggs JD, Bernelli Zazzera F, Cui N. Adaptive attitude tracking with active uncertainty rejection. Journal of Guidance, Control, and Dynamics. 2018;**41**(2):550-558

[4] Thakur D, Srikant S, Akella MR. Adaptive attitude-tracking control of spacecraft with uncertain time-varying inertia parameters. Journal of Guidance, Control, and Dynamics. 2015;**38**(1): 41-52

[5] Sands T. Nonlinear adaptive mathematical system identification. Computation. 2017;**5**(4):1-12

[6] Alvi BA et al. Robust controller design for satellite attitude determination, stabilization and control using LQG/LTR. Wireless Personal Communications. 2015;**85**:329344

[7] Safa A, Baradarannia M, Kharrati H, Khanmohammadi S. Robust attitude tracking control for a rigid spacecraft under input delays and actuator errors. International Journal of Control. 2019; **92**(5):1183-1195

[8] Yang C-D, Sun Y-P. Mixed H_2/H_∞ state-feedback design for microsatellite attitude control. Control Engineering Practice. 2002;**10**:951970

[9] Dasmedir J. Quaternion-based robust satellite attitude tracking control. Balkan Journal of Electrical and Computer Engineering. 2018:53-61

[10] Bai Y, Biggs JD, Wang X, Cui N. Attitude tracking with an adaptive sliding mode response to reaction wheel failure. European Journal of Control. 2018;**42**:67-76

[11] Bemporad A, Morari M. Robust model predictive control: A survey. In: Robustness in Identification and Control. 1999. pp. 207-226

[12] Mayne DQ, Rawlings JB. Model Predictive Control: Theory and Design. Madison, WI: Nob Hill Publishing; 2009

[13] Utkin VI. Variable structure systems with sliding mode. IEEE Transactions on Automatic Control. 1977;**22**(2):212-222

[14] Bartolini G, Ferrara A, Usai E. Chattering avoidance by second-order sliding mode control. IEEE Transactions on Automatic Control. 1998;**43**(2): 241-246

[15] Luo Y, Zhang J, Tang G. Survey of orbital dynamics and control of space rendezvous. Chinese Journal of Aeronautics. 2014;**27**(1):1-11

[16] Kouvaritakis B, Cannon M. Model Predictive Control: Classical, Robust and Stochastic, Advanced Textbooks in Control and Signal Processing. London: Springer; 2015

[17] Utkin VI. Sliding Modes in Optimization and Control Problems. New York: Springer; 1992

[18] Levant A. Sliding order and sliding accuracy in sliding mode control. International Journal of Control. 1993; **58**(6):1247-1263

[19] Utkin VI, Guldner J, Shi J. Sliding Mode Control in Electromechanical Systems. PA, Philadelphia; 1999

[20] Shtessel Y et al. Sliding Mode Control and Observation. New York: Springer New York; 2014

[21] Torres-González V et al. Design of continuous twisting algorithm. Automatica. 2017;**80**:119-126

[22] Behal A, Dixon W, Dawson DM, Xian B. Lyapunov-Based Control of Robotic Systems. Boca Raton, Florida: CRC Press. p. 2010

[23] Leśniewski P, Bartoszewicz A. Hyperbolic tangent based switching reaching law for discrete time sliding mode control of dynamical systems. In: IEEE International Workshop on Recent Advances in Sliding Modes (RASM). Institute of Electrical and Electronics Engineers, Inc; 2015. pp. 1-6

[24] Capello E, Sartori D, Guglieri G, Quagliotti F. Robust assessment for the design of multi-loop proportional integrative derivative autopilot. IET Control Theory and Applications. 2012; **6**(11):1-10

[25] Markley FL, Crassidis JL. Fundamentals of Spacecraft Attitude Determination and Control. Vol. 33. New York: Springer; 2014

[26] Zagrski P. Modeling disturbances influencing an Earth-orbiting satellite. Pomiary Automatyka Robotyka. 2012; **16**:98-103

[27] Cook GE. Satellite drag coefficients. Planetary and Space Science. 1965; **13**(10):929-946

[28] Koppenwallner G. Satellite aerodynamics and determination of thermospheric density and wind. In: AIP Conference Proceedings. Vol. 1333. No. 1. AIP; 2011

[29] Hughes PC. Spacecraft Attitude Dynamics. Chelmsford, Massachusetts, Stati Uniti: Courier Corporation; 2012

[30] Sidi MJ. Spacecraft Dynamics and Control: A Practical Engineering Approach. Vol. 7. Cambridge, UK: Cambridge University Press; 1997

[31] Hur-Diaz S, Wirzburger J, Smith D. Three axis control of the hubble space telescope using two reaction wheels and magnetic torquer bars for science observations. 2008:08-279

[32] Gurrisi C, Seidel R, Dickerson S, Didziulis S, Frantz P, Ferguson K. Space station control moment gyroscope lessons learned. 2010. NASA CP-216272

[33] Mammarella M, Capello E, Park H, Guglieri G, Romano M. Tube-based robust model predictive control for spacecraft proximity operations in the presence of persistent disturbance. Aerospace Science and Technology. 2018;**77**:585-594

[34] Bartlett AC, Hollot CV, Huang L. Root locations of an entire polytope of polynomials: It suffices to check the edges. Mathematics of Control, Signals, and Systems. 1988;**1**:61-71

[35] Barmish BR. New Tools for Robustness of Linear Systems. New York: Macmillan Publishing Company; 1994

[36] Blanchini F, Miani S, editors. Set-Theoretic Methods in Control. Switzerland: Springer International Publishing; 2008

[37] Stoorvogel AA, Ali S. The discrete algebraic Riccati equation and linear matrix inequality. Linear Algebra and its Applications. 1998;**276**(1–3):317-365

[38] Mammarella M, Capello E. A robust MPC-based autopilot for mini UAVs. In: 2018 International Conference on Unmanned Aircraft Systems (ICUAS); Dallas, TX; 2018. pp. 1227-1235

[39] Pilinski EB, Lee AY. Pointing-stability performance of the Cassini spacecraft. Journal of Spacecraft and Rockets. 2009;**46**(5):1007-1015

[40] ESSB-HB-E-003 Working Group. ESA Pointing Error Engineering Handbook. Ref. ESSB-HB-E-003, Issue 1, 19 July 2011

Section 3

Control Moment Gyroscope Actuators

Chapter 10

Investigation of Singularities in a 3/4 CMG Configuration with Mixed Skew Angles

Jonathan W. Lang

Abstract

This paper discusses the singularities that exist within a 3/4 CMG configuration when the CMGs are placed at mixed skew angles. CMGs are typically mounted with the same skew angles and are fixed throughout the spacecraft's lifetime. Changing these skew angles can bring about unique attributes for the spacecraft such as an increased pitch, roll or yaw capability. Mapping out these singularities when each CMG is mounted differently can show an engineer how to maximize these capabilities and enhance a spacecraft's mission completion ability. Using singularity penetration logic, the spacecraft's attitude controls system can pass through these singularities. These singularities would best be avoided to provide optimal control. Finding these limited singularity penetration regions is the focus of this paper. Different mixed skew configurations appear to be more ideal than others for spacecraft that focus on maneuvers about only one axis of rotation.

Keywords: mixed skew angles, 3/4 CMG configuration, singularity penetration, CMG configuration singularity regions

1. Introduction

CMGs have become a staple in the space community as a means to accomplish pointing, tracking, and acquiring. CMGs create a torque by rotating or "gimbaling" the CMGs angular momentum vector. The change in angular momentum is how the CMG produces torque and creates movement for the spacecraft.

The Torque axis (τ) rotation for a CMG is the axis around which the spacecraft maneuver is accomplished. The Torque Axis and the orthogonal relationship between the CMG's gimbal axis (δ) and angular momentum axis (H) can be seen in **Figure 1**. These relationships demonstrate how one may discover the Torque axis and provide the intuition behind singularity generation in this configuration [1].

2. CMG 3/4 configuration model

The 3/4 configuration will have the CMGs positioned in the same configuration around the spacecraft throughout the analysis in this paper and can be seen in **Figure 2**.

The 3/4 configuration [2] is a simple form of spacecraft CMG placement which provides an environment where singularities exist.

IntechOpen

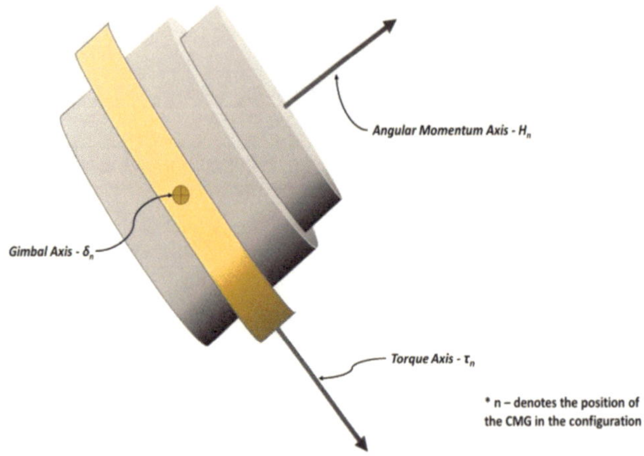

Figure 1.
CMG angular momentum-torque-gimbal axis representation.

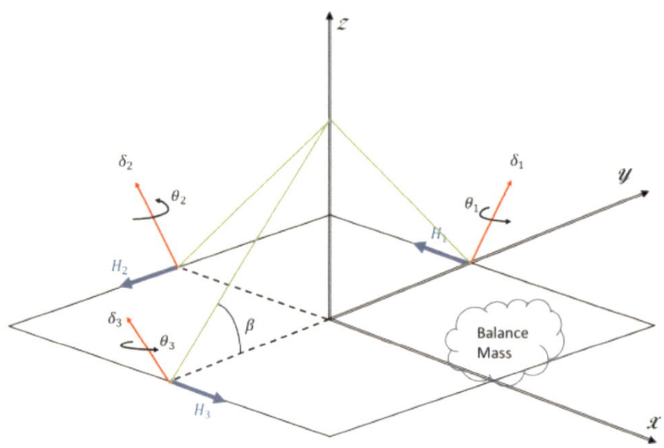

Figure 2.
3/4 configuration of CMGs on a spacecraft.

All modeling in this project will be accomplished through MATLAB—Simulink modeling software. A step size of 0.001 s was found to be the most accurate while using ODE4 Runge-Kutta Integration Solver.

3. CMG angular momentum projected on the body reference axes

The three CMG's each have an angular momentum vector (H_i) which can be seen in **Figure 2** and is projected onto the x, y, z reference frame axes in Eqs. (1)–(3). These three vectors rotate about each of the CMG's gimbal axis (δ_i).

$$H_x = -H_1 \cos(\theta_1) + H_2 \cos(\beta_2)\sin(\theta_2) + H_3 \cos(\theta_3) \quad (1)$$

$$H_y = -H_1 \cos(\beta_1)\sin(\theta_1) - H_2 \cos(\theta_2) + H_3 \cos(\beta_3)\sin(\theta_3) \quad (2)$$

$$H_z = \sin(\beta_1) H_1 \sin(\theta_1) + \sin(\beta_2) H_2 \sin(\theta_2) + \sin(\beta_3) H_3 \sin(\theta_3) \quad (3)$$

A new matrix **[A]** made up of the spatial gradient of Eqs. (1)–(3) can be generated [2]. This new matrix is how singularities can be discovered numerically.

$$\frac{\partial H}{\partial \theta} = \begin{Bmatrix} \partial H_x / \partial \theta_i \\ \partial H_y / \partial \theta_i \\ \partial H_z / \partial \theta_i \end{Bmatrix} = [A]$$

$$[A] = \begin{bmatrix} sin\,(\theta_1) & cos\,(\beta_2)\,cos\,(\theta_2) & -sin\,(\theta_3) \\ -cos\,(\beta_1)\,cos\,(\theta_1) & sin\,(\theta_2) & cos\,(\beta_3)\,cos\,(\theta_3) \\ sin\,(\beta_1)\,cos\,(\theta_1) & sin\,(\beta_2)\,cos\,(\theta_2) & sin\,(\beta_3)\,cos\,(\theta_3) \end{bmatrix} \tag{4}$$

Torque is generated by the time rate of change in the angular momentum, which may be expressed in a chain rule of derivatives and solved for the gimbal rates.

$$T = \dot{H}$$
$$\dot{H} = \frac{\partial H}{\partial \theta}\frac{d\theta}{dt}$$
$$\dot{H} = [A]\dot{\theta}$$
$$\dot{\theta} = [A]^{-1}\dot{H} \tag{5}$$

Now there is a relationship between the time rate of change in the angular momentum $(\dot{H})$ and the time rate of change of gimbal axis rotation $(\dot{\theta})$ found in Eq. (5). The required gimbal axis rotation for a commanded torque may be applied for the appropriate spacecraft maneuver. The one possible danger associated with this is the inversion of the matrix $[\boldsymbol{A}]$.

4. Accuracy of inverting the $[A]$ matrix

Inverting the **[A]** matrix becomes an integral part of determining the creation of the gimbal axis rotation $(\dot{\theta})$ as seen in Eq. (5). Analysis was done on eight different inversion models, the principal one on which all others are compared was the Moore-Penrose matrix inversion [3]. The other seven cases are found in **Table 1**.

The $pinv([\boldsymbol{A}])$ matrix inversion model turns out to be the best suited model as seen in **Table 1**. This seems to be intuitive since the $pinv\ ([\boldsymbol{A}])$ is a reiteration of the Moore-Penrose inversion.

Case	[A] Inversion model	Matrix-Norm difference
1	$[\boldsymbol{A}]^{-1}$	0.872826646563208
2	$inv([\boldsymbol{A}])$	0.872826646563208
3	$pinv([\boldsymbol{A}])$	0
4	$[\boldsymbol{A}]\ eye(size([\boldsymbol{A}]))$	0.872826646560334
5	*LU Decomposition*	0.872826646560693
6	*Analytical Matrix*	6.67204727298e+04
7	*Analytical Formulas*	6.67204727298e+04

Table 1.
Matrix-norm difference between each case and the Moore-Penrose matrix inversion.

5. Applying CMG torque to the spacecraft

Now that the CMG torque has been developed it can now be applied to the spacecraft.

$$\tau = J\dot{\omega} + \omega \times J\omega \tag{6}$$

Using Euler's dynamical Eq. (6), the angular momentum and cross-coupled disturbances can be used to determine the resulting three-axis rotation [4].

6. Singularity penetration logic

Singularity existence has been discussed at length in [2].

Singularities exist when the determinant of the $[\boldsymbol{A}]$ matrix from Eq. (4) approaches zero [2]. Penetration Logic takes advantage of the fact that as the inverse condition number approaches zero the singularity would be met. At a user defined threshold of 1 e-6 penetration logic tells the commanded Gimbal motor rates to repeat the last iteration (before the inverse condition number crossed the user defined threshold and reaches singularity), see **Figure 3**.

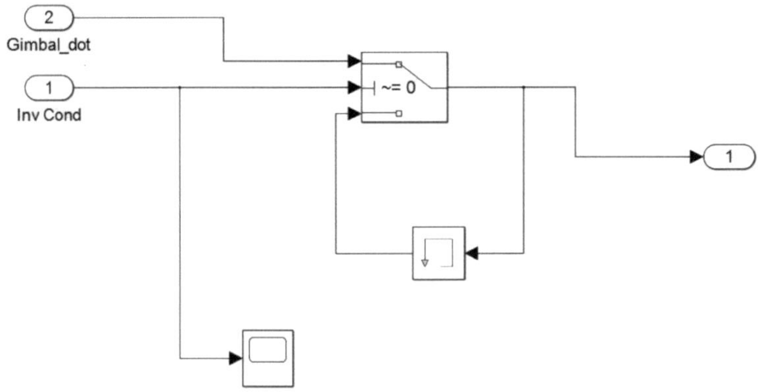

Figure 3.
Singularity penetration with unit delay logic.

7. PDI controller tuning

The PDI controller will be used to generate a control signal from the commanded input signal and the spacecraft feedback signal. The topography for the PDI controller can be seen in **Figure 4**.

The PDI controller accepts both the commanded angle and angular velocity as well as the feedback angle and angular velocity and avoids using the derivative function.

Tuning the gains become the next step in building the controller section of the spacecraft. Three different tuning techniques will be covered:

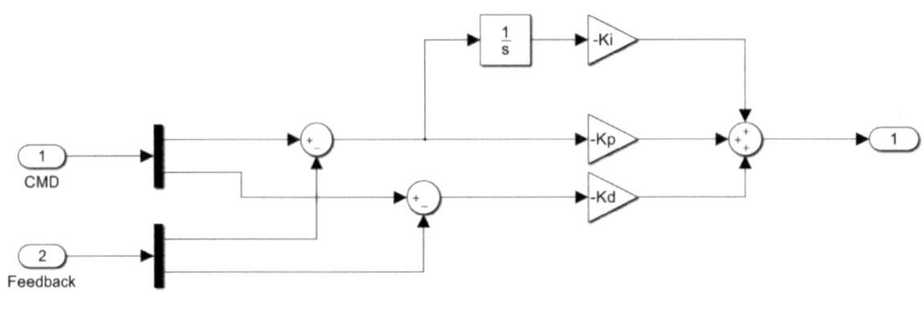

Figure 4.
PDI controller topography.

7.1 Ziegler-Nichols tuning

This tuning technique requires the gain margin (K_u) and bandwidth frequency (ω_b) for the system to be found. Once these two values are discovered the equations in (7) may be used to find the required gains.

$$K_p = 0.6K_u$$
$$K_i = {}^{2K_p}/_{\omega_n}$$
$$K_d = {}^{K_p\omega_n}/_{8} \tag{7}$$

7.2 Manual tuning

Several design criteria must be considered when utilizing the manual tuning technique. Rise time (t_r), damping ratio (ζ), and settling time (t_s) are the design characteristics required and may be used in the equation in (8) to find the new gains.

$$T = {}^{10}/_{\zeta\omega_n}$$
$$K_p = J\left(\omega_n^2 + {}^{2\zeta\omega_n}/_T\right)$$
$$K_i = J\left({}^{\omega_n^2}/_T\right)$$
$$K_d = J\left(2\zeta\omega_n + {}^{1}/_T\right) \tag{8}$$

7.3 Tuning using the linear quadratic regulator function

The LQR function in MATLAB requires the state space form of the control system. Using the form $\dot{x} = Ax + Bu$ and $y = Cx + Du$ the state space form of the torque equation in the inertial frame can be constructed. The state space from in Eq. (9) was developed and used to derive the gains from the LQR function.

$$A = \begin{bmatrix} 0 & 1 \\ 0 & 0 \end{bmatrix}$$
$$B = \begin{bmatrix} 0 \\ J^{-1} \end{bmatrix}$$
$$C = [1 \ \ 0]$$
$$D = [0] \tag{9}$$

The three tuning techniques were completed, and the gain values were calculated. These gains have been compiled and entered in **Table 2** [5].

Using the gains found in **Table 2**, three differing step functions can be found and analyzed. Ziegler-Nichols was found to be the optimal tuning method for the controllers. The LQR function is the worst way of tuning the controller. The overshoot is largest of the three tuning methods and the settling time takes longer than 100 s [5].

Case	Tuning technique	K_p	K_i	K_d
1	Ziegler-Nichols	16.20	0.78	84.65
2	Manual	36.76	1.41	55.67
3	LQR function	1.00	0.10	11.45

Table 2.
Gain solutions for different tuning techniques.

8. Plotting singularity regions with mixed skew angles

A series of mixed skew profiles were selected to perform initial analysis on. Upon investigation, certain profiles were further analyzed to see how they would be advantageous for a spacecraft with a specific maneuver requirement such as maneuver in pitch, roll, or yaw. **Figure 2** demonstrates how these skew angles (β) can change for each of the CMGs in the 3/4 configuration. When these skew angles are altered, new singularity regions are developed and create a hazard for spacecraft control. Five series of singularity plots have been generated to attain an initial picture of how the skew angles affect those singularity regions (**Tables 3–7**).

Profile	CMG 1-β	CMG 2-β	CMG 3-β
1	15	30	60
2	30	15	60
3	30	60	15
4	15	60	30
5	60	15	30
6	60	30	15

Series 1. Using skew angles (β)-15, 30, 60°.

Table 3.
Series 1 CMG β angle configurations.

Profile	CMG 1-β	CMG 2-β	CMG 3-β
1	20	40	80
2	20	80	40
3	40	20	80
4	40	80	20
5	80	20	40
6	80	40	20

Series 2. Using skew angles (β)-20, 40, 80°.

Table 4.
Series 2 CMG β angle configurations.

Profile	CMG 1-β	CMG 2-β	CMG 3-β
1	30	60	90
2	30	90	60
3	60	30	90
4	60	90	30
5	90	30	60
6	90	60	30

Series 3. Using skew angles (β)-30, 60, 90°.

Table 5.
Series 3 CMG β angle configurations.

Profile	CMG 1-β	CMG 2-β	CMG 3-β
1	0	45	90
2	0	90	45
3	45	0	90
4	45	90	0
5	90	0	45
6	90	45	0

Series 4. Using skew angles (β)-0, 45, 90°.

Table 6.
Series 4 CMG β angle configurations.

Profile	CMG 1-β	CMG 2-β	CMG 3-β
1	0	30	60
2	0	60	30
3	30	0	60
4	30	60	0
5	60	3	30
6	60	30	0

Series 5. Using skew angles (β)-0, 30, 60°.

Table 7.
Series 5 CMG β angle configurations.

The singularity plots associated with these series are found in Appendix A of this report. Each of these plots was analyzed to see breadth and depth of singularity surfaces internal to the saturation region of the 3/4 CMG configuration depicted in **Figure 2**.

9. Determining singularity free regions

The five series of singularity plots were observed to determine internal singularity free regions. The objective was to find an internal region in which the spacecraft could maneuver without running into a singularity. Also, if there was a singularity, would the spacecraft be capable of passing through a small amount of singularities in

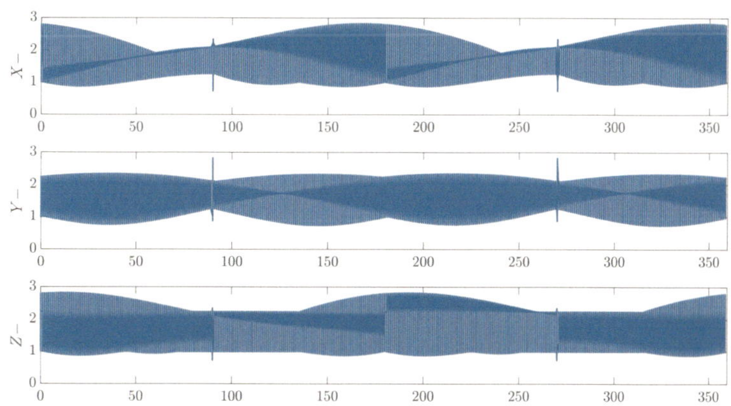

Figure 5.
Singularity regions in the X-, Y- and, Z rotation for 0, 45, and 90° skew angles.

order to continue maneuvering to a commanded rotation. Singularity penetration is a feasible concept however the goal of this paper was to use it sparingly.

The singularity regions found in **Figure 5** demonstrate how with varied skew angles new singularity free regions may be discovered and utilized to a spacecraft designer's advantage. Although the angular momentum plotted in **Figure 5** first interacts with a singularity at a value of 0.73, it only hits the singularity for an instance. Using singularity penetration logic discussed in part VI may allow the spacecraft to operate in an angularity momentum regime at $H = 1.0$ or greater.

The maximum angular momentum achieved by the mixed skew configurations in series 1–5 has been plotted and the data was consolidated into **Table 8**.

Series	CMG 1-β	CMG 2-β	CMG 3-β	Max H*
1	15	30	60	0.54
	30	15	60	0.51
	30	60	15	0.16
	15	60	30	0.16
	60	15	30	0.51
	60	30	15	0.54
2	20	40	80	0.34
	20	80	40	0.17
	40	20	80	0.33
	40	80	20	0.17
	80	20	40	0.33
	80	40	20	0.53
3	30	60	90	0.12
	30	90	60	0.41
	60	30	90	0.51
	60	90	30	0.41
	90	30	60	0.51
	90	60	30	0.25
4	0	45	90	0.73
	0	90	45	0.22
	45	0	90	0.41
	45	90	0	0.22
	90	0	45	0.41
	90	45	0	0.73
5	0	30	60	0.52
	0	60	30	0.39
	30	0	60	0.52
	30	60	0	0.39
	60	0	30	0.52
	60	30	0	0.53

**Maximum Ang. Momentum (H) in singularity free region.*

Table 8.
Series 5 CMG β angle configurations.

10. Conclusion

Mixed skew angles bring a new variety and flexibility in spacecraft design. These new CMG configurations enable engineers to now explore new singularity free regions and push spacecraft to possibly operate with higher levels of momentum.

As seen in **Table 8**, when the opposing CMG configurations have a dramatic change in configuration, the Angular Momentum is typically higher. This is not always the case, as seen in series 3, configurations 1 and 6 (the most dramatic change between CMG 1 and 3 in the series) demonstrate the worst angular momentum possibility. Several of these configurations cater to different requirements of the spacecraft. For example, Series 3 Configuration 5 (ref. Appendix A) may be more suitable for a spacecraft that requires movement about the roll and yaw axes. Series 3 Configuration 2 may be more suitable for pitch and roll spacecraft. Appendix A is meant to be used by spacecraft designers to design a spacecraft suitable to the requirements needed.

Appendix A

Singularity Plots
Series 1:
Configuration 1:

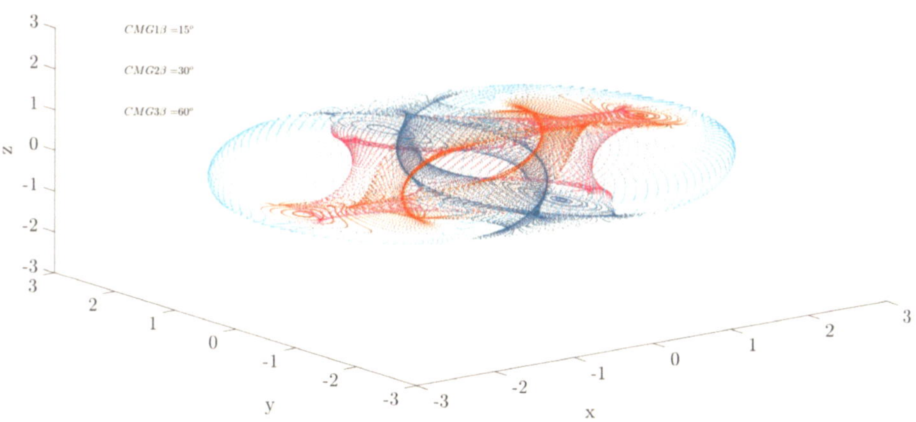

Configuration 2:

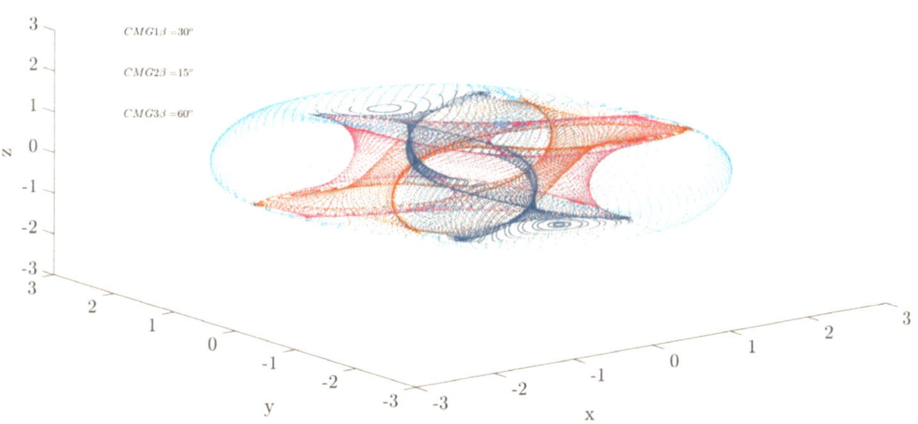

Configuration 3:

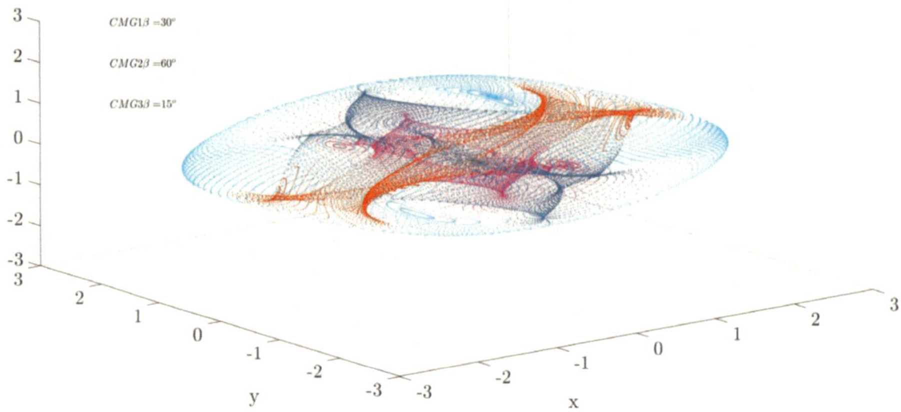

Configuration 4:

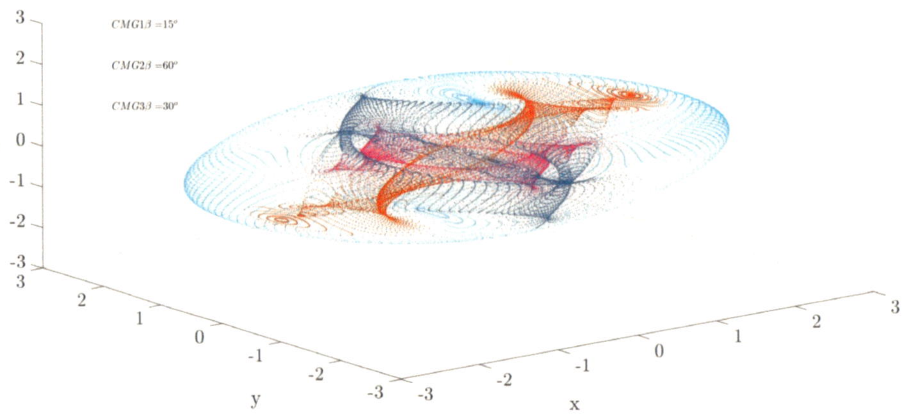

Configuration 5:

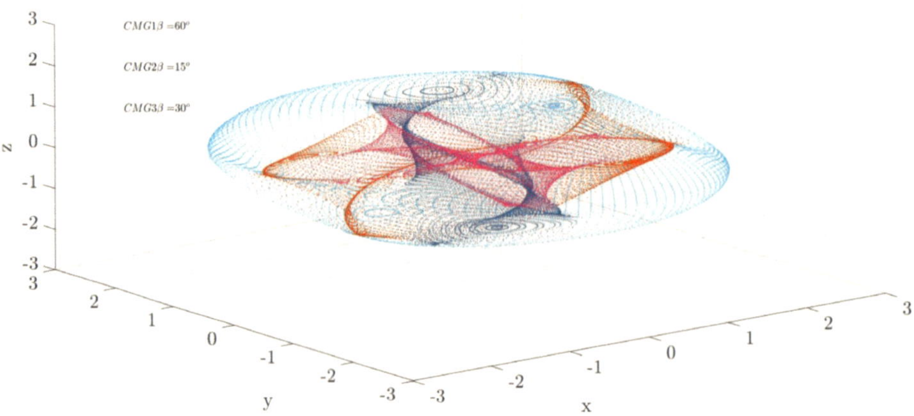

DOI: http://dx.doi.org/10.5772/intechopen.88087

Configuration 6:

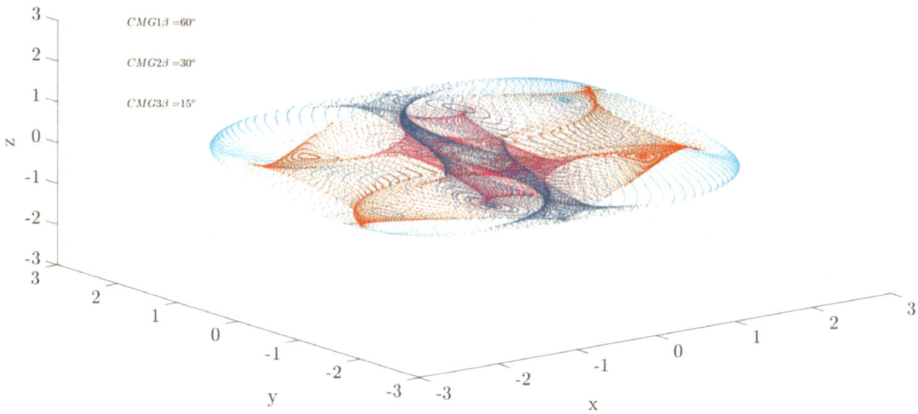

Series 2:
Configuration 1:

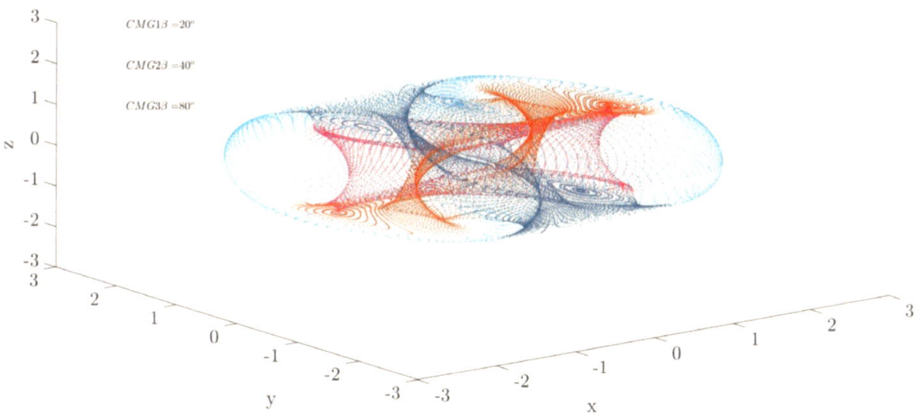

Configuration 2:

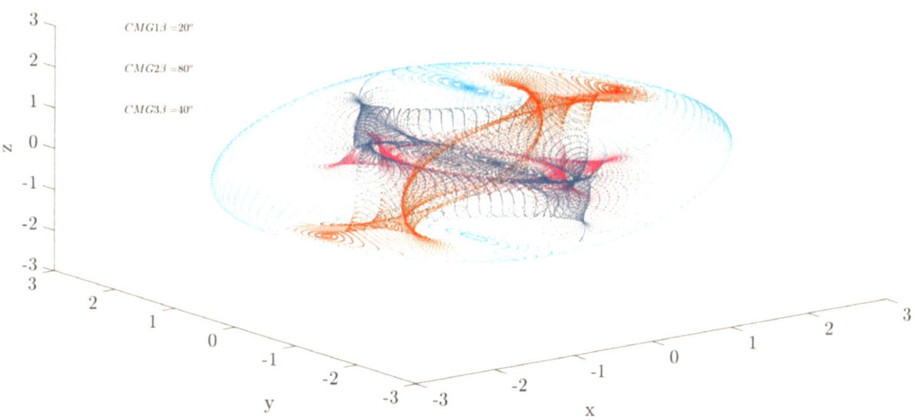

Configuration 3:

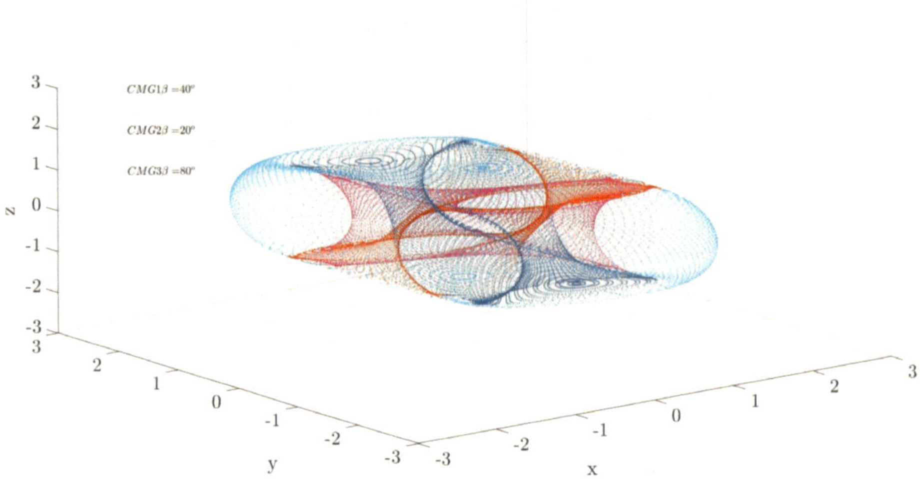

Configuration 4:

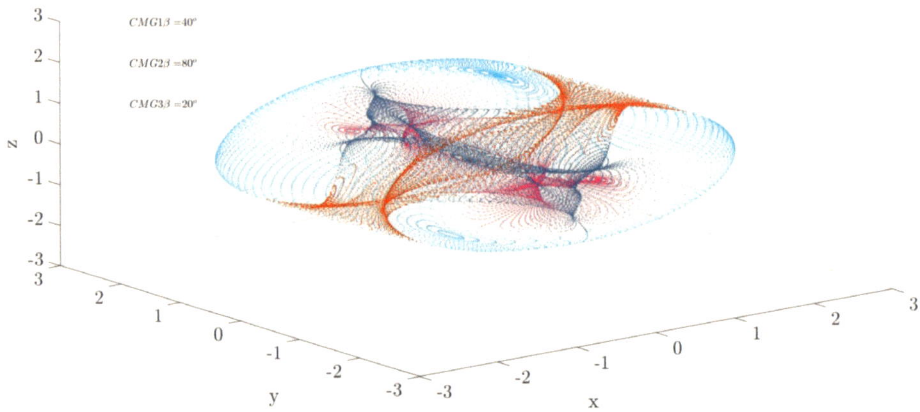

Configuration 5:

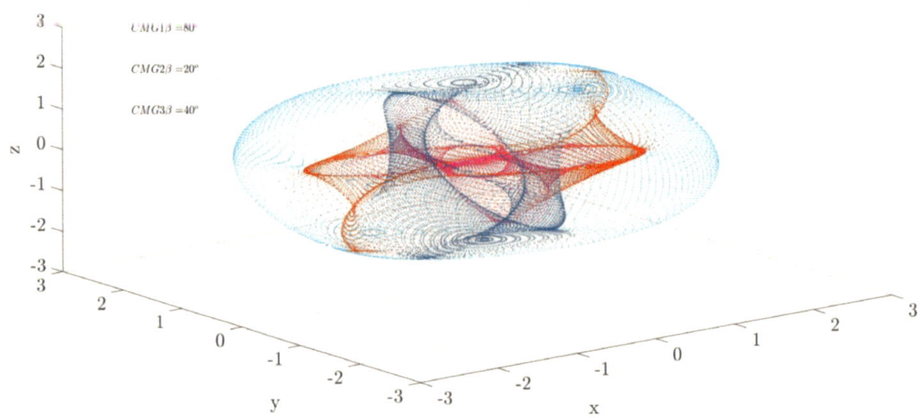

Configuration 6:

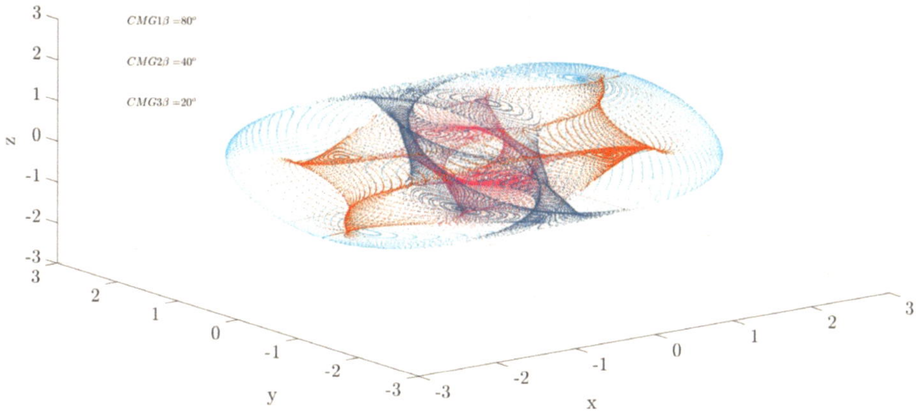

Series 3:
Configuration 1:

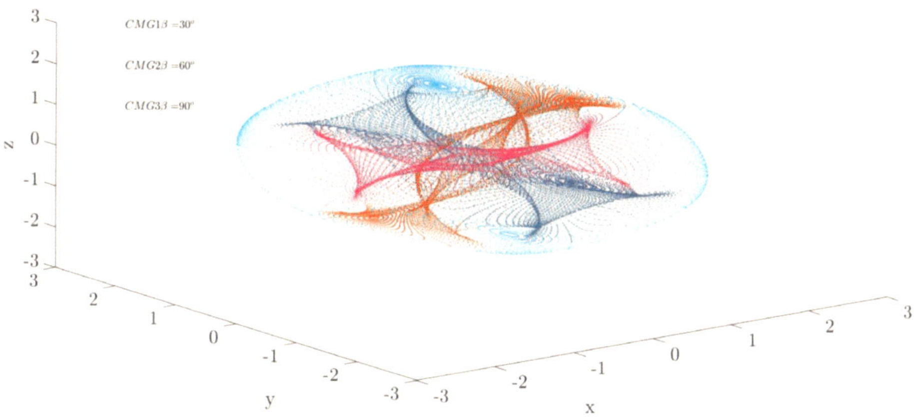

Configuration 2:

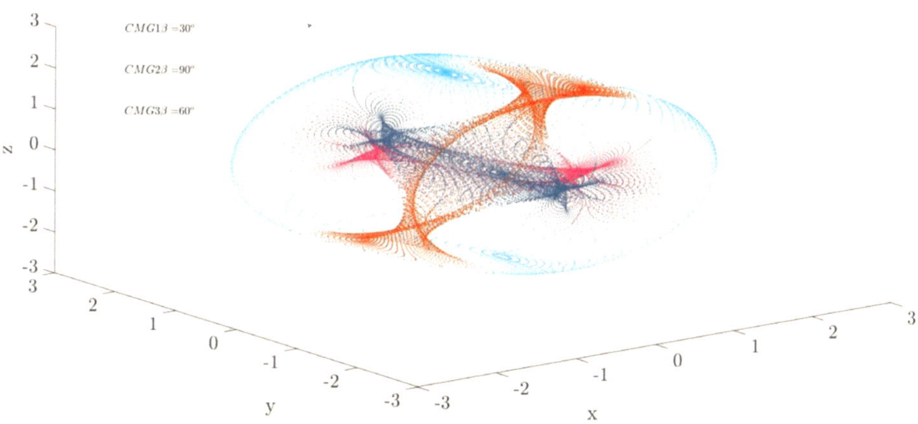

Configuration 3:

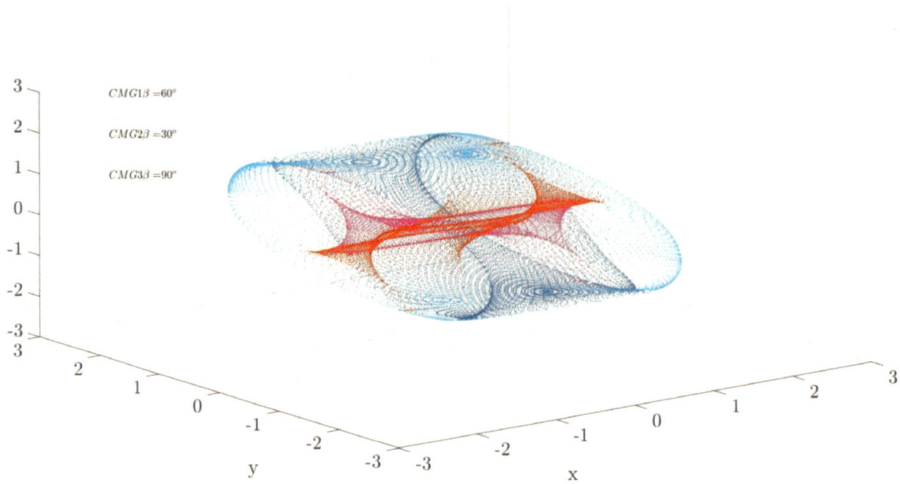

Configuration 4:

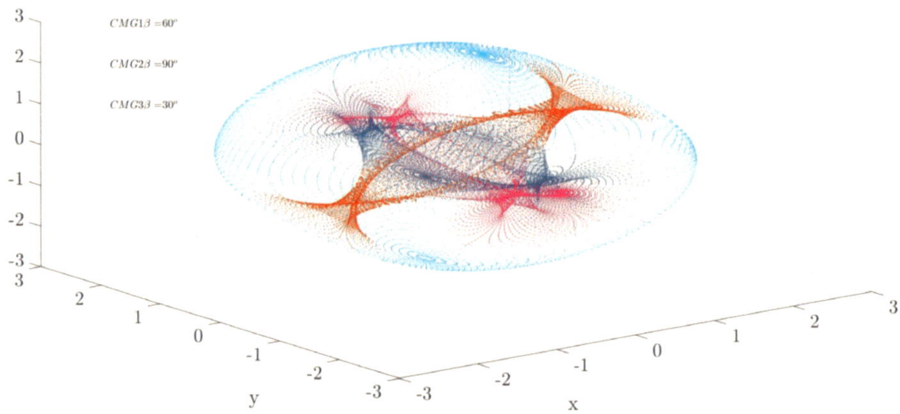

Configuration 5:

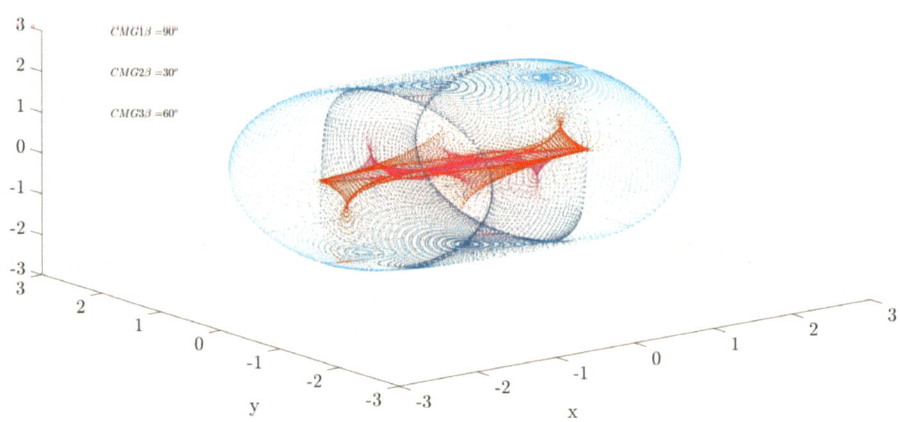

Configuration 6:

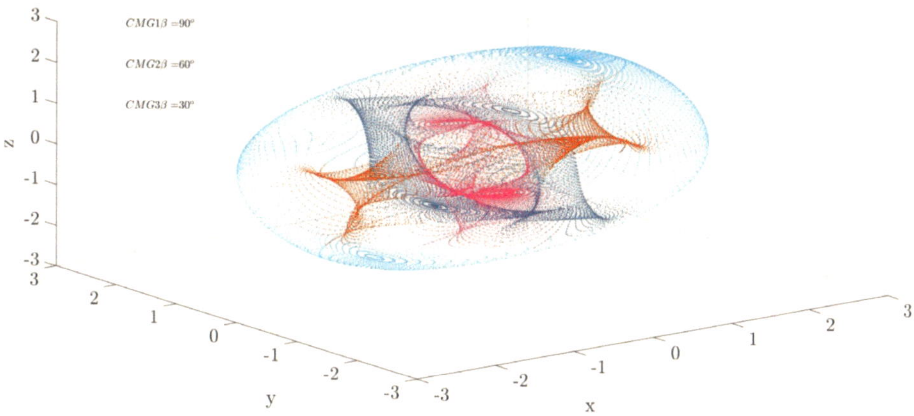

Series 4:
Configuration 1:

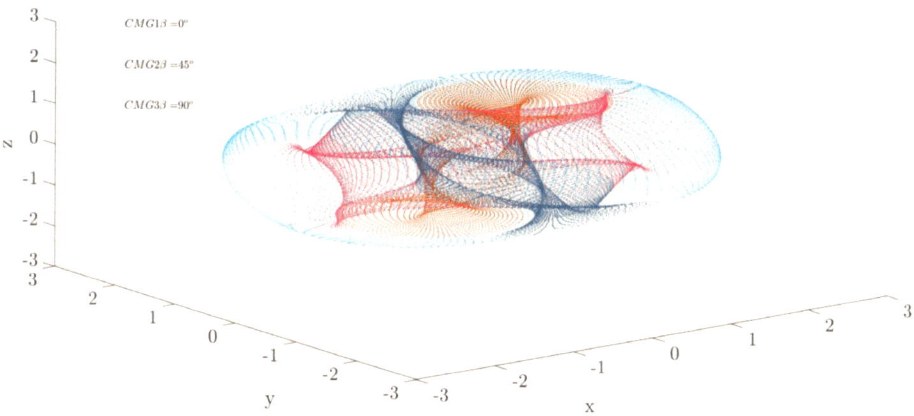

Configuration 2:

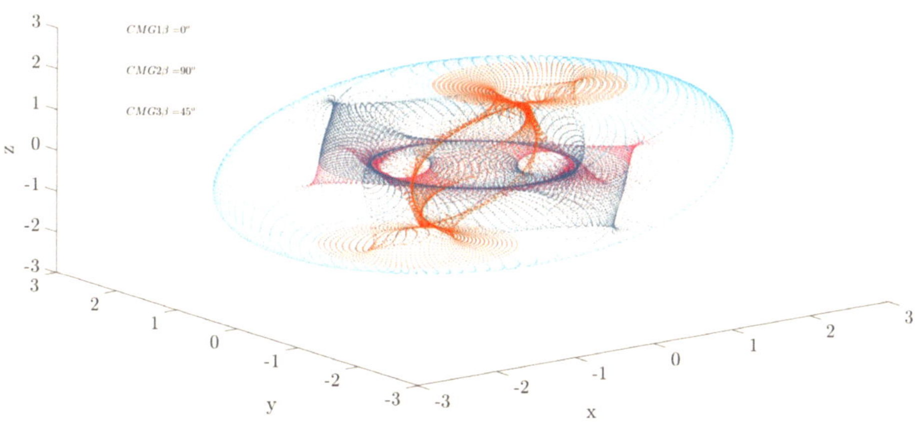

Configuration 3:

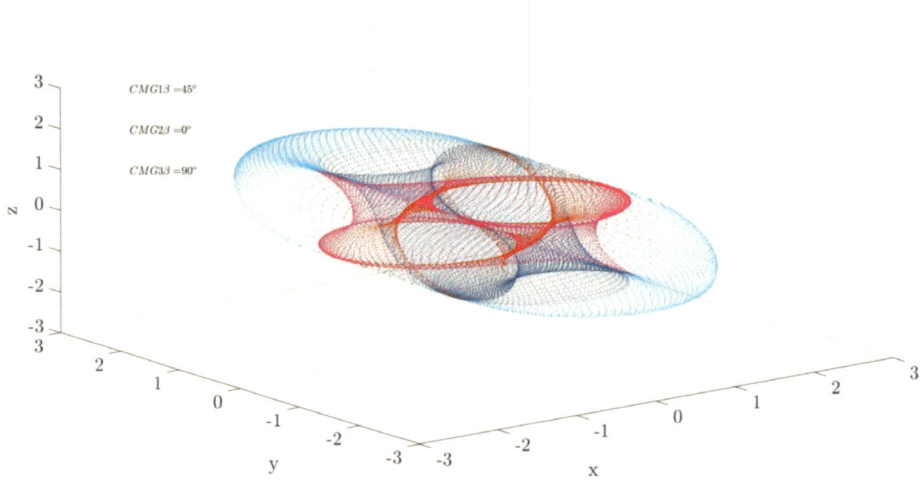

Configuration 4:

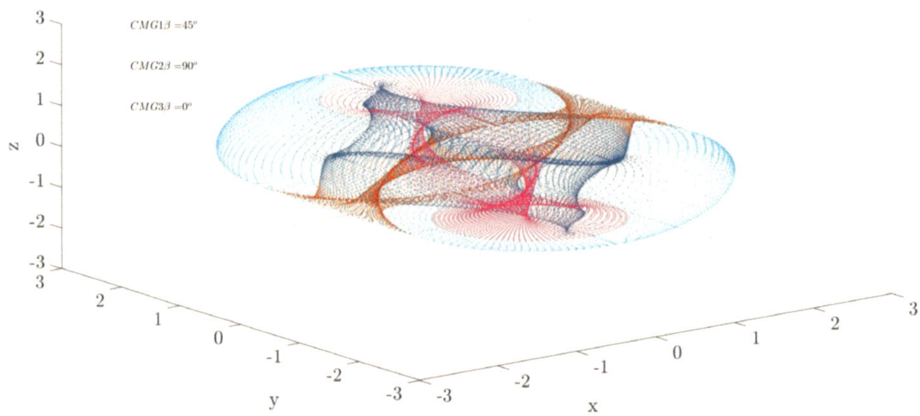

Configuration 5:

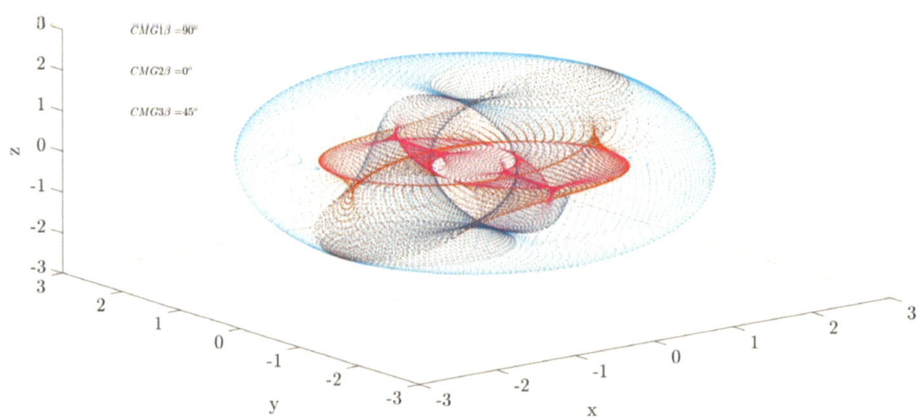

Configuration 6:

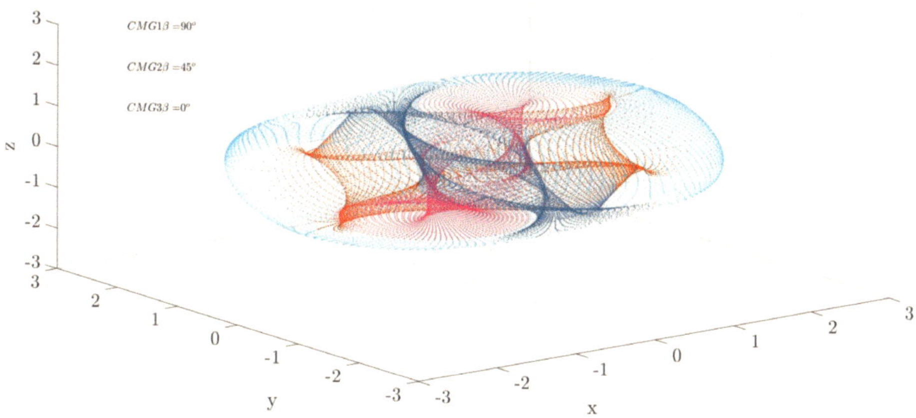

Series 5:
Configuration 1:

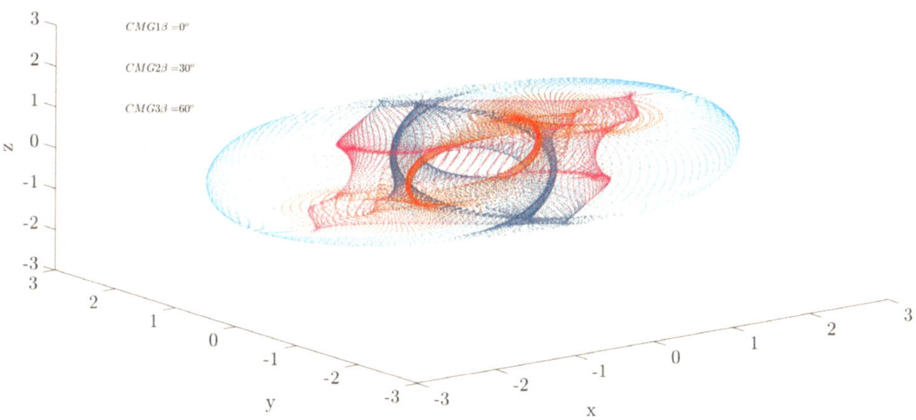

Configuration 2:

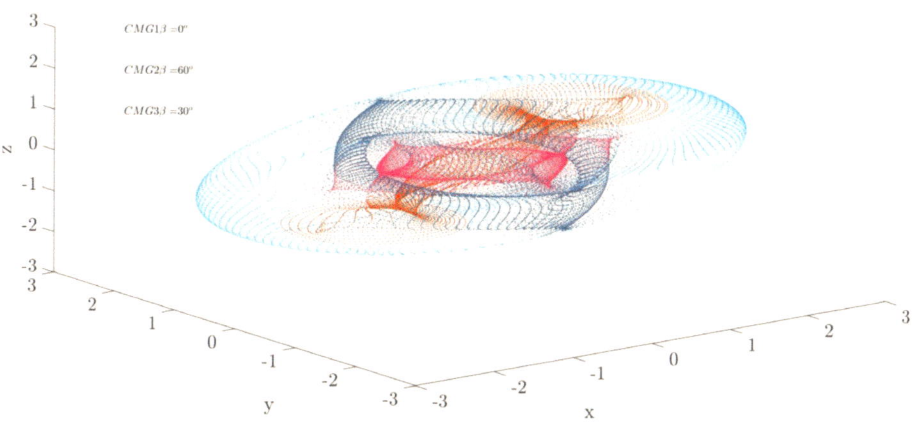

Configuration 3:

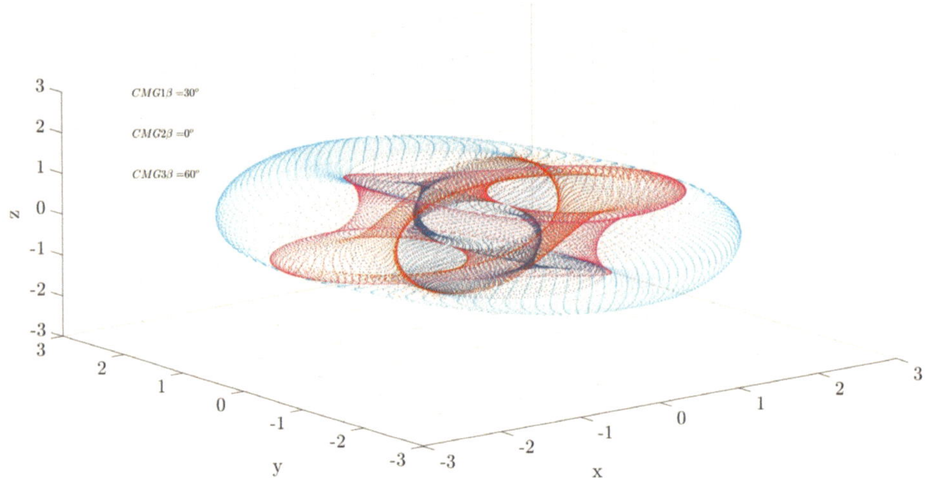

Configuration 4:

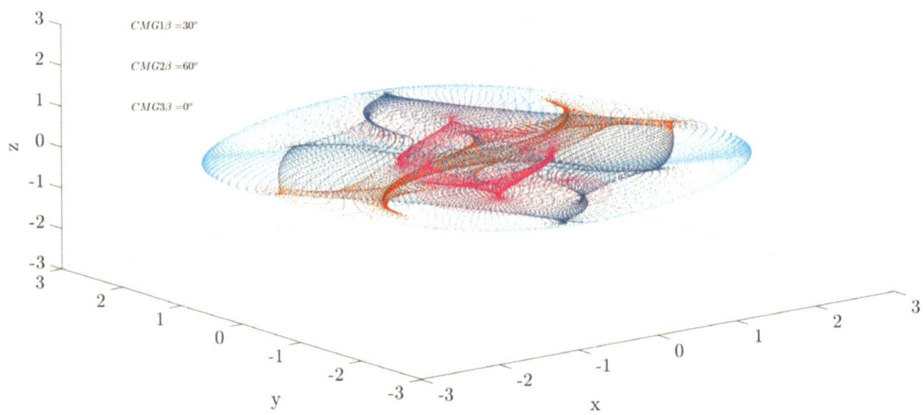

Configuration 5:

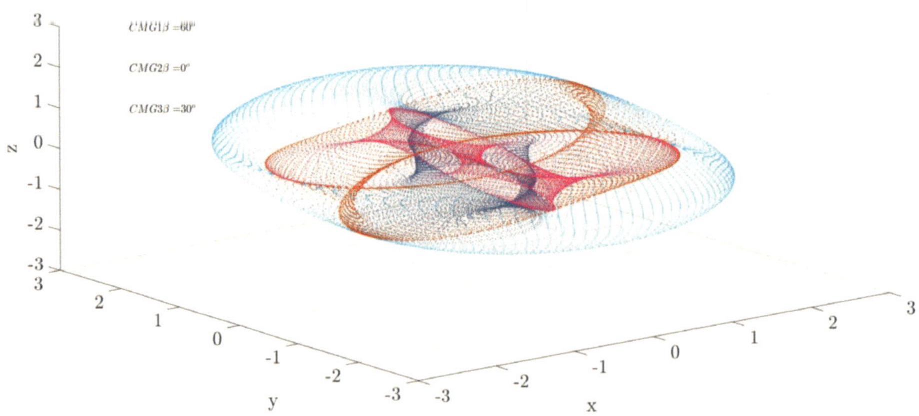

Configuration 6:

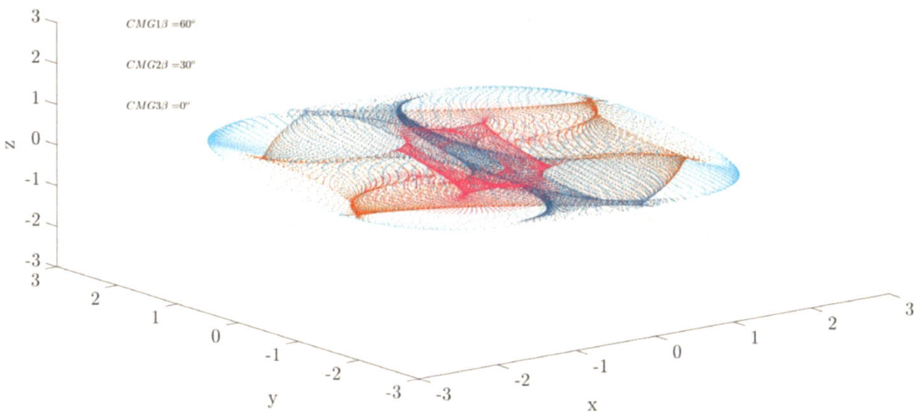

Author details

Jonathan W. Lang
Naval Post Graduate School, United States

*Address all correspondence to: jwlang@nps.edu

IntechOpen

© 2020 The Author(s). Licensee IntechOpen. Distributed under the terms of the Creative Commons Attribution - NonCommercial 4.0 License (https://creativecommons.org/licenses/by-nc/4.0/), which permits use, distribution and reproduction for non-commercial purposes, provided the original is properly cited.

References

[1] Lang J. CMG singularities in a 3/4 configuration. AE3818 Spacecraft Attitude, Determination, and Control Report #1; 2019

[2] Lang J. Singularities in a 3/4 configuration. AE3818 Spacecraft Attitude, Determination, and Control Report #2; 2019

[3] Lang J. Increased CMG capability using singular penetration logic. AE3818 Spacecraft Attitude, Determination, and Control Report #3; 2019

[4] Lang J. Spacecraft kinematics in a 3-2-1 rotation. AE3815 Spacecraft Rotational Mechanics Report #2; 2019

[5] Lang J. Tuning for PID and PDI controllers. AE3818 Spacecraft Attitude, Determination, and Control Report #4; 2019

Chapter 11

Control Moment Gyroscope Skew Angle Variation and Singularity Penetration

Kyle A. Baker

Abstract

This manuscript contains a brief introduction of Control Moment Gyroscopes followed by the mathematical basis for potential singularities and an analysis of how skew angle variability may impact their occurrence. MATLAB is utilized as the primary modeling tool along with WolframAlpha for mathematical derivations of matrix determinants. The results of the modeling efforts show that a uniform skew angle of 90° allows a high maximum angular momentum. Additionally, we attempt to show that having two CMGs at a skew angle of zero could result in similar gains as a uniform 90° configuration and briefly introduce singularity penetration.

Keywords: mixed skew angle, singularity penetration, control moment gyroscope, steering law

1. Introduction

The use of non-tumbling spacecraft, of either the spin stabilized or 3-axis stabilized variety, imposes a requirement for proper attitude control. Control Moment Gyroscopes (CMGs) are pervasive in fulfillment of this requirement not only because of the higher torque profile afforded over a reaction wheels but also due to greater power efficiency gained through their use [1]. These moment exchange devices utilize the principle of conservation of momentum to bypass the fuel and weight limitations of thrusters while still being able to provide a fairly significant slew rate for spacecraft attitude control. This level of performance does not come without its own share of drawbacks: momentum saturation, bearing decay, and most notably singularities all impose operational constraints [2].

CMG steering was implemented into a Simulink model as shown in **Figure 1** in order to analyze these singularities. This model also contained sinusoidal trajectory generation, system dynamics with feedforward control, and a combination of quaternions and direction cosine matrices for the rotational kinematics. The simulations allowed us to observe that varying CMG geometric configuration could have a significant impact on singularities occurrences. With this knowledge in hand one can reason that a spacecraft designer can and should take into account CMG skew variations in order to optimize maneuvering for particular applications. For instance, a communication satellite in geosynchronous orbit and a low earth orbit imaging satellite could end up with very different attitude, pointing, and slew rate problems to solve.

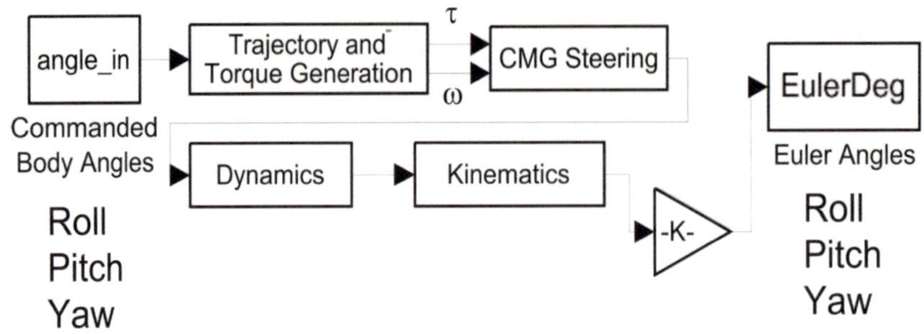

Figure 1.
Simulink model topology.

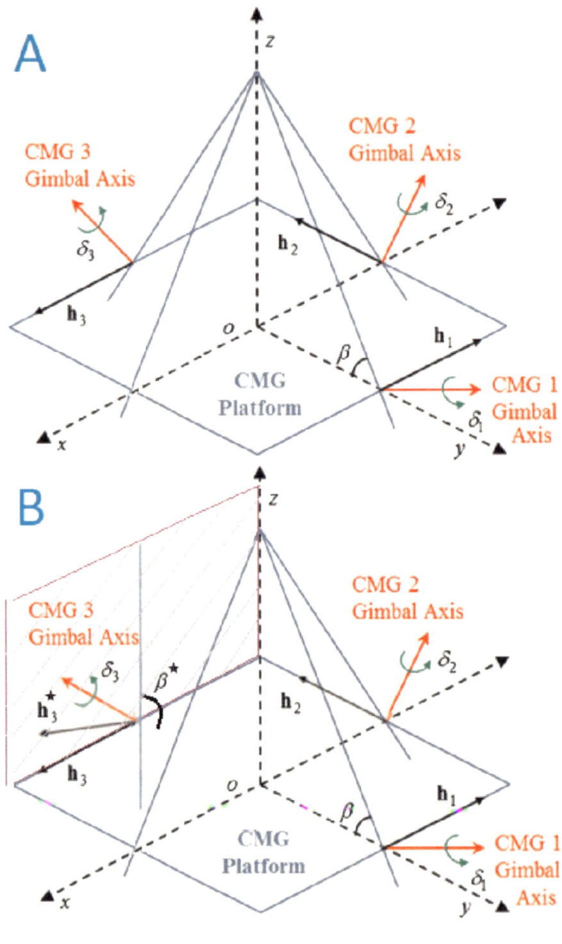

Figure 2.
(A) and (B) CMG arrangement.

A visual representation of CMG placement can be seen in the two depictions of **Figure 2**, which were adapted from drawings in Ref. [3].

Figure 2(A) shows a typical non-redundant skewed pyramid design with a uniform skew angle (β). While typical space applications maintain constant skew angles amongst all the CMGs, it is possible to implement mixed a slew angle design as demonstrated in **Figure 2(B)** where CMG #3 has a different skew angle (a roughly 90° skew angle is shown). These figures also depict another critical variable, gimbal angle. Gimbal angle, denoted by δ in **Figure 2** and θ in the equations of section II, is used to show the rotation that the CMG undertakes in order to impart a torque on the space vehicle. When CMGs are gimbaled, the change in their position alters momentum vector direction (h_1, h_2, and h_3). This in turn changes the torque vector which is perpendicular to both the gimbal axis and the momentum vector. For instance, CMG #3 in depiction (B) will have torque in the shaded plane and it is when a torque is required outside of that plane that a singularity may occur.

These fundamental ideas form the basis for CMG operation and the results shown later in Section III through Section V. We will attempt to show that mixed skew angle CMG configurations may provide some flexibility when dealing with singularities and briefly illustrate some applications and limitations for singularity penetration.

2. Mathematical background

Before going into the results of skew angle variation, let us first set the stage for our future analysis with mathematical derivations used to form the basis for CMG steering logic. For more detailed background information, consult Refs [4,5] for an in depth treatment of kinematics and trajectory generation and Ref. [6] for a complete overview of material found in this section. In order to translate a commanded spacecraft movement into desired torque and CMG gimbal angles rates, we derived the mathematical basis for momentum vectors (h_x, h_y, and h_z) from **Figure 2(A)** and **(B)** as shown in Eq. (1) through Eq. (3).

$$h_x = -\cos\theta_1 + cos\beta sin\theta_2 + cos\,\theta_3 \tag{1}$$

$$h_y = -cos\beta sin\theta_1 - \cos\theta_2 + cos\beta cos\theta_3 \tag{2}$$

$$h_z = cos\beta sin\theta_1 + cos\beta sin\theta_2 + cos\beta sin\theta_3 \tag{3}$$

These momentum vectors were then used to obtain a matrix representing the spatial gradient of torque ($\frac{\partial H}{\partial \theta}$) by taking the partial derivative of each component with respect to gimbal angle. This derived matrix is referred to as the [A] matrix and is shown below in Eq. (4). The rows correspond to h_x, h_y, and h_z components and the columns are the partial derivative of the momentum components with respect to each gimbal angle.

$$H = \frac{\partial H}{\partial \theta} = \begin{bmatrix} sin\,\theta_1 & cos\beta cos\theta_2 & -sin\,\theta_3 \\ -cos\beta cos\theta_1 & sin\,\theta_2 & cos\beta cos\theta_3 \\ sin\beta cos\theta_1 & sin\beta cos\theta_2 & sin\beta cos\theta_3 \end{bmatrix} \tag{4}$$

Torque was found via a change in angular momentum as shown in Eq. (5) and finally a desired gimbal rate to produce this torque was found via Eq. (6).

$$\dot{H} = \frac{\partial H}{\partial \theta}\frac{\partial \theta}{\partial t} = [A]\dot{\theta} \tag{5}$$

$$[A]^{-1}\dot{H} = [A]^{-1}[A]\dot{\theta} = \dot{\theta} \tag{6}$$

Upon closer analysis we can see that there is potential for the $[A]^{-1}$ matrix in Eq. (6) to become singular when applying the matrix inverse per Eq. (7) [7].

$$[A]^{-1} = \frac{1}{det[A]}[Cofactors\ of\ [A]\ matrix] \tag{7}$$

If the determinant of the [A] matrix becomes zero, then we have a situation where $\frac{1}{det[A]}$ is equal to $\frac{1}{0}$ leading to failures in the CMG steering logic. Section III will dig deeper into this singularity analysis.

3. Singularity analysis

This section reemphasizes and builds on work presented in Refs [8, 9]. To begin to analyze when CMG Singularities occur, we must first look into the factors which drive their occurrence. Specifically, we know that when the determinant of the [A] matrix is equal to zero it means that at least one or a combination of multiple CMGs are incapable of producing the required torque. We turned to the WolframAlpha website to solve for the determinant of the [A] matrix from Eq. (4) and produce Eq. (8) [10].

$$\begin{aligned} \det[A] = {} & \sin(\beta)\{2\,\mathbf{cos^2(\beta)}\,\mathbf{cos\,\theta_2}\cos\theta_1\cos\theta_3 - \mathbf{cos\,(\beta)}\,\mathbf{cos\,\theta_2}\sin\theta_1\cos\theta_3 \\ & + \mathbf{cos\,(\beta)}\,\mathbf{cos\,\theta_2}\cos\theta_1\sin\theta_3 + \sin\theta_1\,\mathbf{sin\,\theta_2}\cos\theta_3 + \cos\theta_1\,\mathbf{sin\,\theta_2}\sin\theta_3\} \end{aligned} \tag{8}$$

Combining like terms allows us to factor out a $\cos(\beta)\cos\theta_2$term from the first three lines and apply Eq. (9) for terms with θ_1 and θ_3 . Likewise we could factor a $\sin\theta_2$ term out of the last two lines and apply Eq. (10) to θ_1 and θ_3.

$$\sin(A - B) = \text{cosBsinA} - \text{sinAcosB} \tag{9}$$

$$\sin(A + B) = \text{cosBsinA} + \text{sinAcosB} \tag{10}$$

The final result is Eq. (11) shown below where we have abbreviated sine and cosine functions as “s” and “c” respectively.

$$\det[A] = \sin(\beta)\{s\theta_2[s(\theta_1 + \theta_3))] + c(\beta)c\theta_2[s(\theta_3 - \theta_1) + 2c\theta_1 s\theta_3 c\beta]\} \tag{11}$$

Analyzing Eq. (11) leads us to several cases where singularities can occur as shown in Eqs. (12)–(17). A bolded strike-through represent a condition where the altered terms are equal to zero in order to make the right side of Eq. (11) zero (i.e., singularity conditions).

$$\mathbf{s(\beta)}\{s\theta_2[s(\theta_1 + \theta_3) + c(\beta)c\theta_2[s(\theta_3 - \theta_1) + 2c\theta_1 s\theta_3 c\beta]\} \tag{12}$$

$$s(\beta)\{s\theta_2[\mathbf{s(\theta_1 + \theta_3)} + c(\beta)c\theta_2[\mathbf{s(\theta_3 - \theta_1) + 2c\theta_1 s\theta_3 c\beta}]\} \tag{13}$$

$$s(\beta)\{s\theta_2[s(\theta_1 + \theta_3) + c(\beta)c\theta_2[s(\theta_3 - \theta_1) + 2c\theta_1 s\theta_3 c\beta]\} \tag{14}$$

$$s(\beta)\{\mathbf{s\theta_2}[s(\theta_1 + \theta_3) + \mathbf{c(\beta)c\theta_2}[s(\theta_3 - \theta_1) + 2c\theta_1 s\theta_3 c\beta]\} \tag{15}$$

$$s(\beta)\{s\theta_2[\mathbf{s(\theta_1 + \theta_3)} + \mathbf{c(\beta)c\theta_2}[s(\theta_3 - \theta_1) + 2c\theta_1 s\theta_3 c\beta]\} \tag{16}$$

$$s(\beta)\{\mathbf{s\theta_2}[s(\theta_1 + \theta_3) + c(\beta)c\theta_2[\mathbf{s(\theta_3 - \theta_1) + 2c\theta_1 s\theta_3 c\beta}]\} \tag{17}$$

Note that some of these cases only drive singularities under limited ranges. For instance, Eq. (12) only drives the determinant to zero when sine β is equal to zero; i.e., when β = 0° (i.e., at a single point skew variation). Additionally, Eq. (16) is only

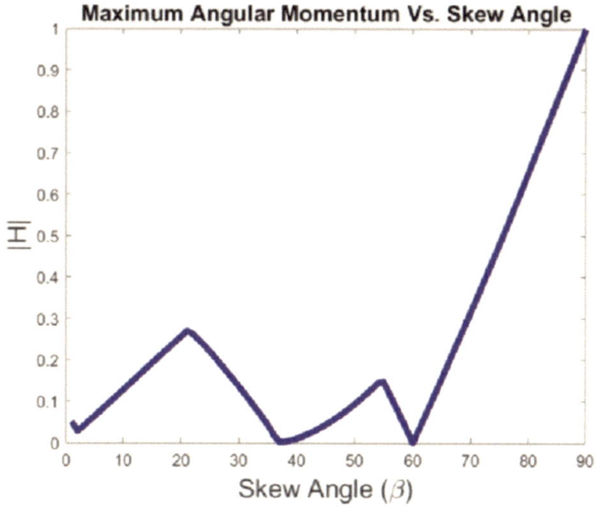

Figure 3.
Maximum angular momentum achieved before encountering singularity (uniform skew angle).

zero when either cosine β or cosine θ_2 is equal to zero; i.e., when either β or θ_2 is equal to 90°.

When the collective effects of Eqs. (12)–(17) are taken into account we obtain **Figure 3** which shows us the maximum magnitude of angular momentum that we can achieve for a given skew angle without running into a singularity. This plot was formed from a discretization of roughly 1 million points for curve smoothing purposes.

Here we found a maximum allowable Momentum which corresponds to $\beta = 90°$. At this point you can achieve one CMG worth of angular momentum without a possibility of singularity occurrence.

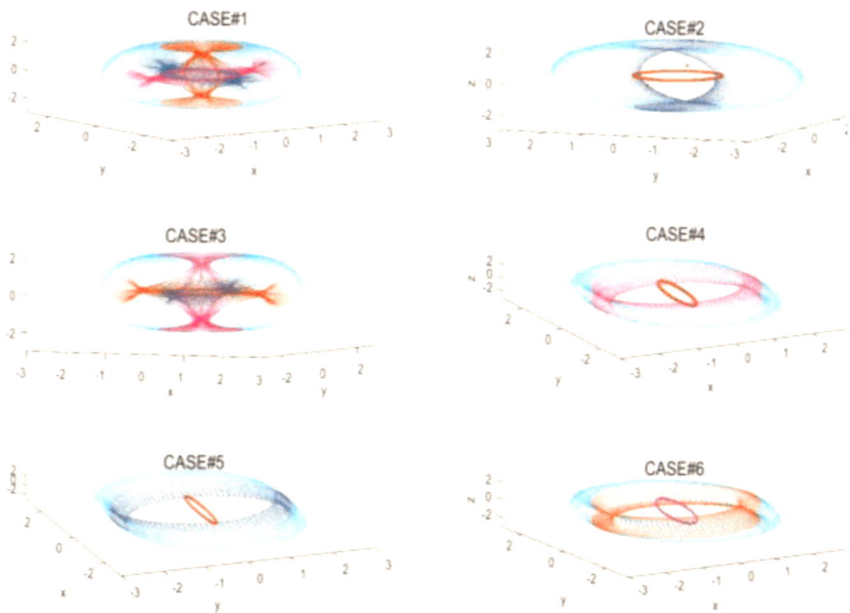

Figure 4.
Singularity surfaces by skew angle.

	CMG#1 skew	CMG#2 skew	CMG#3 skew	Preferred direction
Case 1	0°	90°	90°	N/A
Case 2	90°	0°	90°	Z
Case 3	90°	90°	0°	N/A
Case 4	0°	0°	90°	X
Case 5	0°	90°	0°	Y
Case 6	90°	0°	0°	X

**Skew angle β is shown in degrees.*

Table 1.
Mixed skew angle test cases and preferred direction.

Note that **Figure 3** corresponds to **Figure 2(A)** where all three CMGs have the same skew angle. This graph indicates that when a uniform skew angle of 90° is used, called a box or roof configuration, the spacecraft can exert the highest torque values without experiencing a singularity. Similar results were found in [8] where the researchers indicated that this uniform skew angle of 90° could provide "far superior performance than the commonplace 4 CMG pyramid skewed at 54.73".

Note that it is also possible to operate in a condition similar to **Figure 2(B)** where the CMGs can have non-uniform beta angles. The results of mixed skew angle testing are shown in **Figure 4** with its conditions tabulated in **Table 1**.

From **Figure 4**, you can see that some mixed skew conditions lend themselves to certain applications. For instance, Case 2 allows a similar amount of maximum angular momentum in the X and Y directions but a much greater amount in the Z direction. Both Case 4 and 6 allow more movement in the X direction while Case 5 allows more freedom in the Y direction. Interestingly, note that Case 1 and 3 appear to have no clear advantage like the other four cases. In fact, the only case with two 90° skew angles that has a clear preferred direction is Case 2 while the cases with two 0° skew angles always seem to have a preferred direction. This conclusion is backed up by further analysis conducted in Section IV.

4. Further mixed skew insight

In Section III, we saw that some cases of mixed skew angle CMG configurations could have drastic benefits when moving in a particular direction. This begs the question, what is the maximum angular momentum value that can be achieved in a mixed skew configuration and is there a specific configuration that solves all singularity problems. In order to investigate this we attempted to recreate **Figure 3** but instead of altering a single, uniform skew angle like **Figure 2(A)** we iterated through possible skew angle combinations by increments of five. This involved running a set of about 7000 combinations of skew angle through our MATLAB model. The model analyzed the impact of skew angles on the maximum angular momentum that could be achieved without encountering a singularity. The results of this are plotted in **Figure 5**.

Note that once again the maximum amplitude is one CMG worth of angular momentum but also that there is an almost noticeable pattern to the results. In the results of **Figure 5**, we can see that iterations 201, 301, 401, 501, 601, 701, and 801 all cause spikes in the maximum allowable torque. The skew angle combinations responsible for these spikes can be seen in **Table 2**.

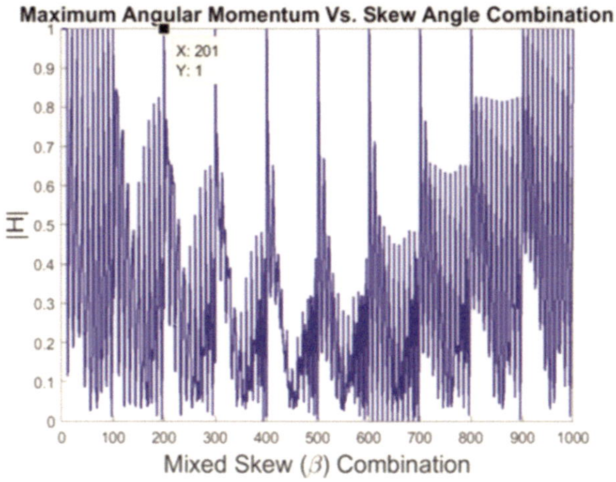

Figure 5.
Maximum angular momentum achieved before encountering singularity (mixed skew angle).

Iteration	CMG#1 skew	CMG#2 skew	CMG#3 skew
201	20	0	0
301	30	0	0
401	40	0	0
501	50	0	0
601	60	0	0
701	70	0	0
801	80	0	0

Table 2.
Mixed skew angle test cases and preferred direction.

These results indicate that having two CMGs with a skew angle of zero results in high allowable angular momentum values. It should also be noted that in **Figure 5** there are smaller peaks that almost reach an angular momentum value of 1. For example, **Figure 5** iteration 704 corresponds to a CMG configuration where CMG#1 is 5°, CMG#2 is 90°, CMG#3 is 0°, allowing a maximum angular momentum of about 0.9. Space is at a premium on a spacecraft and this information could prove useful if certain CMG configuration can be made to fit onboard.

5. Singularity penetration

While the singularities themselves do prove to be a limiting factor it must be noted that there are various techniques available to mitigate their effects. For instance, Refs [6, 11] detail methods of singularity avoidance. A newer technique, illustrated in Ref. [12] espouses a system of singularity penetration called Singularity Penetration with Unit Delay (SPUD). The topology of a circuit for this implementation is depicted in **Figure 6**.

This circuit takes in the $\dot{\theta}$ term as seen in Eqs. (5) and (6). These were implemented within the CMG Steering block of **Figure 1**'s topology. In the topology of **Figure 6**, a singular condition is monitored and the input from the Inverse Condition Number block is sent in as a threshold value of the switching circuit.

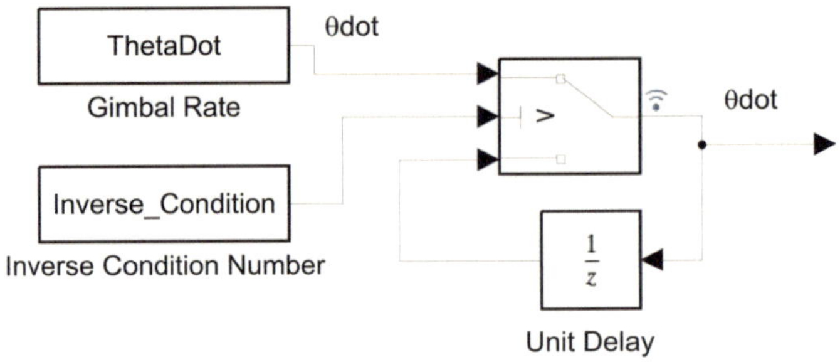

Figure 6.
Singularity penetration with unit delay circuit.

Once a specified threshold is reached the circuit will hold onto the $\dot{\theta}$ value until the singular condition passes. A truly singular condition (e.g., dividing by zero) usually only occurs during one time step, but that one instant is enough to induce extreme errors into a system. We conducted testing under a mildly singular condition where the inverse condition number was only on the order of 1e-6 with the SPUD circuit engaging when it detected the inverse condition number was below 1e-4. The results were not compelling enough to illustrate here due to not being sufficiently singular. However, when a highly singular condition is encountered, SPUD can demonstrate drastic improvements. The results of Ref. [12], tested under a highly singular condition, show a decrease in average Euler Angle tracking error as shown in **Table 3**.

While SPUD provides sizable gains it should be noted that there are factors that must be taken into account when it is utilized. The singular conditions shown in

	Mean error magnitude		
System state	**X-axis**	**Y-axis**	**Z-axis**
SPUD inactive	4.4321	5.1939	4.3712
SPUD active	0.0674	0.1063	0.4682

Table 3.
Euler tracking error reduction.

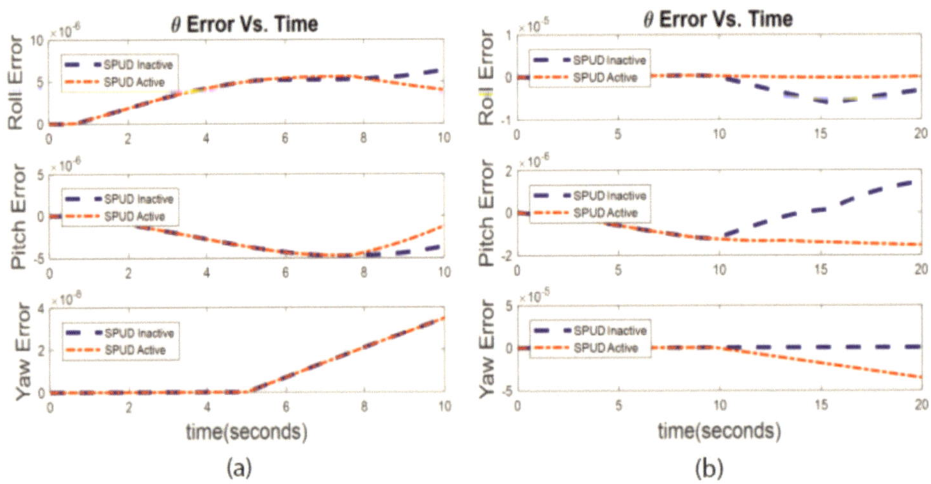

Figure 7.
(A) and (B) SPUD results with variable β.

Eq. (12), are triggered when β is equal to 90°. The circuit of **Figure 1** was implemented with a step size of 0.0001 and a Runge-Kuta Solver with a variable β input and a yaw maneuver of around 60°. The β was implemented with a slope circuit that ran from −90° to +90°. When the slope circuit crossed the 0° mark a minimum inverse condition number on the order of 1e-20 was reached; well below machine epsilon. The results of this simulation are shown in **Figure 7**. SPUD was activated at a threshold of 5e-6 in both cases but the maneuver time was implemented as 10 seconds for **(A)** and 20 seconds for **(B)**.

In depiction (A) SPUD reduced average error by about 1e-3 but in depiction (B) SPUD actually induced more error in the yaw channel (i.e., the maneuvering axis) by over five times. Results where SPUD reduced error in the auxiliary, non-maneuvering roll and pitch axes while increasing error in the maneuver axis of yaw were common when SPUD was utilized with a sliding β angle. The sliding β input forces the SPUD circuit into a prolonged singular condition under which it is less effective. The reason for this is that SPUD is designed to penetrate a singular point encountered in a guidance system instead of operate around a singular point itself. Tools are effective when utilized for their specific purpose and usually less effective when implemented improperly; SPUD is no exception.

6. Future efforts

While it was found that having two CMGs with the same skew angle could help achieve a higher allowable torque, there is still much work to be done. Future efforts will look at a deeper resolution (skew angle increments less than 5 degrees) and try to make mathematical sense of why the results are as seen in **Figure 5**. Additionally, while singularity penetration has proven to be a highly useful alternative to singularity avoidance, further testing should be conducted to analyze when it should be activated. The ideal value of inverse condition number under which SPUD should be activated or ideal rate of change of the condition number are primed for investigation. The key is finding the correct way to identify precursor conditions for an acute singularity occurrence.

Author details

Kyle A. Baker
Department of Mechanical and Aerospace Engineering, Naval Postgraduate School, Monterey, CA, USA

*Address all correspondence to: baker.kas@gmail.com

IntechOpen

© 2020 The Author(s). Licensee IntechOpen. Distributed under the terms of the Creative Commons Attribution - NonCommercial 4.0 License (https://creativecommons.org/licenses/by-nc/4.0/), which permits use, distribution and reproduction for non-commercial purposes, provided the original is properly cited. (cc) BY-NC

Referenes

[1] Votel R, Sinclair D. Comparison of control moment gyroscopes and reaction wheels for earth observing satellites. In: 26th Annual AIAA Conference; 2012

[2] Control Moment Gyroscope. Wikipedia, Wikimedia Foundation. 2018. Available from: https://en.wikipedia.org/wiki/Control_moment_gyroscope

[3] Hao T, Matunaga S. New sliding mode control approach for rapid attitude maneuver using control moment Gyros. Journal of Aerospace Engineering. 2016;**29**(2)

[4] Smeresky B, Rizzo A, Sands T. Kinematics in the information age. Mathematics. 2018;**6**(9):148

[5] Baker K, Cooper M, Heidlauf P, Sands T. Autonomous trajectory generation for deterministic artificial intelligence. Electrical and Electronic Engineering. 2018;**8**(3):59-68

[6] Wie B. Space Vehicle Dynamics and Control. 1st ed. Reston, AIAA: McGraw-Hill; 1998. pp. 310-327

[7] Greenberg M. Advanced Engineering Mathematics. 3rd ed. New Jersey: Prentice Hall; 1998. pp. 508-511

[8] Sands TA, Kim JJ, Agrawal B. 2h singularity-free momentum generation with non-redundant single gimbaled control moment gyroscopes. In: Proceedings of the 45th IEEE Conference on Decision and Control; 2006. pp. 1551-1556

[9] Sands T, Kim JJ, Agrawal BN. Nonredundant single-gimbaled control moment gyroscopes. Journal of Guidance, Control, and Dynamics. 2012; **35**(2):578-587

[10] "Wolfram|Alpha Widgets: "3x3 Determinant Calculator" - Free Mathematics Widget", Wolframalpha.com [Online]. 2019. Available from: https://www.wolframalpha.com/widgets/view.jsp?id=7fcb0a2c0f0f41d9f4454ac2d8ed7ad6

[11] Agrawal BN, Kim JJ, Sands TA. Method and apparatus for singularity avoidance for control moment gyroscope (CMG) systems without using null motion. US Patent 9,567,112; 2017

[12] Sands T, Kim J, Agrawal B. Singularity penetration with unit delay (SPUD). Mathematics. 2018;**6**(2):23

Chapter 12

Single Axis Singularity Mapping for Mixed Skew Angle, Non-Redundant, Single Gimbaled CMG Systems

Eryn A. Culton

Abstract

Control moment gyros are common spacecraft attitude control devices that can be mounted at different orientations within a spacecraft. Some spacecraft need to maximize their maneuverability around a particular axis and, therefore, benefit from particular control moment gyro orientations. This report explains the physics of control moment gyros as attitude control devices and defines a mathematical singularity and its physical manifestation in the spacecraft body. The research continues, analyzing the relation between a control moment gyro's skew angle and its effects on angular momentum magnitude leading to a conclusion defining the best control moment gyro orientations to maximize a spacecraft's yaw maneuverability.

Keywords: rotational mechanics, adaptive control, nonlinear control, control moment gyroscope, momentum exchange, singularity, physics-based control, disturbance decoupling

1. Introduction

Mechanical control has developed over centuries [1–22], expanding original theorems such as Chasle's theorems of motion Phoronomics [23]. With increasing strike capability, advancements in spacecraft technology, and rising political tensions all over the globe, mechanical control has resurfaced as an important research front in order to further current technologies. Opposed nations frequently use satellites on orbit to gather critical intelligence on those around them, a mission that requires precise pointing and an extensive and expansive understanding of the mechanical control envelope provided by the spacecraft's attitude control system. Recent research has been conducted in order to increase the maneuverability of spacecraft with control moment gyroscopes [24–32]. This research takes information and lessons learned from these previous research efforts and builds upon them.

Depending on a spacecraft's mission, it will likely execute a particular kind of attitude maneuver many times during its life span. Characteristic attitude maneuvers should be considered when designing an attitude control system. The type and number of attitude control devices as well as their position within the spacecraft are design choices driven by the physical demands of the attitude maneuvers. These maneuvers should be considered in order to design an attitude control system that ensures the most angular momentum can be generated around that favored axis while also providing maneuverability in other directions.

Constant-speed, single-gimbaled control moment gyros (CMGs) are common spacecraft attitude control devices that, like reaction wheels, are momentum exchange devices that operate on the law of conservation of momentum in an undisturbed system. Unlike reaction wheels, CMGs do not change their rotational velocity to alter the spacecraft's attitude but, rather, change their direction. Although this ability allows CMGs to uniquely control spacecraft attitude, it also poses challenges: CMGs can only provide torque in a plane orthogonal to their gimbal axis. When a desired torque orthogonal to this plane is commanded, the CMG encounters a mathematical singularity and attitude control is lost.

The locations of these singularities can be plotted 3-dimensionally in order to gain an understanding of the singularity free angular momentum available to command. These singularity maps change based upon the CMG's skew angle within the spacecraft and can be optimized to maximize the singularity free, angular momentum space about a particular axis.

2. Theory

It is necessary to understand how CMGs are commanded and how they physically affect the spacecraft in order to understand how a mathematical singularity causes a spacecraft to lose control. Like any actuator system, a command is entered and a trajectory is generated to reach the commanded position from the initial position; applied to a CMG, a specific rotation is the command and Eq. (1) through Eq. (3) are the equations used to generate the attitude maneuver trajectory [33].

$$\theta = A\sin(\omega t) \tag{1}$$

$$\omega = A\omega\cos(\omega t) \tag{2}$$

$$\dot{\omega} = -A\omega^2\sin(\omega t) \tag{3}$$

where θ is the gimbal angle, ω is the gimbal rotational velocity, and $\dot{\omega}$ is the gimbal rotational acceleration. To send a command to the CMG actuators, the trajectory is plugged into a feedforward controller that calculates the commanded torque required to set the spacecraft on the created trajectory. The best method of calculating the commanded torque is to use the Newton-Euler equation written in the body frame, represented as Eq. (4).

$$\tau = J\dot{\omega} + \omega \times J\omega \tag{4}$$

The feedforward uses Eq. (5), an adapted version of Eq. (4), to calculate this torque command. Eq. (5) is the nonlinear feedforward control equation based off of the Newton-Euler equation written in the body frame. Since Eq. (5) directly describes the physics of the system, it is the best feedforward control to use.

$$u_{ff} = \hat{J}\dot{\omega}_d + \omega_d \times \hat{J}\omega_d \tag{5}$$

where $\hat{J}$ is the "best guess" spacecraft moment of inertia matrix, $\dot{\omega}_d$ is the desired rotational acceleration, and ω_d is the desired rotation rate. Using this idealized feed forward control eliminates phase lag in the system.

At this point in the system topology, the torque command is converted to a voltage or current and sent directly to the actuators. The actuators move and torque is exerted on the spacecraft as described by Eq. (6).

$$J\dot{\omega} = -J_{CMG}|\dot{\boldsymbol{\omega}}_{CMG}|\hat{\dot{\boldsymbol{\omega}}}_{CMG} \tag{6}$$

where $\dot{\omega}$ is the spacecraft's rotational acceleration, JCMG is the CMG moment of inertia, and $\dot{\omega}_{CMG}$ is the CMG angular acceleration. As the direction of the CMG angular momentum changes, the spacecraft's rotation changes on the other side of Eq. (6). In order to predict how changing the direction of the CMG angular momentum affects the spacecraft, the CMG system orientation must be understood and the angular momentum vectors must be resolved into the three body axes.

For analysis purposes, a simplified, non-redundant, single gimbaled CMG system will be used. This system will consist of three CMG's as pictured in **Figure 1**. To note, the CMG skew angle is defined as the angle between a vertical line parallel to the Z axis at each CMG location and the Z axis; in other words, the gimbal axis would be pointing out from the spacecraft in the x-y plane when β = 0° or would be pointing straight up when β = 90°. In **Figure 1**, β is annotated at its equivalent angle. Also, each angular momentum vector is drawn at its initial position, θ = 0°.

Figure 1 provides a visual aid in generating a set of three equations that resolve the angular momentum of each CMG into the x, y, and z axes. These equations are described in Eqs. (7), (8), and (9).

$$\boldsymbol{h}_x = (\cos\theta_3 - \cos\theta_1 + \sin\beta_2 \sin\theta_2)|\boldsymbol{H}| \tag{7}$$

$$\boldsymbol{h}_y = (\sin\beta_3 \sin\theta_3 - \sin\beta_1 \sin\theta_1 - \cos\theta_2)|\boldsymbol{H}| \tag{8}$$

$$\boldsymbol{h}_z = (\cos\beta_1 \sin\theta_1 + \cos\beta_2 \sin\theta_2 + \cos\beta_3 \sin\theta_3)|\boldsymbol{H}| \tag{9}$$

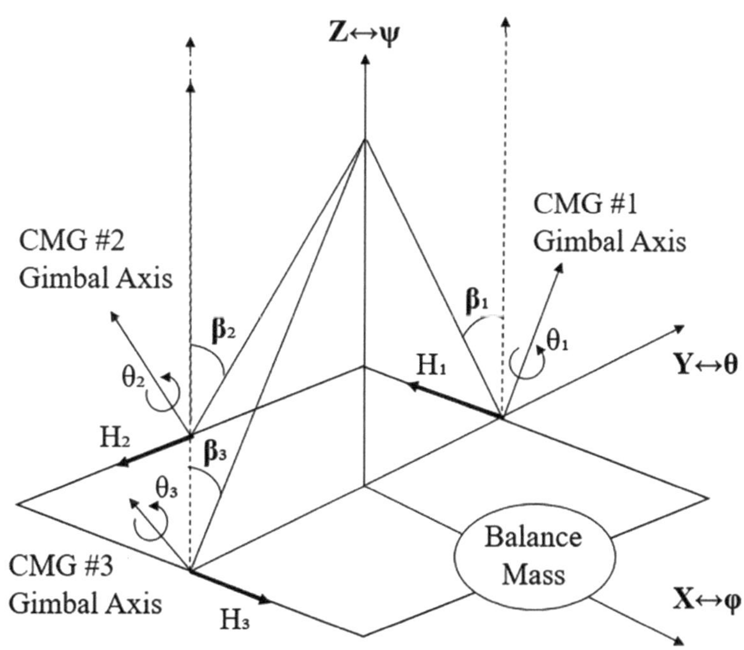

Figure 1.
¾ CMG system [34].

where h is angular momentum about a particular axis, β is the skew angle of each CMG, θ is the angle the momentum vector has rotated about the CMG gimbal axis, and H is the maximum angular momentum a single CMG can produce.

The desired torque given from the system described in Eqs. (7), (8), and (9) can be written as Eq. (10), where the desired torque is equal to the partial derivative of angular momentum with respect to the gimbal angle multiplied by the time derivative of the gimbal angle.

$$\tau = \frac{\partial H}{\partial \theta}\frac{d\theta}{dt} \tag{10}$$

The partial derivative of angular momentum with respect to the gimbal angle is found by taking the spatial gradient of Eqs. (7), (8), and (9) which produces a Jacobian matrix, the A matrix. The A matrix describes the components of torque provided by each CMG in each axis; this is represented in Eq. (11).

$$[A] = \frac{\partial \boldsymbol{H}}{\partial \theta_i} = \begin{bmatrix} \sin\theta_1 & \sin\beta_2\cos\theta_2 & -\sin\theta_3 \\ -\sin\beta_1\cos\theta_1 & \sin\theta_2 & \sin\beta_3\cos\theta_3 \\ \cos\beta_1\cos\theta_1 & \cos\beta_2\cos\theta_2 & \cos\beta_3\cos\theta_3 \end{bmatrix} \tag{11}$$

Given the A matrix's definition, Eq. (10) can be written inversely to find the commanded gimbal rotation rates as Eq.(12) where the inverse of A is equal to the reciprocal of the determinant of A multiplied by its cofactor [35].

$$\dot{\boldsymbol{\theta}} = \frac{1}{\det(A)} CoF(A)\boldsymbol{\tau} \tag{12}$$

Eq. (12) encounters a mathematical singularity when the determinant of A equals zero; within the control system, the computer will continually attempt to calculate one over zero and, in the process, send the absurdly large results as torque commands to the CMGs. The CMG actuators follow the randomly large commands and the spacecraft loses attitude control. Physically, this kind of singularity is hit when a particular combination of gimbal angles is reached and the CMG cannot produce torque in the desired direction. These combinations of gimbal angles are defined by the determinant of the A matrix. For the CMG system in **Figure 1** when all skew angles could be different, the determinant of A is evaluated in Eq. (13).

$$\begin{aligned} \det[A] = {} & \sin\theta_1(\sin\theta_2\cos\beta_3\cos\theta_3 - \sin\beta_3\cos\theta_3\cos\beta_2\cos\theta_2) \\ & + \sin\beta_2\cos\theta_2(-\sin\beta_1\cos\theta_1\cos\beta_3\cos\theta_3 - \cos\beta_1\cos\theta_1\sin\beta_3\cos\theta_3) \\ & - \sin\theta_3(-\sin\beta_1\cos\theta_1\cos\beta_2\cos\theta_2 - \sin\theta_2\cos\beta_1\cos\theta_1) \end{aligned} \tag{13}$$

There are a multitude of cases when Eq. (13) is equal to zero, causing a singularity. Within each of these cases, at any chosen combination of skew angles, there are numerous different gimbal angle combinations resulting in a singularity; each of these gimbal and skew angle combinations produces a certain angular momentum in the x, y, and z directions as calculated by Eqs. (7), (8), and (9) respectively. For a particular skew angle combination, there is a gimbal angle combination such that a singularity is hit with the smallest achievable angular momentum: this becomes the maximum angular momentum the entire CMG system can reach before encountering a singularity at that particular skew angle combination set up.

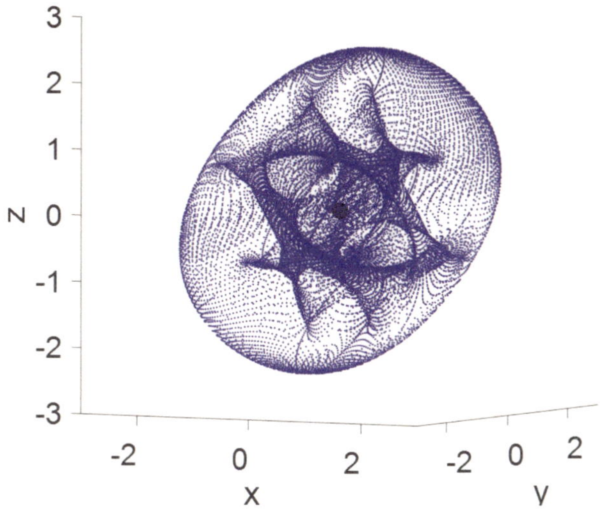

Figure 2.
Restricted angular momentum sphere within entire command space [34].

Although this reduction in the commandable angular momentum has been applied to many spacecraft on orbit, it is extremely limiting. **Figure 2** illustrates this reduction with the black sphere representing the singularity free maximum angular momentum space while the space enclosed with the blue surface represents all valid angular momentum commands. Furthermore, the outer blue surface defines the angular momentum saturation limit for its particular CMG setup. In **Figure 2**, the CMG set up includes three CMGs at equivalent skew angles of 56°.

In an attempt to remove this limit, Sands created a mechanism with which to penetrate this smallest angular momentum and expand the commandable angular momentum to everything up until saturation [32, 36, 37]. This mechanism is called singularity penetration with unit delay (SPUD) [32] and pierces the inner singularity surfaces by sending the CMG actuators valid control commands while the system passes through a singularity. This mechanism is critical in order to reach the maximum angular momentum at a particular axis.

3. Analysis

Defining the maximum angular momentum achievable without encountering a singularity for a CMG system over all possible skew angle combinations can be calculated via two methods: numerically or analytically. To numerically define this surface, the skew angle combinations are discretized and the associated minimum angular momentum is calculated numerically. To analytically define the same surface, each case that makes the determinant of A equal to zero is identified. The equation defining each case is then evaluated for its minimum angular momentum over all gimbal angle combinations for every skew angle. The minimum angular momentum data for all cases is then plotted on a single graph and the minimum angular momentum out of each case is taken as the maximum angular momentum achievable for that skew angle combination.

For this research, numerically calculating the maximum angular momentum without reaching a singularity for each discretized skew angle was chosen over the analytical method because the numeric solution creates a conservative model. The conservative nature of the numeric solution was determined by comparing a numerically calculated and analytically determined maximum angular momentum plot when all skew angles were equivalent. To compare these methods, however, a discretization size for the numeric solution had to be chosen. Three numeric solutions were plotted with discretizations of 0.1, 1, and 2°. One degree was chosen because using a smaller discretization, such as 0.1°, introduced noise into the plots while using a larger discretization, such as 2°, missed critical data points leading to important singularity locations. The 1° discretization plotted a smooth singularity location line while not skipping any important values. The plots using 0.1 and 2° are pictured in **Figure 3** while the 1° discretization is plotted in **Figure 4** with the analytic solution derived and created in Sands' dissertation [36].

Table 1 describes the mean error and standard deviation between the numerically obtained and analytically obtained data in **Figure 4**.

The numeric results vary from the analytic angular momentum values for most skew angles from 1 to 55° as can be seen in **Figure 4** and **Table 1**. After 55° however, both the numeric and analytic data is equivalent; **Figure 4** shows they plot along the same line while **Table 1** confirms the mean error and standard deviation between the values are both approximately zero. Although the numerically obtained results differ from the analytic values before 55°, the numeric results claim a lower possible

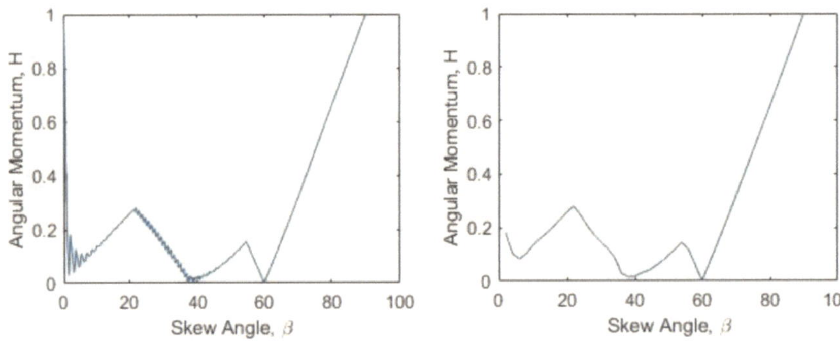

Figure 3.
0.1° discretization (left) versus 2° discretization (right) for numerically determined maximum angular momentum [34].

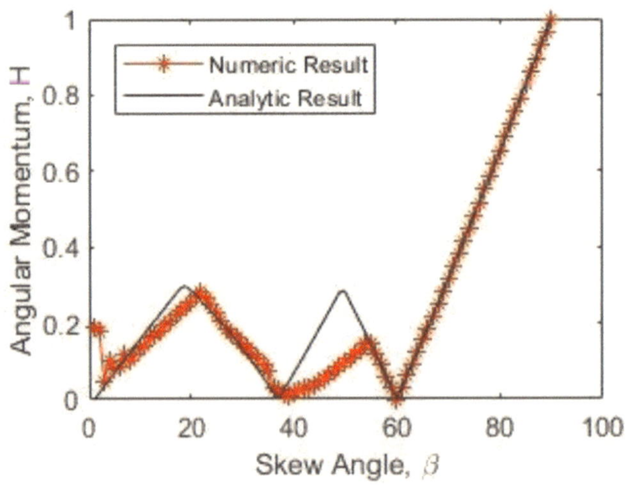

Figure 4.
Numeric versus analytic determination of maximum angular momentum [34].

Data points	μ	σ
1–37	0.0333	0.0388
38–60	0.0811	0.0707
61–90	5.51e-5	1.15e-4
Total	0.0344	0.0530

Table 1.
Mean error and standard deviation between numeric and analytic data [34].

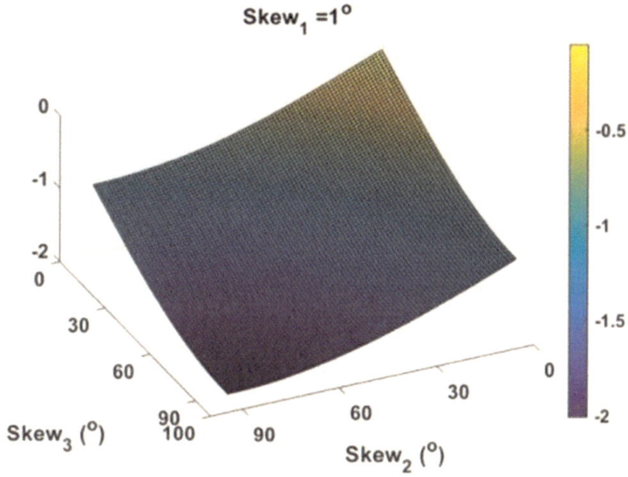

Figure 5.
3-D maximum angular momentum on the z axis for $\beta_1 = 1°$, $\beta_2 = \beta_3 =$ free *[34].*

angular momentum is possible before reaching a singularity. Using these data points would provide a buffer between where the singularities are expected to be versus where they actually are, protecting the attitude control system from hitting a singularity. Because this buffer is on the "safe" side, the maximum angular momentum without hitting a singularity for a CMG system with different skew angles was determined numerically.

Figure 4 plotted the maximum angular momentum in any direction for a non-redundant CMG system with equivalent skew angles. In order to design an attitude control system for a spacecraft with a characteristic maneuver, a similar figure can be produced plotting only the maximum angular momentum in that favored axis. This research aims to characterize skew angle combination effects on maximum angular momentum around the spacecraft's z axis, in other words, mixed skew angle effects on yaw maneuverability. To analyze this relationship, the maximum achievable angular momentum about the z axis was calculated for different skew angle combinations using the numerical method used to produce **Figure 4**. When creating the plots in **Figures 5** and **6**, the actual angular momentum values were plotted instead of strictly their magnitude; as a result, the plots are negative.

In order to plot the maximum achievable angular momentum about the z axis for all skew angle combinations, a four dimensional plot would be needed. Since this is not achievable, skew angle one was held constant while skew angles two and three were varied from 0 to 90°. Three dimensional plots were created as can be seen in **Figure 5**. However, due to the difficulty of orienting each graph to show the angular momentum magnitude, a color bar was employed instead. This allowed the same data to plot in two dimensions as can be seen in **Figure 6**.

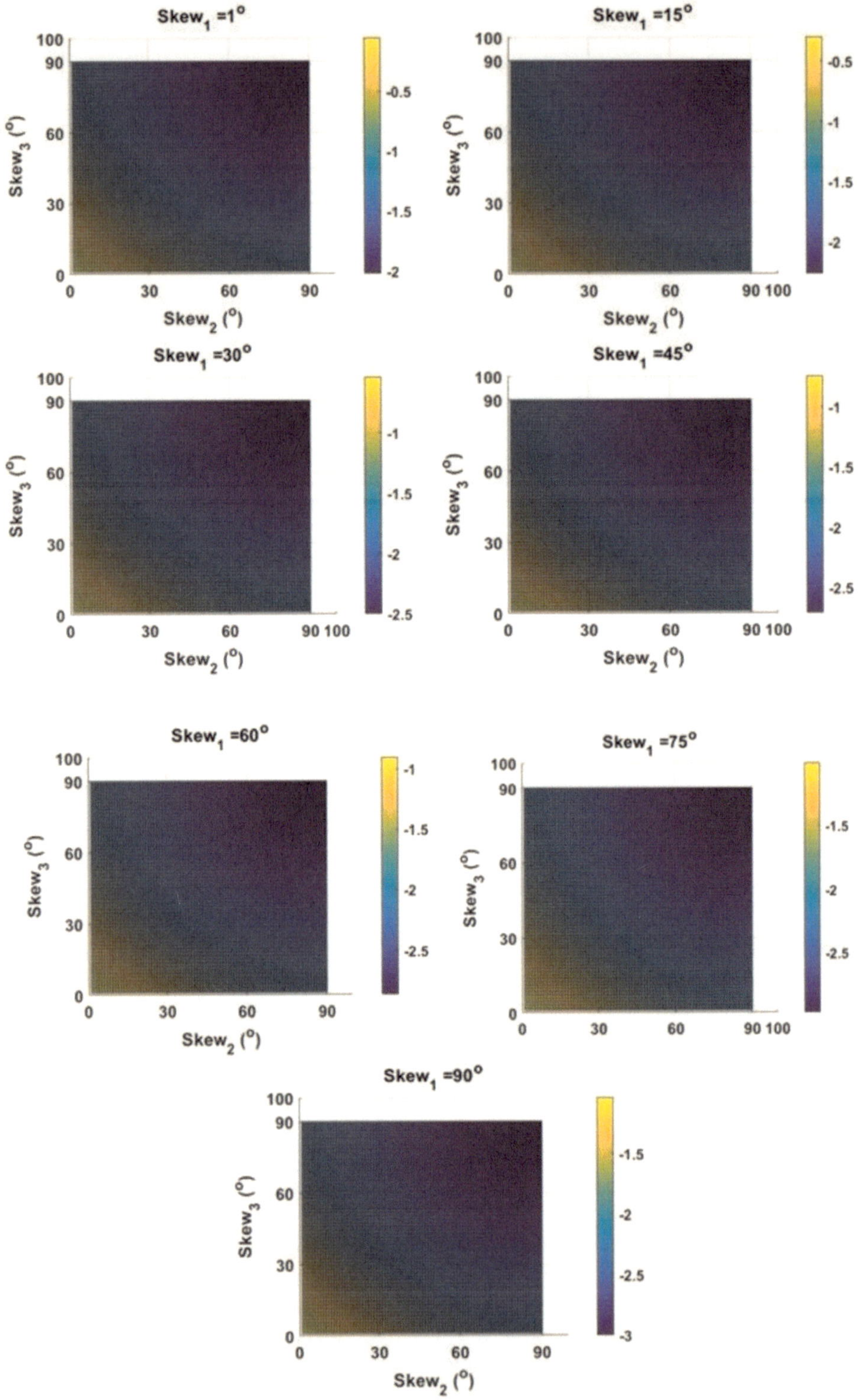

Figure 6.
Maximum angular momentum for β_1 = 1°,15°, 30°, 45°, 60°, 75°, 90°, $\beta_2 = \beta_3$ = free [34].

|H|	β_1 (°)	β_2 (°)	β_3 (°)
2.017	1	90	90
2.259	15	90	90
2.5	30	90	90
2.707	45	90	90
2.866	60	90	90
2.966	75	90	90
3	90	90	90

Table 2.
Maximum yaw maneuverability skew angle combinations [34].

Figure 6 illustrates the same trend for all β_1: maximum achievable angular momentum is smallest when both β_2 and β_3 are close to 0° and largest when both β_2 and β_3 are equal to 90°. Additionally, the magnitude of achievable angular momentum increases with β_1. For small β_1, such as 1°, the maximum angular momentum when β_2 and β_3 are close to 0° is 0|**H**| while for large β_1, such as 90°, the maximum angular momentum when β_2 and β_3 are close to 0° is 1|**H**|. **Table 2** lists the maximum angular momentum and associated skew angles for each plot in **Figure 6**.

Plotting the singularity maps for the skew angle combinations listed in **Table 2** visualizes the commandable angular momentum on the z axis. These mixed skew angle combinations produce the singularity maps pictured in **Figure 7**.

Within **Figure 7**, the highlighted blue surface in each plot contains the singularity defining the maximum achievable angular momentum about the z axis. For skew angle combinations with β_1 lower than 45° and β_2 and β_3 equal to 90°, the saturation limit on the z axis is defined by one of the inner singularity surfaces. For β_1 larger than 45° and β_2 and β_3 equal to 90°, the saturation limit is defined by the outer singularity surface. As long as β_1 is larger than 0°, there are no singularities exactly on the z axis before the saturation limit because there are at least two CMG's capable of exerting maximum angular momentum in the z direction. Since angular momentum can be commanded in that direction regardless of the orientation of the third CMG, there is no singularity until the saturation limit.

4. Conclusion

Drawing from the key points of this research, it is clear that different skew angles create drastically different singularity plots. These singularity plots map out the unattainable torque commands for a particular CMG system, ultimately defining the attitude envelope a spacecraft can achieve within a defined amount of time. As a result of this important relationship, CMG skew angles should be carefully chosen when designing a spacecraft attitude control system.

When designing a non-redundant CMG attitude control system for a spacecraft that needs to maximize its yaw maneuverability, a CMG system with all skew angles equal to 90° would maximize the commandable angular momentum about the z axis as **Figure 6**, **Table 2**, and **Figure 7** all show. The next greatest combination would be to set two of the skew angles equal to 90° and the third skew angle equal to something greater than zero in order to avoid a singularity at the origin.

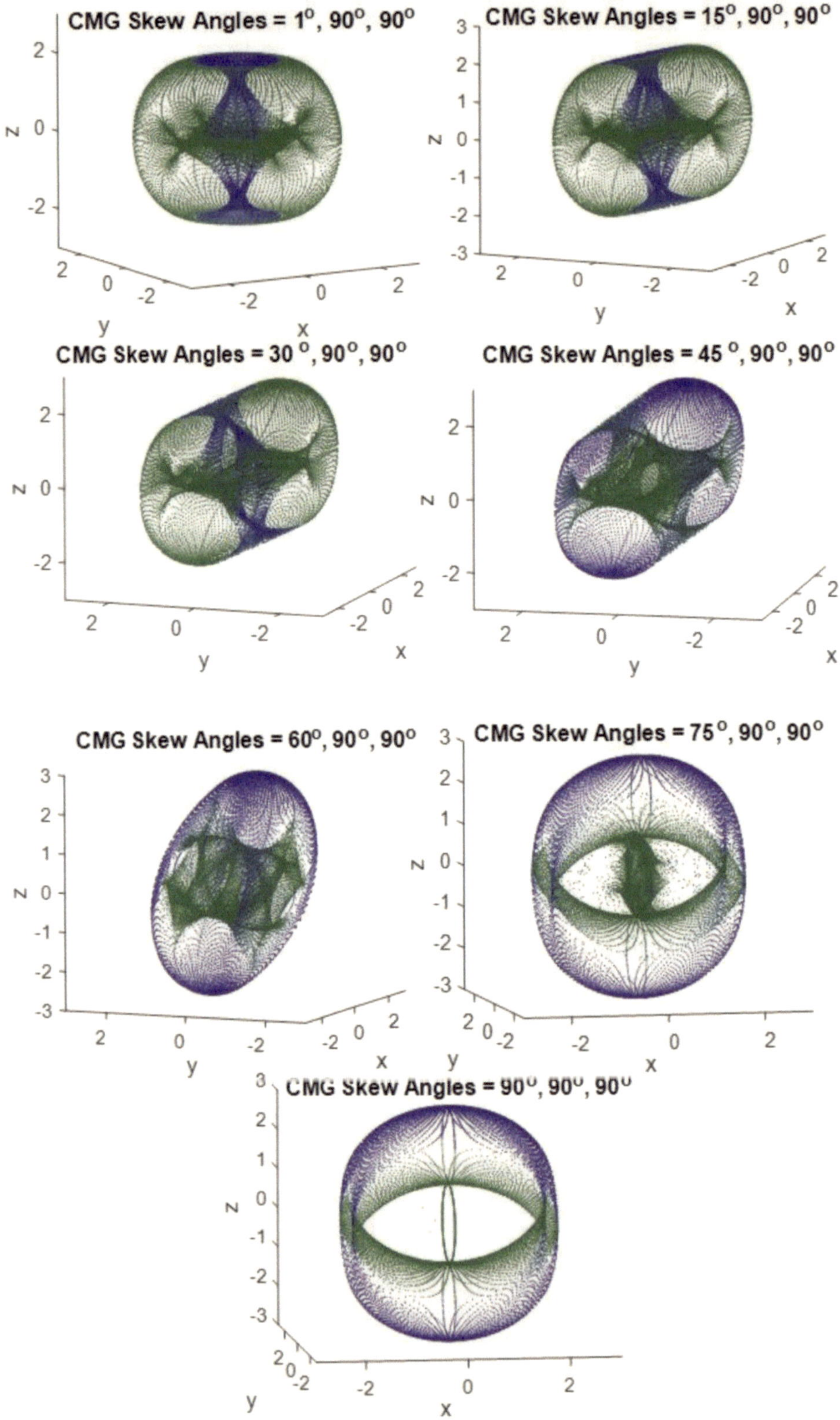

Figure 7.
Singularity maps [34].

Author details

Eryn A. Culton
Naval Postgraduate School, United States Navy, Monterey, CA, USA

*Address all correspondence to: eaculton@nps.edu

IntechOpen

© 2020 The Author(s). Licensee IntechOpen. Distributed under the terms of the Creative Commons Attribution - NonCommercial 4.0 License (https://creativecommons.org/licenses/by-nc/4.0/), which permits use, distribution and reproduction for non-commercial purposes, provided the original is properly cited.

References

[1] Merz J. A History of European Thought in the Nineteenth Century. London: Blackwood; 1903. p. 5

[2] Whittaker E. A Treatise on the Analytical Dynamics of Particles and Rigid Bodies. Cambridge, UK: Cambridge University Press; 1904

[3] Church IP. Mechanics of Engineering. New York: Wiley; 1908. p. 111

[4] Wright T. Elements of Mechanics Including Kinematics, Kinetics, and Statics, with Applications. New York: Nostrand; 1909

[5] Eduard Study (D.H. Delphenich translator), Foundations and goals of analytical kinematics. Sitzber. d. Berl. math. Ges. 1913;**13**:36-60. Available from: http://neo-classical-physics.info/uploads/3/4/3/6/34363841/study-analytical_kinematics.pdf [Accessed: 14 April 2017]

[6] Gray A. A Treatise on Gyrostatics and Rotational Motion. London: MacMillan; 1918 (published 2007 as ISBN 978-1-4212-5592-7)

[7] Rose ME. Elementary Theory of Angular Momentum. New York, NY: John Wiley & Sons; 1957 (published 1995), ISBN 978-0-486-68480-2

[8] Kane T. Analytical Elements of Mechanics Volume 1. New York and London: Academic Press; 1959

[9] Kane T. Analytical Elements of Mechanics Volume 2 Dynamics. New York and London: Academic Press; 1961

[10] Thompson W. Space Dynamics. New York: Wiley and Sons; 1961

[11] Greenwood D. Principles of Dynamics. Englewood Cliffs: Prentice-Hall; 1965 (reprinted in 1988 as 2nd ed.), ISBN: 9780137089741

[12] Fung AC, Zimmermun BG. Digital Simulation of Rotational Kinematics. NASA Technical Report NASA TN D-5302. Washington, DC; 1969. Available from: https://ntrs.nasa.gov/archive/nasa/casi.ntrs.nasa.gov/19690029793.pdf

[13] Henderson DM. Euler angles, quaternions, and transformation matrices—Working relationships. NASA Technical Report NASA-TM-74839. McDonnell Douglas Technical Services Co. Inc.; July 1977. Available from: https://ntrs.nasa.gov/archive/nasa/casi.ntrs.nasa.gov/19770024290.pdf

[14] Henderson DM. Euler angles, quaternions, and transformation matrices for space shuttle analysis. McDonnell Douglas Technical Services Co. Inc., Houston Astronautics Division as NASA Design Note 1.4-8-020, 9 June 1977. Available from: https://ntrs.nasa.gov/archive/nasa/casi.ntrs.nasa.gov/19770019231.pdf

[15] Goldstein H. Classical Mechanics. 2nd ed. Massachusetts: Addison-Wesley; 1981

[16] Kane T, Levinson D. Dynamics: Theory and Application. Boston: McGraw-Hill; 1985

[17] Huges P. Spacecraft Attitude Dynamics. New York: Wiley and Sons; 1986

[18] Wiesel W. Spaceflight Dynamics. 2nd ed. Boston: Irwin McGraw-Hill; 1989. p. 1997

[19] Wie B. Space Vehicle Dynamics and Control. Virginia: AIAA; 1998

[20] Slabaugh GG. Computing Euler angles from a rotation matrix. 1999; **6**(2000):39-63. Available from: http://www.close-range.com/docspacecraf tomputing_Euler_angles_from_a_rotation_matrix.pdf

[21] Vallado D. Fundamentals of Astrodynamics and Applications. 2nd ed. El Segundo: Microcosm Press; 2001

[22] Roithmayr C, Hodges D. Dynamics: Theory and Application of Kane's Method. New York: Cambridge; 2016

[23] Chasles M. Note sur les propriétés générales du système de deux corps semblables entr'eux. Bulletin des Sciences Mathématiques, Astronomiques, Physiques et Chemiques (in French). 1830;**14**: 321-326

[24] Sands T, Lu D, Chu J, Cheng B. Developments in angular momentum exchange. International Journal of Aerospace Sciences. 2018;**6**(1):1-7. DOI: 10.5923/j.aerospace.20180601.01

[25] Sands T, Kim J, Agrawal B. 2H singularity free momentum generation with non-redundant control moment gyroscopes. Proc. IEEE CDC; 2006. pp. 1551-1556. DOI: 10.1109/CDC.2006.377310

[26] Sands T. Fine pointing of military spacecraft [dissertation]. Monterey, CA, USA: Naval Postgraduate School; 2007

[27] Kim J, Sands T, Agrawal B. Acquisition, tracking, and pointing technology development for bifocal relay mirror spacecraft. Proceedings of SPIE. 2007. p. 6569. DOI: 10.1117/12.720694

[28] Sands T, Kim J, Agrawal B. Control moment gyroscope singularity reduction via decoupled control. Proc. IEEE SEC. 2009. pp. 1551-1556. DOI: 10.1109/SECON.2009.5174111

[29] Sands T, Kim J, Agrawal B. Nonredundant single-gimbaled control moment gyroscopes. Journal of Guidance, Control, and Dynamics. 2012; **35**:578-587. DOI: 10.2514/1.53538

[30] Sands T, Kim J, Agrawal B. Experiments in control of rotational mechanics. International Journal Automatic Control Intelligent Systems. 2016;**2**:9-22. ISSN: 2381-7534

[31] Agrawal B, Kim J, Sands T. Method and apparatus for singularity avoidance for control moment gyroscope (CMG) systems without using null motion. U.S. Patent 9567112 B1, Feb 14, 2017

[32] Sands T, Kim J, Agrawal B. Singularity penetration with unit delay (SPUD). Mathematics. 2018;**6**:23-38. DOI: 10.3390/math6020023

[33] Baker K, Cooper M, Heidlauf P, Sands T. Autonomous trajectory generation for deterministic artificial intelligence. Electrical and Electronic Engineering. 2018;**8**(3):59-68

[34] Culton E, Sands T. Single axis singularity mapping for mixed skew angle, non-redundant, single gimbaled CMG systems. In: Unpublished manuscript submitted to Advancements in Spacecraft Attitude Control (working title). London, UK: InTech; 2019 Print ISBN: 978-1-78984-802-1. Project in draft form, Due to Publisher April 2019

[35] Sands T, Lu D, Chu J, Cheng B. Developments in angular momentum exchange. International Journal of Aerosol Science. 2018;**6**(1):7

[36] Sands T. Fine pointing of military spacecraft [dissertation]. CA: MAE, NPS; 2010

[37] Agrawal BN, Kim JJ, Sands TA. Method and apparatus for singularity avoidance for control moment gyroscope (CMG) systems without using null motion. US Patent 9,567,112; 2017

Chapter 13

Momentum Space Analysis for Mixed Skew Angle Arrays

Zachary A. Lewis

Abstract

This report explores non-redundant constant-speed single-gimballed control moment gyroscope mechanics for spacecraft maneuvers. A 3/4 control moment gyroscope pyramid design was examined. This report focuses on maximum available momentum for mixed skew angle configurations. A method to maximize off-axis momentum is developed.

Keywords: control moment gyroscopes, mixed skew angle, momentum, singularity

1. Introduction

This manuscript describes the principles of control moment gyroscope (CMG) steering and attitude control with a specific focus on mixed CMG skew angles. CMGs are momentum exchange devices commonly used for spacecraft maneuvers. The studied configuration was a 3/4 CMG pyramid scheme with a balance mass configuration as shown in **Figure 1**. Rotation about the x-axis is denoted as roll, y-axis pitch, and z-axis yaw, respectively.

1.1 Research introduction

This paper explores the maximum available momentum for a mixed skew angle, non-redundant, constant-speed, single-gimballed CMG configuration. Mixed skew angles provide the opportunity to bias CMG momentum in a particular direction. This may enhance maneuverability of a spacecraft and improve performance for various mission sets.

This study assumes a spacecraft has a proper attitude control system that includes some form of singularity avoidance or penetration logic; this is necessary to utilize the full momentum of a CMG in the studied configuration.

1.2 System introduction

A Simulink model was used to develop, test, and simulate a spacecraft attitude control system. The system topology is shown in **Figure 2**.

The attitude control system, shown in **Figure 2**, begins with a user input attitude maneuver, the system then generates a trajectory for the maneuver, and the trajectory is fed to a controller and actuator to generate desired torque. The dynamics describes the kinematics of the spacecraft motion. For more information on the spacecraft kinematics, see [1]. Sensors and observers are used to determine the state

IntechOpen

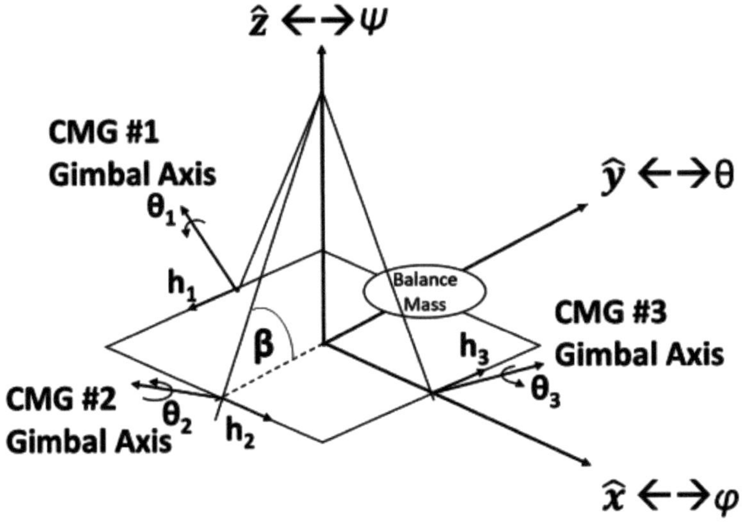

Figure 1.
CMG orientation schematic.

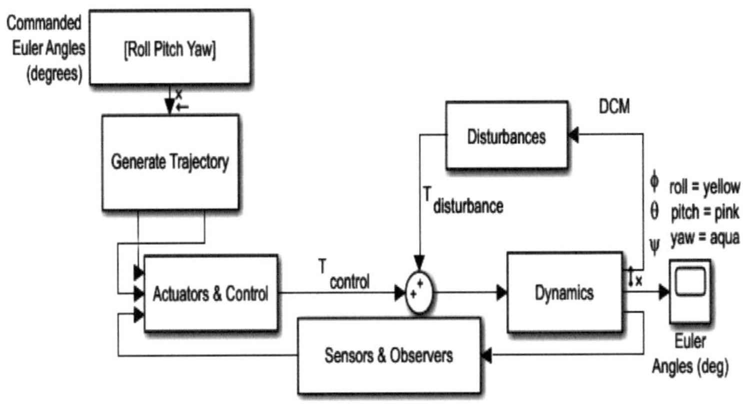

Figure 2.
Attitude control system topology.

of the spacecraft and the information is fed back to the controllers. This process continues until the spacecraft completes the maneuver.

2. Theory

2.1 Trajectory generation

Given a desired maneuver, trajectory generation is the first phase of attitude control. The trajectory provides a path and input for the controllers. It is unrealistic for a spacecraft to instantly move from one position to the desired end state; therefore, a trajectory must be generated that provides time for the spacecraft to maneuver.

A sine curve can be used as a rule of thumb trajectory. It provides time for the spacecraft to accelerate, reach a maximum rate, and then gradually decelerate to land at the desired endpoint. More advanced trajectory generation techniques can be found in [2].

2.2 Torque generation

CMGs, depicted in **Figure 3**, are torque devices used to maneuver spacecraft. They operate by rotating a mass to establish angular momentum (**h**). In order to generate torque, a constant speed, single-gimbaled CMG changes the direction of the angular momentum.

Shown in **Figure 3** is the angular momentum vector, **h**, the gimbal axis, **g**, (used to rotate the CMG and change the direction of the angular momentum), and the torque vector. The **h** vector is determined by the direction of the spinning mass. **g** is fixed by gimbal motor placement and is determined by the direction of rotation, θ. The resulting torque vector, **T**, is found using the right hand rule with **g** and **h**.

The Euler's momentum exchange torque equation is shown in Eq. (1).

$$T = J\dot{\omega} + \omega \, \mathrm{X} \, J\omega \tag{1}$$

In Eq. (1), J is the system moment of inertia and ω is the system angular velocity. The total system angular momentum, H_s, is defined as

$$H_s = J_{s/c}\omega_{s/c} + J_{CMG}\omega_{CMG} = J_{s/c}\omega_{s/c} + \mathbf{h} \tag{2}$$

where $J_{s/c}$ and J_{CMG} are the moment of inertia for the spacecraft and CMG, respectively, and **h** is the CMG angular momentum. Substituting Eq. (2) into Eq. (1) yields

$$T_{ext} = \dot{H}_s + \omega \, \mathrm{X} H_s. \tag{3}$$

Applying the Law of Momentum Conservation, the torque a CMG applies will result in an opposite effect on the spacecraft. The application of this concept is shown in Eq. (4)

$$-u = \dot{\mathbf{h}} + \omega \, \mathrm{X} \, \mathbf{h} \tag{4}$$

where u is the system control torque input.

With the configuration shown in **Figure 1**, β depicts the CMG skew angle, θ is the rotation of the CMG about the gimbal axis, and **h** is the CMG angular momentum vector. The **h** vector, shown in Eq. (4), is determined by projecting each CMG's angular momentum onto the x, y, and z axes, respectively, as shown in Eq. (5). The derivative, $\dot{\mathbf{h}}$, vector is shown in Eqs. (6) and (7).

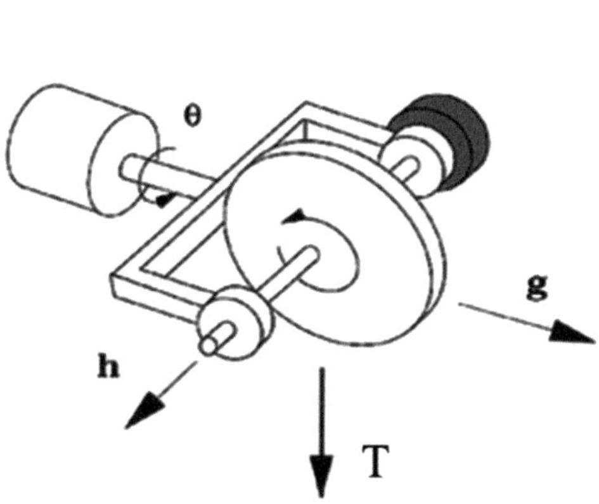

Figure 3.
Simplified schematic of a CMG [3].

$$\mathbf{h} = \begin{bmatrix} h_x \\ h_y \\ h_z \end{bmatrix} = \begin{bmatrix} \cos(\beta_1)\sin(\theta_1) & \cos(\theta_2) & \cos(\beta_3)\sin(\theta_3) \\ -\cos(\theta_1) & \cos(\beta_2)\sin(\theta_2) & \cos(\theta_3) \\ \sin(\beta_1)\sin(\theta_1) & \sin(\beta_2)\sin(\theta_2) & \sin(\beta_3)\sin(\theta_3) \end{bmatrix} \begin{Bmatrix} \hat{x} \\ \hat{y} \\ \hat{z} \end{Bmatrix}. \tag{5}$$

$$\dot{\mathbf{h}} = \frac{d\mathbf{h}}{dt} = \frac{\partial h}{\partial \theta_i}\frac{\partial \theta}{\partial t} = [A]\dot{\theta} \tag{6}$$

$$[A] = \begin{bmatrix} \cos(\beta_1)\cos(\theta_1) & -\sin(\theta_2) & \cos(\beta_3)\cos(\theta_3) \\ \sin(\theta_1) & \cos(\beta_2)\cos(\theta_2) & -\sin(\theta_3) \\ \sin(\beta_1)\cos(\theta_1) & \sin(\beta_2)\cos(\theta_2) & \sin(\beta_3)\cos(\theta_3) \end{bmatrix} \tag{7}$$

2.3 Controller torque generation

In a real system, there must be some form of commanded torque generation. Controllers take the state as an input and generate the torque command, u. There are two primary forms of controllers, feedforward and feedback.

2.3.1 Feedforward controller

The feedforward controller takes the state as an input and generates a torque command. A physics-based feedforward controller, pioneered by Lorenz, is the ideal form of a controller and is far more accurate than a linearized controller [4–6]. The feedforward controller for the system analyzed is shown in Eq. (8).

$$u_{FF} = \hat{J}\dot{\omega_d} + \omega_d X \hat{J}\omega_d \tag{8}$$

By equating coefficients between Eqs. (8) and (1), it is clear that the feedforward controller is based exactly on the dynamics of a rotational system where $\hat{J}$ is the best guess of the spacecraft's moment of inertia and ω_d is the desired gimbal rate.

Feedforward control is effective because it is based exactly on the dynamics of the spacecraft. In addition, proper application of the feedforward controller can eliminate phase lag [7]. However, the basic feedforward, as shown in Eq. (8), has no ability to account for imperfections in the maneuver. Various advances in adaptive feedforward control have been developed to advance the ability of a physics-based feedforward controller and can be found in [8–13].

2.3.2 PID feedback controller

The feedback controller is a method to create a control command based on an error signal between the current and desired state. This proves incredibly useful to correct for imperfections in a maneuver. Imperfections or disturbances can be a result of various things such as an inaccurate estimate of *J*, vibrations of a non-rigid body, aerodynamic drag, and more. Various advances and approaches for disturbance rejection are discussed in [14–17].

A PID controller creates a control command based on a linearized dynamic model with proportional, integral, and derivative gains as shown in Eq. (9).

$$u_{FB} = K_p e(t) + K_I \int_0^t e(t')dt' + K_D \frac{de(t)}{dt} \tag{9}$$

where, K_p, K_I, and K_D are the proportional, integral, and derivative gains, respectively, and e is the error signal. The gains are tuned in order to minimize error and create dynamic stability for the system.

There are various tuning methods utilized for PID controllers. Some common methods are iterative/experimental, Zeigler, and Liner Quadratic Regulator (LQR).

A first order approach to tuning is simply to choose gains through iteration. This experimental approach is simply to vary gains until the system performs within the acceptable parameters.

A second approach, Zeigler tuning, was developed empirically over time and can be considered a basic rule-of-thumb approach. Zeigler tuning requires first to determine the ultimate gain that creates stable oscillations. The ultimate gains, along with the oscillation period, are then used to calculate the gains for a PID controller. The equation for calculating Zeigler PID gains is given in Eq. (10).

$$\begin{aligned} K_p &= 0.6K_u \\ K_I &= \frac{K_P}{0.5T_u} \\ K_D &= 0.125K_pT_u \end{aligned} \tag{10}$$

LQR tuning is theoretically the optimal tuning method for linearized systems. It utilizes an optimization algorithm such that an engineer can input a linear state-space model and the optimal, minimum-cost, gains will be given [18].

2.3.3 PDI feedback controller

The PDI controller is a variant of the PID controller. David Luenberger developed an approach to avoid differentiating within the controller to reduce error. Rather than differentiating to provide derivative gain as shown in Eq. (9), the observed derivative state is fed to the controller after PI command calculation, the derivative state and gain are then summed after an integrator. The approach induces derivative action while avoiding differentiation within the controller. In addition, the PDI controller utilizes R.D. Lorenz' approach to eliminate virtual zero references in a cascaded topology. For more information on Luenberger and Lorenz' PID variations, see [17].

2.4 Actuators

The actuator takes the torque command and converts it into a gimbal rate for the CMGs to apply the desired torque. Rearranging Eq. (6) and substituting Eq. (7) yields

$$\dot{\theta} = [A]^{-1}\dot{\mathbf{h}} \tag{11}$$

Therefore, for a given desired torque, $\dot{\mathbf{h}}$ Eq. (11) can be used to command a rotation rate for all CMGs to provide the desired torque. Advanced in the application of Eq. (11), also known as the inverse steering law, can be found in [19]. In a true actuator circuit the desired gimbal rate is then converted into either currents or voltages based on the gimbal motor specifications.

As shown in Eq. (11), $[A]^{-1}$ is an integral component to the commanded torque calculation. However, Eq. (12) shows that $[A]^{-1}$ is also the source of numerical singularities. For example, if $\beta_1 = \beta_2 = \beta_3 = 90°$,

$$[A]^{-1} = \begin{bmatrix} \frac{s(\theta_3)}{(c\theta_1 s\theta_3 + s\theta_1 c\theta_3)t(\theta_2)} & \frac{s(\theta_3)}{c\theta_1 s\theta_3 + s\theta_1 c\theta_3} & \frac{s(\theta_3)}{c\theta_1 s\theta_3 + s\theta_1 c\theta_3} \\ \frac{-1}{s(\theta_2)} & 0 & 0 \\ \frac{s(\theta_1)}{(c\theta_1 s\theta_3 + s\theta_1 c\theta_3)t(\theta_2)} & \frac{-c(\theta_1)}{c\theta_1 s\theta_3 + s\theta_1 c\theta_3} & \frac{\sin(\theta_1)}{c\theta_1 s\theta_3 + s\theta_1 c\theta_3} \end{bmatrix} \quad (12)$$

where s, c, and t are sin, cos, and tan, respectively. There are numerous CMG orientations in which the denominator for a given term in [A] may be zero—this results in a numerical singularity.

2.5 Sensors, observers, and filters

Sensors are used to measure the actual state of the spacecraft. Ideally, a spacecraft would have enough sensors to provide full state feedback for the controllers. However, the number of sensors required to provide full state feedback may be cost or space prohibitive. Regardless, sensors are not ideal and induce noise into a system.

As a result, generally a state sensor will be used in conjunction with filters and observers. Filters, such as Lowpass or Kalman filters, are used to remove noise from the sensed state signal. Observers, as duals to the controller, take the state and differentiate to provide full state feedback. A number of space system identification algorithms can be found in [20].

2.6 Singularity generation

There are various definitions of a singularity for a non-redundant CMG system. Physically, singularities are situations in which at least a single CMG torque vector is perpendicular to the commanded torque direction. As a result, the commanded torque cannot be applied and, in general, control systems attempt to calculate and send infinite results to the CMGs and lose control of the spacecraft.

Numerically, singularities occur when

$$\det([A]) = 0. \quad (13)$$

In general, for the configuration shown in **Figure 1**,

$$\det([A]) = c(\beta_1)c(\theta_1)c(\theta_2) * [c(\beta_2)s(\beta_3)c(\theta_3) + s(\beta_2)s(\theta_3)] + s(\theta_2) * [s(\beta_1)c(\theta_1)s(\theta_3) + s(\theta_1)s(\beta_3)c(\theta_3)] + c(\theta_2)c(\beta_3)c(\theta_3) * [s(\theta_1)s(\beta_2) - s(\beta_1)c(\theta_1)c(\beta_2)] \quad (14)$$

Eqs. (7), (13), and (14) show that singularities are dependent on both β and θ. For any spacecraft configuration, β is fixed upon CMG installation and singularities become a function of θ alone. Therefore, it is required to clearly define the effect of β on singularities and available CMG momentum.

It is essential to note that singularities are instantaneous positions—this means that an infinitesimal shift in angle will result in a position theoretically capable of producing torque. However, though positions like this are theoretically capable of producing torque, the required commanded $\dot{\theta}$ could be significantly large. In this case the CMG would not be singular, however the behavior would remain singularity-like.

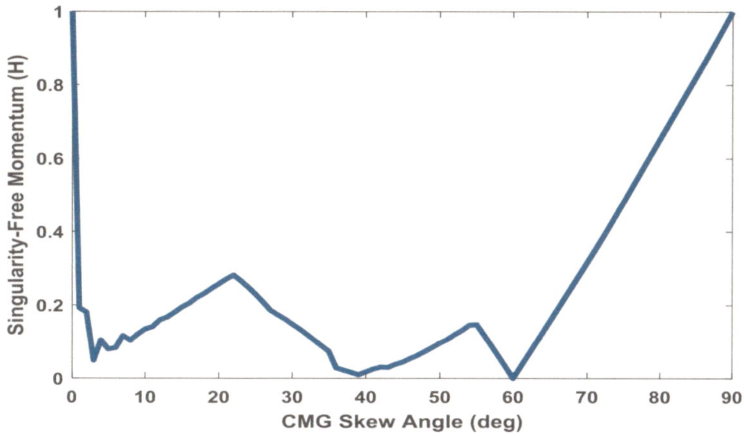

Figure 4.
Singularity free momentum.

2.6.1 Singularity penetration

As shown in Eq. (12), CMGs are coupled in all three axes [21]. Consequently, if one CMG is singular, the resulting $[A]^{-1}$ would technically be singular. However, in many situations, even if one CMG is singular resulting in a singular $[A]^{-1}$, the other CMGs are capable of providing useable torque.

With this in mind, to effectively maneuver a spacecraft using CMGs, singularities and singularity-like regions near singularities must be characterized and avoided or penetrated. The industry standard is to operate only in totally singular-free regions, however the number of singularities for any CMG system is often large and as a result, the control space for CMG maneuvers is highly restrictive. Without singularity controller logic, for example, a benchmark 54.73° CMG skewed array has a minimum singularity free momentum of 0.15 H (shown in **Figure 4**) where H is maximum momentum of one CMG [22].

One approach for singularity control logic is singularity avoidance as shown in [23]. The newest, state-of-the-art solution for CMG singularities is singularity penetration with unit delay (SPUD). SPUD was invented to expand the CMG operating area and protect the spacecraft from losing control near singularities. SPUD's basic

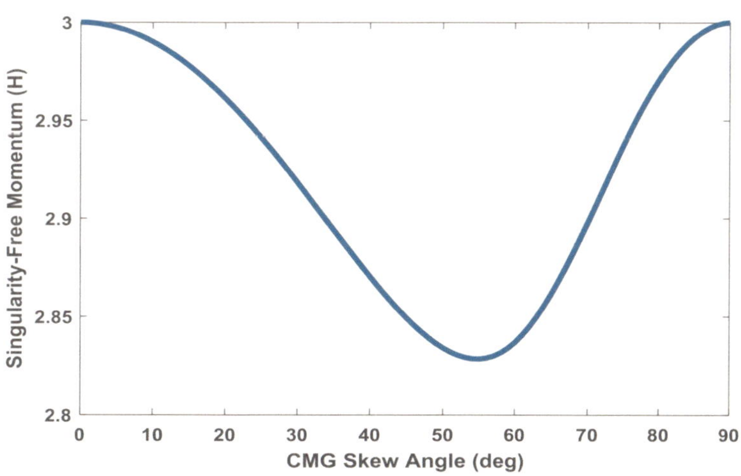

Figure 5.
Homogeneous skew angle maximum momentum.

premise is to recognize when the CMG is approaching a singularity and hold the last useful command until the CMG passes through the singularity [24].

2.7 Maximum CMG momentum

With the assumption that sufficient singularity avoidance or penetration logic is in place, a spacecraft attitude control system has the opportunity to utilize the entirety of a CMG's momentum.

Figure 5 shows the maximum available CMG momentum for a homogeneous skew angle CMG array. However, there are many possible configurations in which the skew angle for all CMGs may not be able to be equal. There is a lack of characterization of available CMG momentum for mixed skew angle CMG arrays.

3. Results and discussion

MATLAB code and a Simulink model, based on the CMG fundamentals described, were utilized as tools to analyze the maximum available momentum for mixed skew angle CMG arrays. For the momentum analysis, CMG #1 was held constant while CMG #2 and #3 were iterated from 0° to 90° with a step size of 1°. Then, CMG #1 was stepped 5° and the process repeated until CMG #1 cycled from 0° to 90°.

3.1 Discretization analysis

A discretization analysis was conducted to ensure accurate results were being recorded. A step size analysis for the homogeneous skew angle array was conducted and is shown in **Table 1**.

It was clear that as step size decreased, the standard deviation decreased. As the step size approached 0.1°, the standard deviation began to reach asymptotic behavior. However, as the step size approached 0.1°, computational time increased exponentially.

The analysis in **Table 1** was conducted iterating one skew angle from 0° to 90°. With the knowledge that the mixed skew angle study would increase the iteration dimensionality from one to two while holding CMG #1 constant, the number of iterations would increase by the power of two. Taking into account computation time, a 1° step size was chosen for the study.

3.2 Maximum singular momentum results

The maximum available momentum for a non-redundant mixed skew array is shown in **Figure 6**.

Step size (°)	Standard deviation	Computation time (s)
0.1	0.2541	182.09
0.5	0.2592	39.86
1	0.2652	19.20
2	0.2815	10.10
5	0.3250	4.35

Table 1.
Discretization analysis.

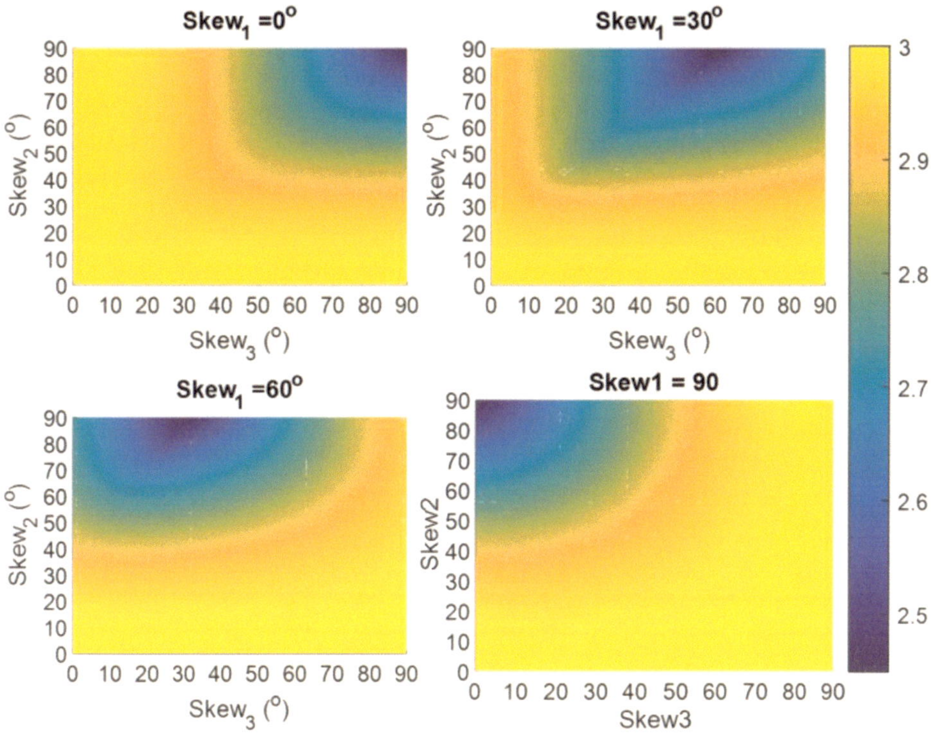

Figure 6.
Maximum available momentum.

CMG 1-2-3 β angles (°)	Maximum momentum (H)
0-0-*free*	3
0-*free*-0	3
30-0-*free*	3
60-0-*free*	3
90-0-*free*	3
90-*free*-90	3
0-90-90	2.449
30-90-60	2.449
60-90-30	2.449
90-90-0	2.449

Table 2.
Significant CMG skew angles and momentum.

For each image in **Figure 6**, the horizontal axis shows β_3 from 0° to 90° and the vertical shows β_2 from 0° to 90°. The top left image shows β_1 at 0°, top right 30°, bottom left 60°, and bottom right 90°. **Table 2** shows the significant skew angle combinations and resulting maximum available momentum.

Table 2 shows that if β_2 is zero, any combination of β_1 and β_3 will result in the maximum available momentum being 3 H. In addition, when β_2 is 90, the complimentary angle between β_1 and β_3 results in the lowest maximum available momentum for all mixed skew combinations.

3.3 Momentum space rotation

As depicted in **Figure 1**, CMG #1 and #3 are opposite one another while CMG #2 is offset by 90°. This orientation is critical for momentum space rotation. As shown in **Figure 6**, when β_1 and β_3 are 90°, β_2 does not affect the maximum available momentum. An investigation of the effect of β_2 on the array momentum space was conducted.

It is known from [25] that certain combinations of 0° and 90° mixed skew angles can shift the momentum space for an array without affecting the shape of the singularity surface. **Figures 7–9** show the momentum space and singularities for a

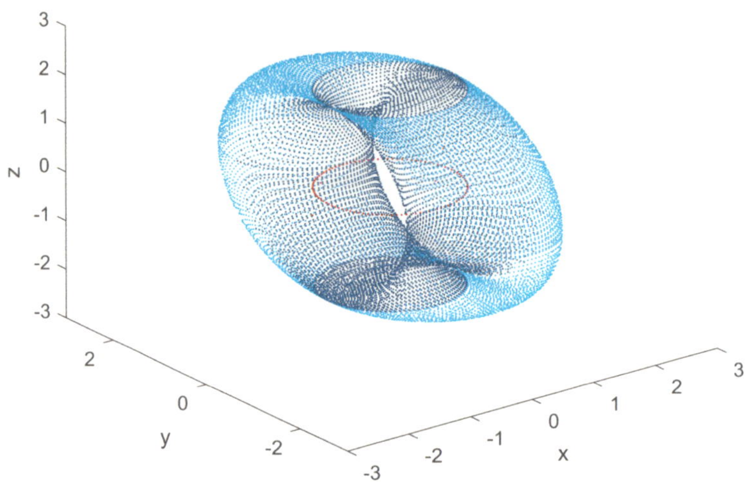

Figure 7.
CMG skew 1-2-3 of 90-0-90 degrees respectively.

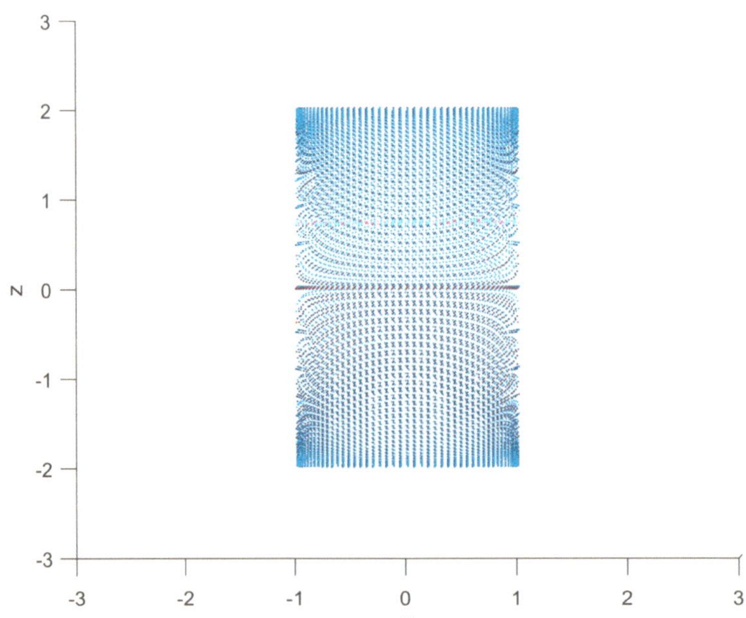

Figure 8.
CMG skew 90-0-90 in the x-z plane.

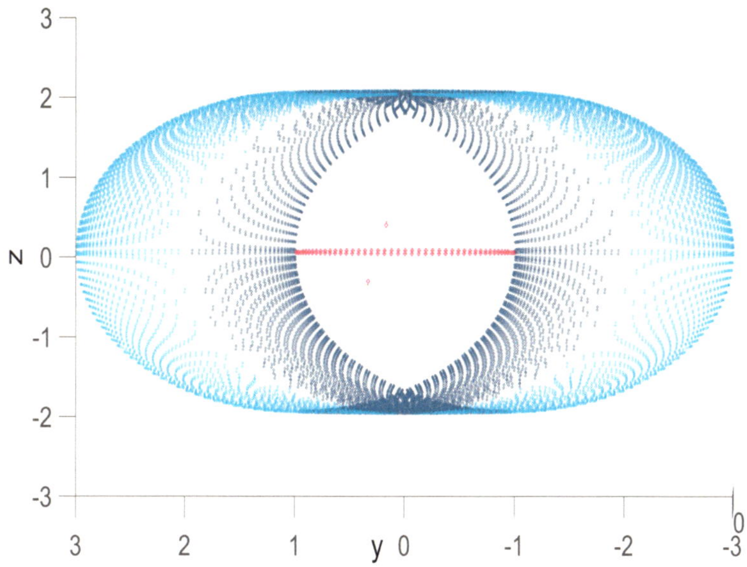

Figure 9.
CMG skew 90-0-90 in the y-z plane.

90-0-90 CMG array configuration. It has a donut shape with the maximum available momentum of 1, 3, and 2 H for x, y, and z, respectively.

Changing β_2 to 30° maintains the same singularity surface shape, however, the donut rotates 30° up in the y-z plane as shown in **Figure 10**.

With a 90-30-90 configuration, rather than having 3 H available in the y axis, 3 H is available in the axis 30° up from the y-z plane. This momentum space tilt occurs for all values of β_2 when β_1 and β_3 are 90°.

β_2 alone can point the direction of maximum available momentum. This behavior is purely a function of geometry. **Figure 1** and Eq. (7) show that when β_1 and β_3 are 90°, CMG #1 and #3 can put their full momentum anywhere in the y-z plane, therefore by changing the skew of CMG #2, the axis of 3 H momentum can shift

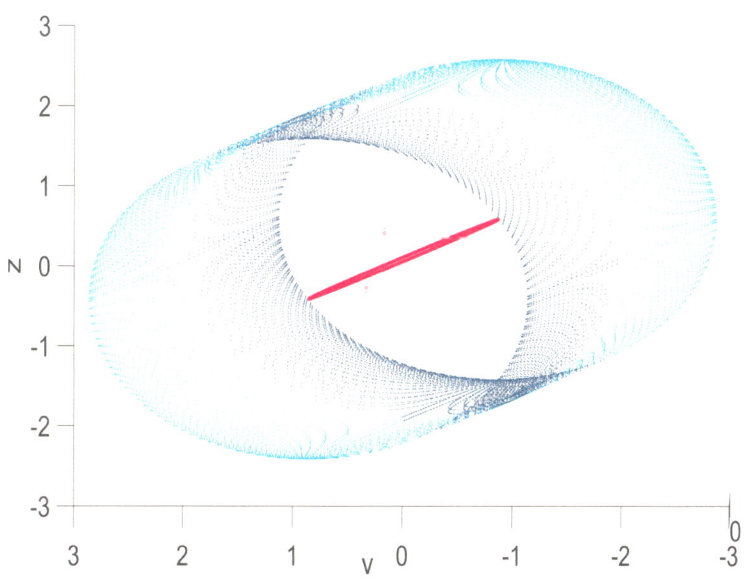

Figure 10.
CMG skew 90-30-90 in the y-z plane.

from the y axis up towards the z axis without affecting the momentum surface shape.

Shifting the angle momentum axis occurs for any combination of skew angles. However, for other mixed skew angles, as shown in **Figure 6**, the ability to provide a full 3 H is no longer preserved and the singularity surface changed. Physically, as β_1 and β_3 decrease from 90, they no longer have the ability to provide their full momentum in the y-z plane. As an example, the resulting change in available momentum for a 60-X-30 configuration is shown in **Figure 11** and **Figure 12**.

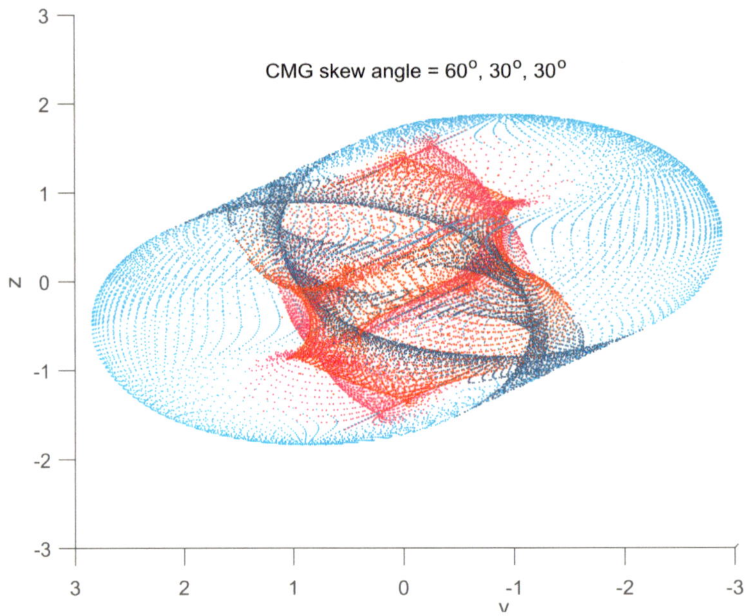

Figure 11.
CMG skew 60-30-30 singularity surface.

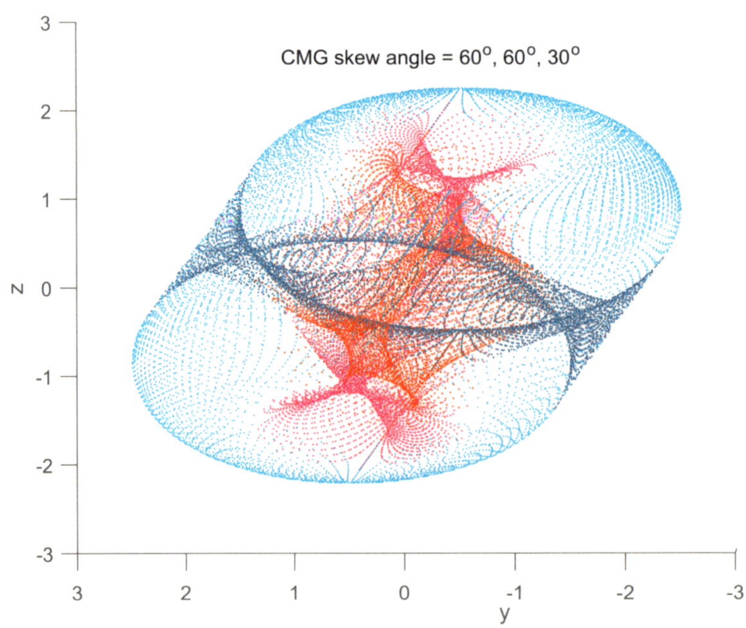

Figure 12.
CMG skew 60-60-30 singularity surface.

As β_2 increases from 30° to 60°, the maximum available momentum shifts up from 30° to 60°, but the magnitude of momentum in that axis shrinks.

3.4 Off-axis maneuver recommendation

For spacecraft that depend on off axis rotations, mixed skew angles can provide the benefit of tilted momentum space to maximize the CMGs' ability to rotate about that axis. A 90-X-90 configuration (where X is the angle of the desired rotation axis) is recommended because it allows the momentum space to tilt without changing shape or sacrificing available momentum. To achieve an off-axis 3 H momentum space, the CMGs must be oriented as shown in **Figure 1** such that two CMGs are located on the x axis and one is located on the y axis where the desired axis of rotation is angled in the y-z plane.

4. Conclusion

CMGs are a useful momentum exchange tool to maneuver spacecraft. There are many configurations of mixed skew angle arrays that allow for a maximum 3 H available momentum. If β_2 is 0°, any combination of β_1 and β_3 will allow a full 3 H to be available in a given direction. For a 90-X-90 configuration, β_2 can be used to tilt the maximum momentum space in the y-z plane without altering the shape of the momentum space. This configuration could be useful for spacecraft requiring off axis rotations.

Author details

Zachary A. Lewis
Naval Postgraduate School, California, USA

*Address all correspondence to: zachary.lewis@nps.edu

IntechOpen

© 2020 The Author(s). Licensee IntechOpen. Distributed under the terms of the Creative Commons Attribution - NonCommercial 4.0 License (https://creativecommons.org/licenses/by-nc/4.0/), which permits use, distribution and reproduction for non-commercial purposes, provided the original is properly cited. CC BY-NC

References

[1] Smeresky B, Rizzo A, Sands T. Kinematics in the information age. Mathematics. 2018;**6**(9):148

[2] Baker K, Cooper M, Heidlauf P, Sands T. Autonomous trajectory generation for deterministic artificial intelligence. Electrical and Electronic Engineering. 2018;**8**(3):59-68

[3] Schallig SAV, Chu QP, Rhee SW, Kampen EV. Maximum null motion algorithm for single gimbal control moment gyroscopes. Advances in Aerospace Guidance, Navigation and Control. 2017:689-707

[4] Cooper M, Heidlauf P, Sands T. Controlling chaos—Forced van der pol equation. Mathematics. 2017;**5**(4):70

[5] Sands T, Lorenz R. Physics-based automated control of spacecraft. In: AIAA SPACE; 2009. p. 6625

[6] Sands T, "Physics-based control methods," chapter in Advancements in Spacecraft Systems and Orbit Determination, London, UK: InTech; 2012, pp. 29-54

[7] Sands T. Elimination of phase lag using feedforward: Phase lag elimination at all frequencies for full state estimation of spacecraft attitude. Physics Journal. 2017;**3**(1):1-12

[8] Nakatani S, Sands T. Battle-damage tolerant automatic controls. Electrical and Electronic Engineering. 2018; **8**(1):23

[9] Nakatani S, Sands T. Simulation of spacecraft damage tolerance and adaptive controls. IEEE Aerospace. 2014;**1-16**:2014

[10] Nakatani S, Sands T. Nonlinear adaptive control: Autonomous damage recovery in space. International Journal of Automation, Control and Intelligent Systems. 2016;**2**:23-36

[11] Sands T, Kim JJ, Agrawal BN. Spacecraft fine tracking pointing using adaptive control. In: Proceedings of 52th International Astronautical Congress; 2007

[12] Sands T, Kim JJ, Agrawal B. Spacecraft adaptive control evaluation. In: Infotech@ Aerospace; 2012. p. 2476

[13] Sands T, Kim JJ, Agrawal BN. Improved hamiltonian adaptive control of spacecraft. In: 2009 IEEE Aerospace Conference. pp. 1-10

[14] Lobo K, Lang J, Starks A, Sands T. Analysis of deterministic artificial intelligence for inertia modifications and orbital disturbances. International Journal of Control Science and Engineering. 2018;**8**(3):53-62

[15] Sands T, Bollino K, Kaminer I, Healey A. Autonomous minimum safe distance maintenance from submersed obstacles in ocean currents. Journal of Marine Science and Engineering. 2018; **6**(3):98

[16] Sands T, Kevin Bollino, "Autonomous underwater vehicle guidance, navigation, and control," chapter in Autonomous Vehicles (Working Title), London, UK: InTech; 2019

[17] Sands T. Improved magnetic levitation via online disturbance decoupling. Physik Journal. 2015;**1**: 272-280

[18] Trammell L, Microstar Laboratories, Inc. Ziegler-Nichols Tuning Rules for PID [Internet]. Available from: http://www.mstarlabs.com/control/znrule.html [Accessed: 24 March 2019]

[19] Sands T, Lu D, Chu J, Cheng B. Developments in angular momentum exchange. International Journal of Aerospace Sciences. 2018;**6**(1):7

[20] Sands T. Space system identification algorithms. Journal of Space Exploration. 2017;**6**:138

[21] Sands TA, Kim JJ, Agrawal B. Control moment gyroscope singularity reduction via decoupled control. In: IEEE SEC 2009; 2009. pp. 388-391

[22] Sands TA, Kim JJ, Agrawal B. 2h singularity-free momentum generation with non-redundant single gimbaled control moment gyroscopes. In: Proceedings of the 45th IEEE Conference on Decision and Control; 2006. pp. 1551-1556

[23] Agrawal BN, Kim JJ, Sands TA. Method and apparatus for singularity avoidance for control moment gyroscope (CMG) systems without using null motion. US Patent 9,567,112; 2017

[24] Sands T, Kim J, Agrawal B. Singularity penetration with unit delay (SPUD). Mathematics. 2018;**6**(2):23

[25] Sands T, Kim JJ, Agrawal BN. Nonredundant single-gimbaled control moment gyroscopes. Journal of Guidance, Control, and Dynamics. 2012; **35**(2):578-587

Chapter 14

Mixed Skew Angle Singularity Plotting for Non-Redundant Single Gimbal CMG Array

Joshua A. Ten Eyck

Abstract

Control moment gyroscopes (CMGs) are utilized on spacecraft to control torsional movements. This chapter displays how a singularity map of a non-redundant single gimbal CMG array is created, analyzes the maximum singularity free momentums versus skew angle for three symmetric skew angles, and then analyzes for mixed skew angles. The objective of this paper is to demonstrate mixed skew angle values that produce large singularity free regions.

Keywords: rotational mechanics, adaptive control, deterministic artificial intelligence, nonlinear control, control moment gyroscope, momentum exchange, singularity, voice coil, armature resistance, physics-based control, disturbance decoupling

1. Introduction

Singularities are locations where a CMG is unable to achieve the desired gimbal rate or angular momentum and saturates the specific CMG, causing large transient commands to all CMGs and total loss of attitude control. By determining the locations of each singularity, a three-dimensional map can be produced. From this map, the singularities can be avoided or passed though utilizing singularity penetration with unit delay (SPUD) [1]. By modifying the skew angle, the amount of available singularity free angular momentum changes with respect to reaching a singularity.

2. Theory

Understanding how the three CMGs operate and provide torque to the spacecraft is necessary before determining the locations of singularities. The desired rotation is input into the trajectory generator, and the trajectory is fed to the controller which calculates the control designed to achieve the desired maneuver. This is necessary for the actuators to produce the desired torque since it is impossible for the rotational body to change from the initial state to the final state instantaneously. The trajectory generation produces the following three equations, Eqs. (1)–(3), to produce the maneuver as per Ref. [1]:

$$\theta_d = A sin \omega t \# \quad (1)$$

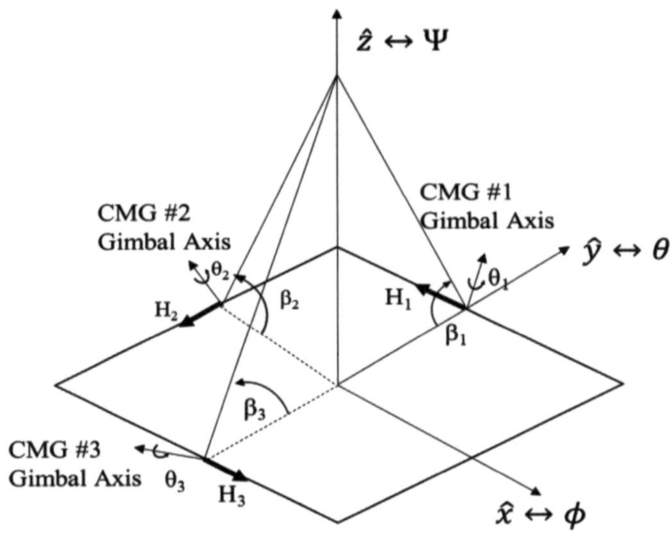

Figure 1.
¾ skew angle array [1].

$$\omega_d = A\omega cos\omega t\# \tag{2}$$

$$\dot{\omega_d} = -A\omega^2 sin\omega t\# \tag{3}$$

where θ is the gimbal angle, ω is the gimbal rate, and $\dot{\omega}$ is the gimbal acceleration. Next, a feedforward loop is implemented to eliminate phase lag. Instead of waiting for an error to be produced by the feedback loop, the feedforward starts the spacecraft on the desired track before the feedback calculates an error. In the field of deterministic artificial intelligence, this is known as a self-awareness statement as the CMG would know that it is a rotational body subject to the physics described in Eq. (4) with some expected inertia, J. The ideal feedforward control inputting the desired dynamics as displayed in Eq. (5):

$$\sum T = \hat{J}\dot{\omega} + \omega x \hat{J}\omega\# \tag{4}$$

$$\hat{J}\dot{\omega}_d + \omega_d x \hat{J}\omega_d = u_{ff}\# \tag{5}$$

The u_{ff} value is then fed to the actuators—the CMGs, reaction wheels, or thrusters—as a voltage or current command to produce the desired torque. Because the circuit is assumed to have a unity gain, no additional treatment was necessary to achieve the desired torques from the voltages or currents. The torque applied to the spacecraft is equal and opposite to the torque produced by the CMG array. The rate of change of the angular momentum is equal to the torque produced by the CMG, as per Eq. (6). The CMG array's angular momentum can be broken down into the different rotational directions.

$$T_{desired} = -\dot{H}\# \tag{6}$$

Figure 1 depicts the magnitude and direction of the angular momentum, H, gimbal angle, θ, skew angle, β, and gimbal axis for the three CMGs: H1, H2, H3, θ1, θ2, θ3, β1, β2, β3, respectively. The angular momentum will maintain the same magnitude but can rotate around each gimbal axis to change its direction.

If the CMG is commanded to create a torque in a direction that cannot be obtained by rotating the H vector around the gimbal axis, then that point will produce a singularity for that CMG. From **Figure 1**, the angular momentum can be expressed in directional components as seen in Eq. (7):

$$\begin{aligned} h_x &= \cos\theta_3 - \cos\theta_1 + \cos\beta_2 \sin\theta_2 \\ h_y &= -\cos\beta_1 \sin\theta_1 - \cos\theta_2 + \cos\beta_3 \sin\theta_3 \\ h_z &= \sin\beta_1 \sin\theta_1 + \sin\beta_2 \sin\theta_2 + \sin\beta_3 \sin\theta_3 \end{aligned} \tag{7}$$

Breaking down Eq. (6), the CMG torque into its separate components the CMG torque, $\dot{H}$, is the product of the gimbal angle rate and A matrix:

$$T_{desired} = -\dot{H} \quad where \quad \dot{H} = \frac{\partial H}{\partial \theta}\frac{\partial \theta}{\partial t} = \frac{\partial H}{\partial \theta_i}\dot{\theta} = [A]\dot{\theta} \tag{8}$$

$\frac{\partial \theta}{\partial t}$ is the gimbal angle rate and $\frac{\partial H}{\partial \theta}$ is defined as the A matrix per Eq. (9):

$$\frac{\partial H}{\partial \theta} = A = \begin{bmatrix} \sin\theta_1 & \cos\beta_2\cos\theta_2 & -\sin\theta_3 \\ -\cos\beta_1\cos\theta_1 & \sin\theta_2 & \cos\beta_3\cos\theta_3 \\ \sin\beta_1\cos\theta_1 & \sin\beta_2\cos\theta_2 & \sin\beta_3\cos\theta_3 \end{bmatrix} \tag{9}$$

The A matrix is the key to being able to control the system [3]. Locations where the A matrix is not invertible are rank deficient, meaning there are not enough independent vectors to achieve the desired torque. These locations can be determined by setting the determinant of the A matrix equal to zero as per Eq. (10):

$$A^{-1} = \frac{1}{\det[A]}[CoF]\# \tag{10}$$

where CoF is the cofactors matrix. As the det[A] approaches zero, the inverse of A goes to infinity, therefore the matrix is not invertible. In order to calculate these locations, the determinant of the A matrix is necessary as per Eq. (11). For the initial determination of the maximum singularity free region, the beta angles were set equal:

$$\begin{aligned} \det[A] = sin\beta(&2\cos\theta_1\cos\theta_2\cos\theta_3\cos^2\beta - \sin\theta_1\cos\theta_2\cos\theta_3 cos\beta + \cos\theta_1\sin\theta_2\sin\theta_3 \\ &+ \cos\theta_1\cos\theta_2\sin\theta_3 cos\beta + \sin\theta_1\sin\theta_2\cos\theta_3) \end{aligned} \tag{11}$$

Each singularity occurs at the location where the determinant of the matrix is equal to zero, in other words when Eq. (11) is set equal to zero. At these locations there is a singular inversion of the matrix causing at least one of the CMGs to try to produce an infinite gimbal rate. Because the three CMGs are coupled by the matrix inversion equation, a large transient command of the CMGs is created until the singularity has passed.

3. Results

There are two different approaches to produce a minimum magnitude of nonsingular momentum versus skew angle plot: analytically and numerically. The analytical method requires evaluation of the determinant of the A matrix. From this, several different cases are determined that will cause the determinant of the A matrix to be equal to zero, causing singularities. For each case, the equation causing

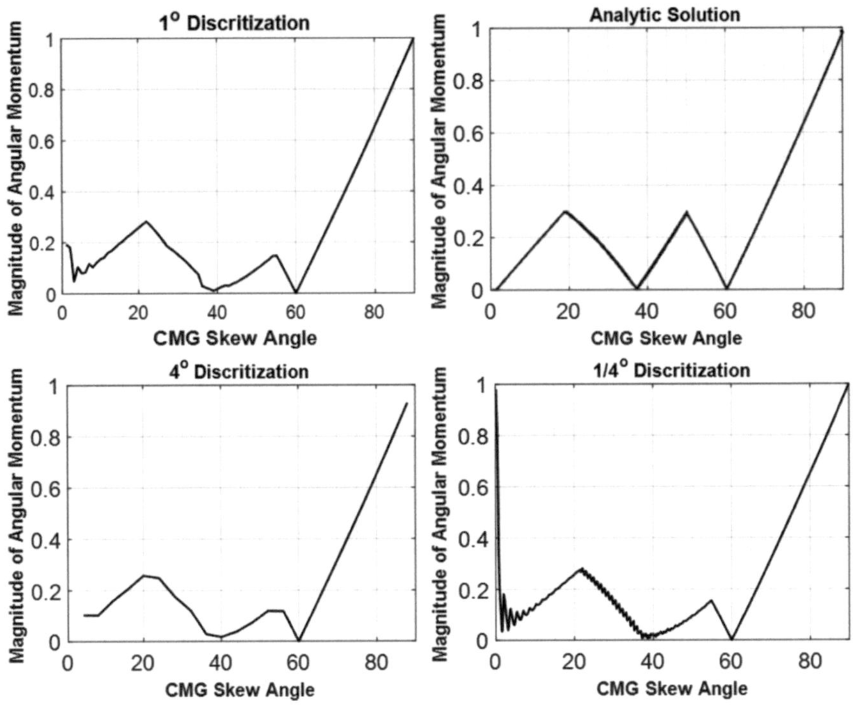

Figure 2.
Plots of analytic and numerical maximum singularity free momentum versus skew angle for ¾ array [1].

the singularities to occur was then minimized over all skew angles to determine the minimum angular momentum of that case for each skew angle. Each case was then plotted and overlaid. The curve creating the minimum values for the entire system determines the maximum value for angular momentum at any specific skew angle for that array to not reach a singularity. This plot is shown in **Figure 2**.

The second approach is numerically. In this case, the beta angles are discretized and the angular momentum is determined at each singular point then for each case the minimum singular value is determined at each discretized beta angle. These values are then plotted against the beta discretization.

Figure 2 compares the analytic solution [4], to the numerical solution. Before comparing the two different solutions, different discretizations of the skew angles were analyzed. Values of discretizations analyzed were 4, 2, 1, ½ and ¼°. **Figure 2** displays three of these plots in comparison to the analytic solution.

If the discretization was too small, excessive noise was introduced, as is visible in the ¼° discretization plot. If the discretization was too large, the trend line is smooth but removes the instantaneous changes happening in between the step sizes. Due to the decrease in the noise, yet maintaining the fidelity of the model, the 1° discretization was used.

Comparing the 1° discretization to the analytic solution, both follow a similar trend: there are peaks around 20, 50 and 90°. However, the exact location and magnitude differs. An overlay of the two plots is located in **Figure 3**. The analytic solution has higher peaks that occur at smaller angles. For example, the analytic solution has a peak of 0.3H at 49.5° whereas the numerical solution has a peak of only 0.15H at 54°. In order to evaluate the differences between the two plots, the standard deviation, σ, and mean, μ, of the error between the two were taken for three separate regions as well as all together. The plot was broken into regions as the numerical result behaved differently from the 1 to 38° range then it did in the 39–60°, or the 61–90°. The values are tabulated in **Table 1**.

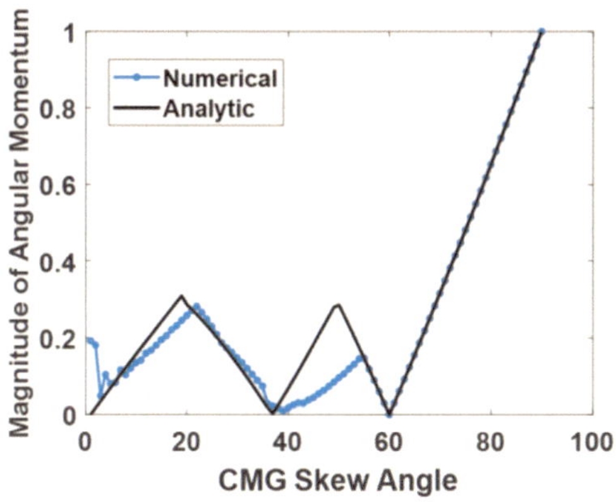

Figure 3.
Overlay of 1° discretization and analytic result [1].

	1–38°	39–60°	61–90°	Total
σ	0.033	0.084	3.69×10^{-5}	0.053
μ	0.039	0.071	6.86×10^{-5}	0.035

Table 1.
Analysis of numerical result error.

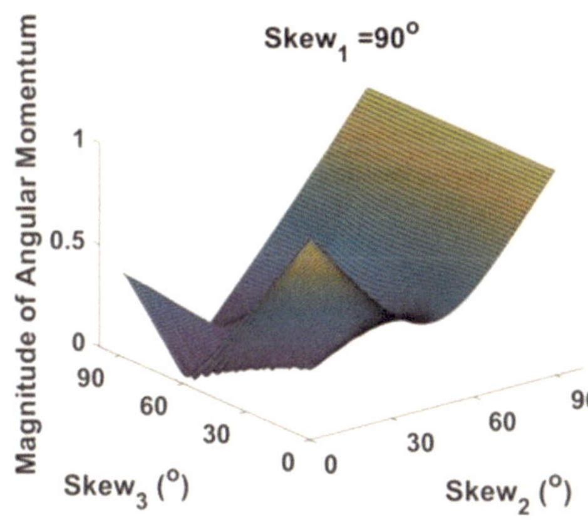

Figure 4.
Numerical maximum singularity free momentum of $\beta 1 = 90°$, $\beta 2$ and $\beta 3$ free [5].

The majority of the error incurred over the 1–38° range, occurred at between 1 and 5°. Numerically, the angular momentum at a skew angle of zero would hit a singularity causing the value to go to infinity, whereas analytically the value should be zero. The large deviation is caused by the definition of the numerical solution. Because the singularities are discretized, there will be locations where singularities occur between the steps of the numerical solution. The second region, 39–60°, acquires the most error; the peak is shifted to a larger skew angle but with a smaller maximum angular momentum. Therefore, because the values of the maximum singularity free angular momentum were usually smaller for the numerical approach than the analytic, the numerical approach serves as a conservative representation of the maximum angular momentum. The final region, 61–90° had very little error as the two lines were nearly coincided.

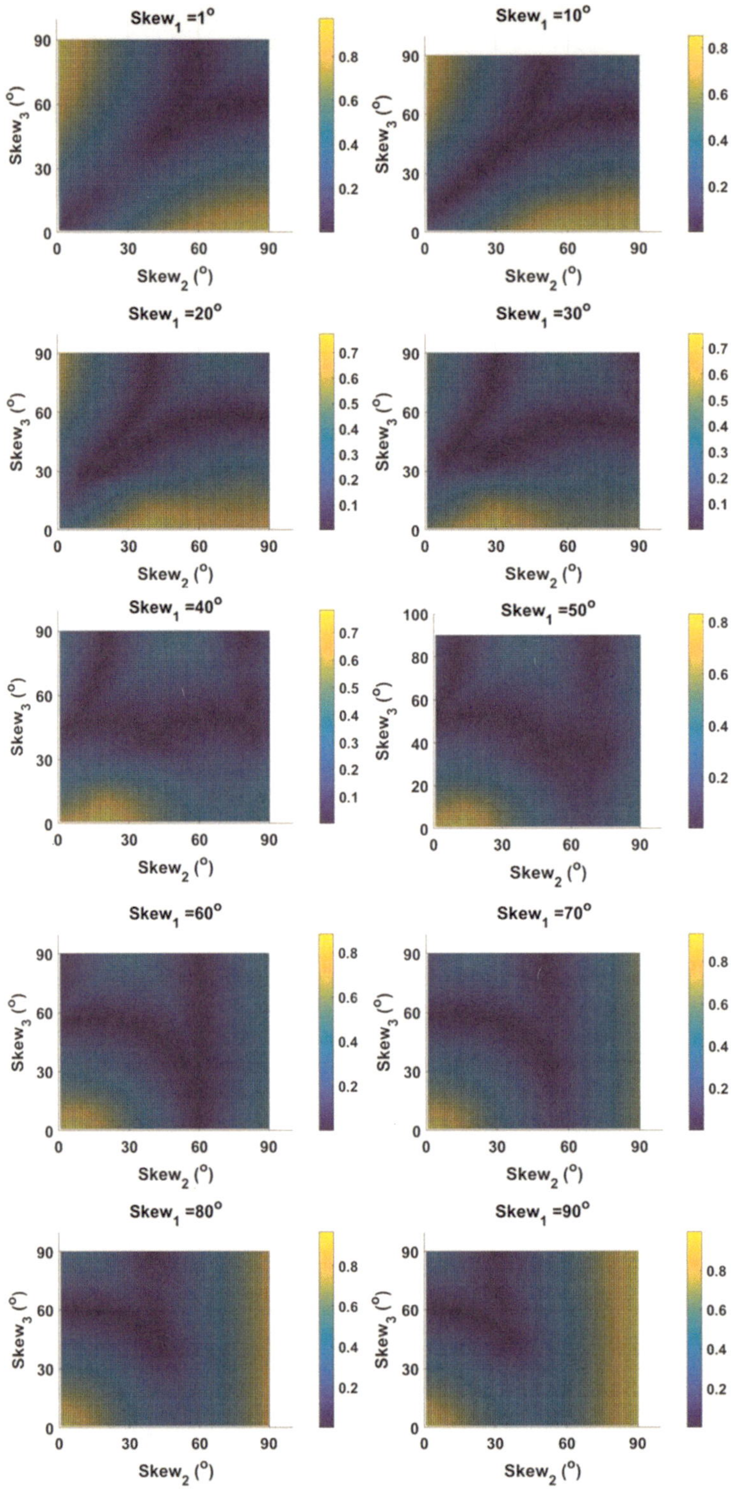

Figure 5.
2-D plots of maximum singularity free angular momentums for mixed skew angles [5].

Figure 2 describes the maximum magnitude of angular momentum when all three CMGs hold the same skew angle. However, by enabling mixed skew angles, the determinant of the A matrix, expands as per Eq. (11).

$$\begin{aligned}\det[A] = {} & \sin\beta_1\cos\beta_2\cos\beta_3\cos\theta_1\cos\theta_2\cos\theta_3 + \sin\beta_1\cos\theta_1\sin\theta_2\sin\theta_3 \\ & + \cos\beta_1\cos\beta_2\sin\beta_3\cos\theta_1\cos\theta_2\cos\theta_3 - \sin\beta_2\cos\beta_3\sin\theta_1\cos\theta_2\cos\theta_3 \\ & + \sin\beta_3\sin\theta_1\sin\theta_2\cos\theta_3 + \cos\beta_1\sin\beta_2\cos\theta_1\cos\theta_2\sin\theta_3\end{aligned}$$

The analytic solution was not used to display the maximum singularity free regions for mixed skew angles. By adding two more degrees of freedom, plotting the maximum singularity free momentums against the mixed beta angles would ideally be produced in four dimensions. Instead, three-dimensional plotting was used with keeping one of the beta angles held constant. **Figure 4** displays the numerical results of holding β1 at 90° while iterating β2, and β3:

Three-dimensional plotting has its limitation; in order to better demonstrate the locations of large singularity free momentum ranges, the plots were compressed to a two-dimensional representation with a color bar demonstrating the maximum value of singularity free angular momentum for each plot, as per **Figure 5**.

Figure 4 displays several areas of interest; locations where yellow and orange occur on the plot or locations with large singularity free locations, additionally locations where the color is dark blue represent small singularity free locations. From **Figure 3**, having all three skew angles set to 90° provided the greatest singularity free region, however **Figure 5** displays several other angles producing similar results. **Table 2** displays several of the large singularity free skew angles.

The skew angle arrays displayed in **Table 2** are portrayed graphically in **Figure 6** where the black coloration is the inner wall of singularities and the blue are the remaining singularities for varying angular momentum. Each plot has a void of singularities in their respective centers; these singularity free regions enable maneuvers that do not exceed the maximum angular momentum of the void to work without requiring singularity penetration.

At almost every skew angle, there is at least one location where the achievable angular momentum, without hitting a singularity, is >0.75H. Therefore, if the CMG array in a spacecraft is limited to a specific angle for one of the three CMGs, the array will still be able to operate in a large singularity free field if the other two are designed with the remaining angles in mind [2]. Similarly, setting one skew angle and arbitrarily picking the remaining skews could lead to small singularity free regions. The mixed skew angles of 20, 38, 1° provides an angular momentum region of 0.78H however, as depicted in **Figure 7**, by changing one angle from 1 to 45° led to drastic changes in the singularity free region.

Angular momentum	Skew 1	Skew 2	Skew 3
1.00	90	90	free
0.97	80	1	1
0.93	70	3	1
0.89	60	7	1
0.83	50	12	1
0.79	40	18	1
0.76	30	26	1
0.78	20	38	1
0.85	10	54	1
1.00	1	90	1

Table 2.
Singularity free maximum angular momentum values for mixed skew angles.

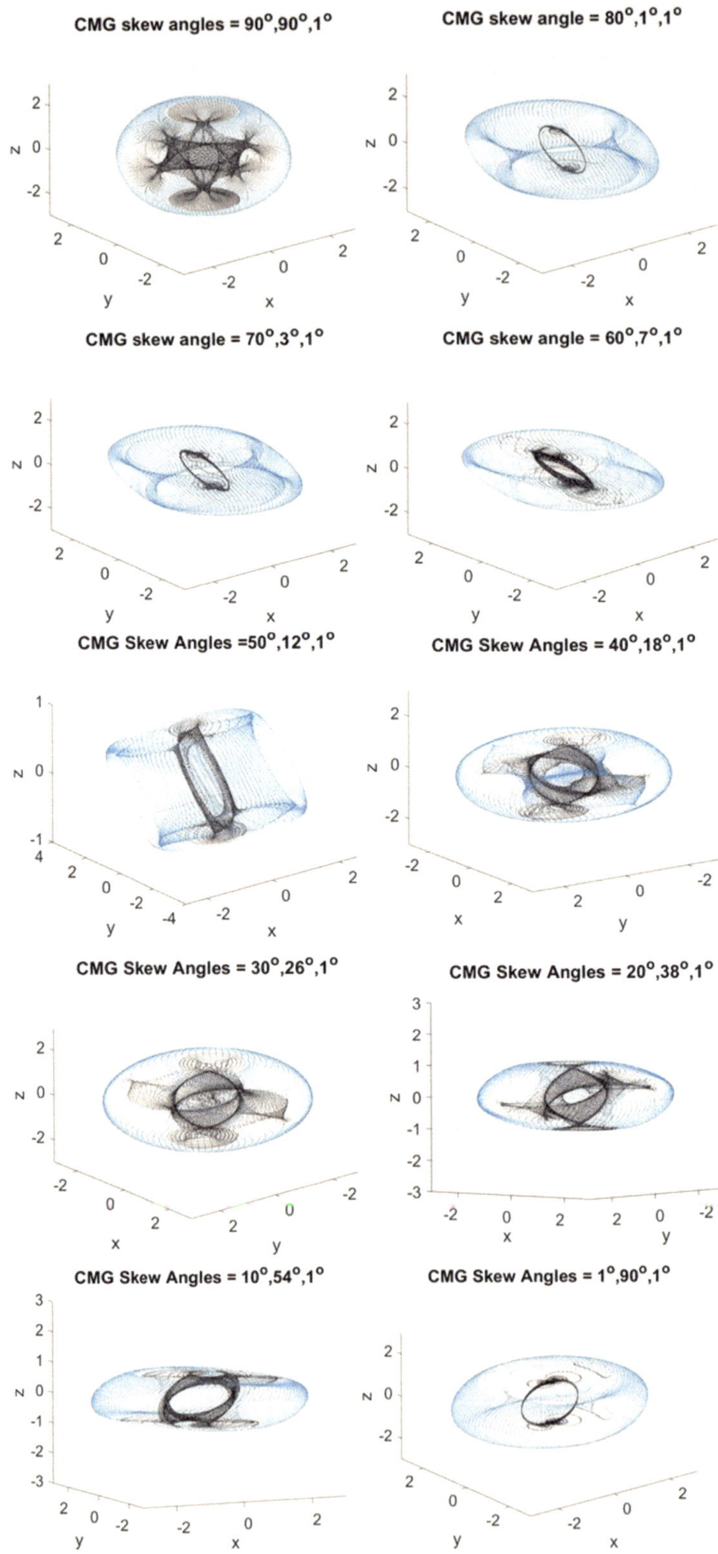

Figure 6.
Singularity hypersurfaces for mixed skew angles [5].

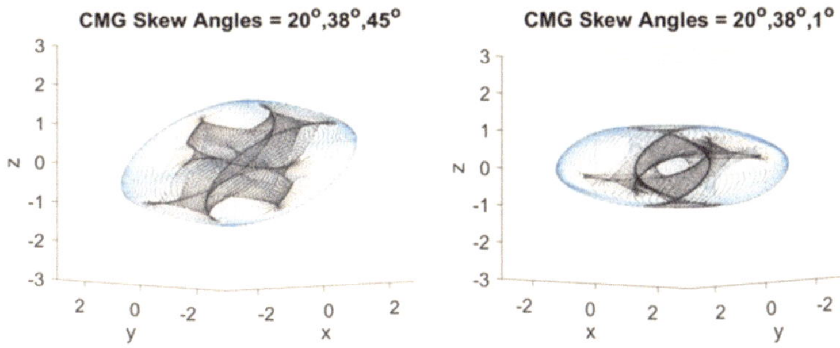

Figure 7.
Singularity hypersurface for β = 20, 38, 45° compared to β = 20, 38, 1° [5].

The maximum singularity free angular momentum drastically decreased from 0.78 to 0.02H when the third CMG skew angle became 45°. Therefore, the orientation of each CMG can severely affect the capability of the spacecrafts ability to control maneuvers.

4. Conclusion

CMG singularity maps provide a great deal of understanding of how a CMG array will be able to operate. A non-redundant array displays the minimum necessary CMGs to provide all degrees of freedom to operate a spacecraft. The maximum singularity free momentum depictions reveal that certain skew angles provide larger regimes to operate within and demonstrate the necessity to pick the skew angle to be able to achieve the desired torques. By investigating mixed skew angle plots of maximum singularity free angular momentum regions, several mixed angles were determined to display large regions where the array could operate without hitting singularities. This future work will enable a better understanding of the control and capabilities of the non-redundant CMG array.

Acknowledgements

This work was supported in part by the Naval Postgraduate School.

Author details

Joshua A. Ten Eyck
Astronautical Engineering, Naval Postgraduate School, Monterey, California, USA

*Address all correspondence to: joshua.teneyck@nps.edu

IntechOpen

© 2020 The Author(s). Licensee IntechOpen. Distributed under the terms of the Creative Commons Attribution - NonCommercial 4.0 License (https://creativecommons.org/licenses/by-nc/4.0/), which permits use, distribution and reproduction for non-commercial purposes, provided the original is properly cited. BY-NC

References

[1] Baker K, Cooper M, Heidlauf P, Sands T. Autonomous trajectory generation for deterministic artificial intelligence. Electrical and Electronic Engineering. 2018;**8**(3):59-68

[2] Sands T. Fine pointing of military spacecraft [dissertation]. CA: MAE, NPS; 2007

[3] Wie B. Space Vehicle Dynamics and Control. 2nd ed. Virginia, USA: AIAA; 2008

[4] Sands TA, Kim JJ, Agrawal B. 2H singularity-free momentum generation with non-redundant single gimbaled control moment gyroscopes. In: Proceedings of the 45th IEEE Conference on Decision and Control; 2006. pp. 1551-1556

[5] Ten Eyck J, Sands T, editors. Advancements in Spacecraft Attitude Control (working title). London, UK: InTech; 2019. Unpublished manuscript submitted. Print ISBN: 978-1-78984-802-1. Project in draft form, due to publisher April 2019